Mechanics for Physicists

An Introduction, Including Special Relativity

Mechanics for Physicists

An Introduction, Including Special Relativity

Torsten Fliessbach

University of Siegen, Germany

World Scientific

NEW JERSEY • LONDON • SINGAPORE • BEIJING • SHANGHAI • HONG KONG • TAIPEI • CHENNAI • TOKYO

Published by

World Scientific Publishing Co. Pte. Ltd.
5 Toh Tuck Link, Singapore 596224
USA office: 27 Warren Street, Suite 401-402, Hackensack, NJ 07601
UK office: 57 Shelton Street, Covent Garden, London WC2H 9HE

British Library Cataloguing-in-Publication Data
A catalogue record for this book is available from the British Library.

This book is the English translation to the 8th Edition of the original German language version, published by Springer-Verlag GmbH Germany, a part of Springer Nature (Copyright © 2020), under the title "Mechanik: Lehrbuch zur Theoretischen Physik I".

MECHANICS FOR PHYSICISTS
An Introduction, including Special Relativity

ISBN 978-981-12-8457-1 (hardcover)
ISBN 978-981-12-8458-8 (ebook for institutions)
ISBN 978-981-12-8459-5 (ebook for individuals)

For any available supplementary material, please visit
https://www.worldscientific.com/worldscibooks/10.1142/13631#t=suppl

Typeset by Stallion Press
Email: enquiries@stallionpress.com

Preface

This book entitled *Mechanics for Physicists: An Introduction, including Special Relativity* is the English version of my German textbook *Mechanik* (8th edition) [1]. This book is the first one of a series of textbooks [1–4] which developed from repeatedly held lectures. In German universities, the course on Theoretical Mechanics is usually offered to physics students in their third semester.

The presentation is on an average university level in theoretical physics. The approach is rather intuitive instead of deductive. Formal derivations and proofs are carried out without emphasis on mathematical rigor. The material is divided into chapters which roughly correspond to a one- to two-hour lecture. Each chapter starts with a short abstract and is devoted to a specific subject. As far as possible, each chapter is kept self-contained.

As usual, we start with Newton's laws and simple applications (Part I). Thereby the decisive role of the reference systems is emphasized. A broad space is then dedicated to Lagrange formalism and its applications. The Lagrange equations are introduced especially in a clear-cut way (Part II). In preparation of Hamilton's principle and the Noether theorem, an introduction to the variational calculus is given (Part III), always illustrated by examples (brachistochrone, catenary, geodesic lines).

Parts IV–VI are dedicated to standard applications (Kepler problem, oscillations, rotation of a rigid body). Due to our emphasis on applications (normally done with the Lagrange formalism), the alternative Hamilton formalism is treated in a compact way only (Part VII). Still all necessary prerequisites for quantum mechanics are provided.

Part VIII presents a short introduction to continuum mechanics. One could write several books about this wide field. Every physicist should, however, know some basic applications like string vibrations, beam bending, and fluid motion (including sound waves). For a continuum, one needs a Lagrange density instead of a Lagrange function. In Chapter 33, we compare the Lagrange densities of mechanics (for example, for the string vibrations) with some Lagrange densities in other fields of physics (electrodynamics, quantum mechanics). This elementary discussion introduces the reader to the concept of field theories.

The title of this book says "introducing special relativity". This means that we present a thorough introduction to the theory of Special Relativity (Part IX). This starts with a clear-cut introduction of the basic physical idea, which replaces Galilei's relativity principle with that of Einstein (and consequently the Galilei transformations by that of Lorentz). Then we investigate the most important applications (length contraction, time dilation, production of heavy particles, and the twin paradoxon). An appendix addresses the question of the relation between Newton's force and the Minkowski force. For this relation, one finds different expressions in various textbooks.

All over the world, one finds different conventions for the typesetting of formulas or mathematical expressions. This book follows the European style that numbers are written as upright (roman) letters, for example, Euler's number $\mathrm{e} = 2.71\ldots$, the imaginary unit $\mathrm{i} = \sqrt{-1}$, or the number $\mathrm{\pi} = 3.14\ldots$ In contrast to this, physical quantities (like the elementary charge e or the time t) are written in italic (slanted fonts). Physical quantities consist of a number times a (physical) unit (like $t = 3\,\mathrm{s}$, units like s for second are written in roman). If the physical quantity is a vector, then it is written in bold italic. In the US, one normally finds $\mathrm{e} = 2.71\ldots$ (roman), but $\pi = 3.14\ldots$ (italic), and physical quantities are written in italic except for capital Greek letters (roman) or vectors (bold roman). The derivative of a function $f(t)$ may be found as df/dt or as $\mathrm{d}f/\mathrm{d}t$.

The d in $\mathrm{d}t$ is written in roman to point out the mathematical limit (d stands for differential). Alternatively (from a more physical point of view), one may consider df/dt just as a fraction with a sufficiently small interval dt; then dt is a physical quantity and is written in italic. This book uses the form df/dt.

I want to thank Dr. Andrea Wolf (World Scientific) for her initiative that led to this book. I am deeply thankful to Hans Walliser (coauthor of [5]) for a careful reading of the manuscript and for pointing out numerous corrections. Last but not least, many readers of various German editions helped improving my book.

Preface of the 8th German Edition

This book is part of a lecture elaboration [1–4] of the cycle Theoretical Physics I–IV. It gives the material for my lecture on Theoretical Mechanics. This lecture is frequently offered for physics students in the third semester.

The presentation is on an average course level in theoretical physics. The approach is rather intuitive instead of deductive; formal derivations and proofs are carried out without emphasis on mathematical rigor.

Closely following the text, but also partly to its continuation and supplementation, more than 80 exercises are provided. These exercises fulfill their purpose if they are worked out by the students as independently as possible. This work should be done before reading the corresponding sample solutions, which are offered in the *Arbeitsbuch zur Theoretischen Physik* [5]. In addition to the solutions, the workbook contains a compact repetition of the material of the textbooks [1–4].

The scope of this book goes somewhat beyond the material that is usually covered during one semester in a physics course at a German university. The material is divided into chapters, which correspond on average to a lecture lasting one to two hours. Of course, different chapters build on each other. However, the individual chapters are more or less self-contained. On the one hand, this makes it possible to select chapters for a specific course (for example, in a Bachelor's program), in which the material is to be more limited. On the other hand, the student can more easily select a subject of particular interest.

There are many good representations of mechanics that are suitable for in-depth studies. I will list only a few books that I have preferably consulted myself and which are occasionally cited in the text. As a standard work, I would like to mention first the *Classical mechanics* by Goldstein [6]. For the introductory chapters, a similar approach has been chosen as in the *Theoretische Mechanik* by Stephani and Kluge [7]. Finally, volume 1 of the course of Landau-Lifschitz [8] should be mentioned. For relativistic mechanics, I refer to the introductory chapters of Weinberg's book [9] on General Relativity.

Compared to the previous edition of this book, some errors have been eliminated, and in numerous places minor additions and improvements have been made. Newly added is Appendix A with the title *Newtonian and Minkowski force.* It deals with the question of the correct relation between the Newtonian force and the Minkowski force. For this relation, one finds different statements in the literature. The appendix gives the formal derivation of the correct relation and discusses its practical and logical relevance.

I would like to thank the numerous readers of the previous editions for their valuable hints. Error messages, comments, and hints are always welcome, for example, via the contact link on my homepage www2.uni-siegen.de/~flieba/. On this homepage, you will also find correction lists.

References

[1] T. Fliessbach, *Mechanics for Physicists: An Introduction, including Special Relativity*, 1st edition, World Scientific Publishing Company, Singapore, 2023 (this book); T. Fließbach, *Mechanik*, 8th edition, Springer Spektrum, Heidelberg, 2020 (German original of this book).

[2] T. Fließbach, *Elektrodynamik*, 7th edition, Springer Spektrum, Heidelberg, 2022.

[3] T. Fließbach, *Quantenmechanik*, 6th edition, Springer Spektrum, Heidelberg, 2018.

[4] T. Fließbach, *Statistische Physik*, 6th edition, Springer Spektrum, Heidelberg, 2018.

[5] T. Fließbach and H. Walliser, *Arbeitsbuch zur Theoretischen Physik — Repetitorium und Übungsbuch*, 4th edition, Spektrum Akademischer Verlag, Heidelberg, 2020.

[6] H. Goldstein, *Classical Mechanics*, various editions, also available in the internet.

[7] H. Stephani, G. Kluge, *Theoretische Mechanik*, Spektrum Akademischer Verlag, Heidelberg, 1995.

[8] L. D. Landau, E. M. Lifschitz, *Mechanics*, various editions, also available in the internet.

[9] S. Weinberg, *Gravitation and Cosmology*, John Wiley & Sons, New York, 1972, also available in the internet.

Contents

Preface v

Introduction xxi

Part I: Elementary Newtonian Mechanics **1**

Chapter 1. Trajectory **3**

Time Measurement 5
Euclidean Space 6
Coordinates 7
Velocity and Acceleration 8
Exercises 10

Chapter 2. Newton's Laws **11**

Applications 16
One-dimensional Motion 17
Exercises 19

Chapter 3. Conservation Laws **21**

Momentum Conservation 21
Conservation of Angular Momentum 21
Energy Conservation 23
Potential 25
Exercises 28

Chapter 4. System of Mass Points **31**

Momentum 32
Angular Momentum 33
Energy 34
Determination of Potentials 36
Exercises 38

Chapter 5. Inertial Frames **39**

Galilean Transformation 40
Second Law 44
Passive and Active Transformation 46
Invariance of Newton's Laws 47
Validity of the Galilean Transformation 49

Chapter 6. Accelerated Reference Frames **51**

Linearly Accelerated Reference Frame 51
Rotating Reference System 53
Exercises 59

Part II: Lagrangian Formalism **61**

Chapter 7. Lagrange's Equations of First Kind **63**

Constraints 63
Constraint Forces 66
General Case 67
Conservation Laws 70

Chapter 8. Applications I **73**

General Procedure 73
Inclined Plane 75
Mass Point on the Rotating Rod 79
Atwood's Falling Machine 81
Exercises 82

Chapter 9. Lagrange's Equations Second Kind **85**

Generalized Coordinates 85
Elimination of Constraint Forces 87
Lagrange Function 88
Constants of Motion 92
Electromagnetic Forces 94
Friction Forces 95
Exercises 96

Chapter 10. Applications II **99**

Inclined Plane 100
Mass Point on a Rotating Rod 100
Atwood's Falling Machine 101
Double Pendulum 102
Curvilinear Coordinates 103
Mass Point on a Circular Cone 104
Exercises 108

Chapter 11. Space–Time Symmetries **111**

Introduction 111
Homogeneity of Time 115
Homogeneity of Space 116
Isotropy of Space 117
Relativity of Space–Time 119
Exercises 120

Part III: Variational Principles **121**

Chapter 12. Variation without Constraint **123**

Problem 123
Euler–Lagrange Equation 125
Solving the Euler–Lagrange Equation 128
Generalizations 129
Exercises 132

Chapter 13. Variation with Constraint **133**

Problem 133
Lagrange Multiplier 135
Isoperimetric Constraint 137
Holonomic Constraint 140
Applications 142
Exercises 145

Chapter 14. Hamilton's Principle **147**

Lagrange Equations of Second Kind 147
Lagrange Equations of First Kind 148
Notes on Hamilton's Principle 149
Gauge Transformation 151
Exercises 153

Chapter 15. Noether's Theorem **155**

Extended Noether Theorem 161
Exercises 163

Part IV: Central Potential **167**

Chapter 16. Two-Body Problem **169**

Reduction to One-Body Problem 170
Reduction to the Radial Equation 171
Conservation of Energy 173
Solution 174
Discussion of Motion 175
Exercises 179

Chapter 17. Kepler Problem **181**

Kepler's Laws of Planetary Motion 183
Deterministic Many-Body Problem 185
Hyperbolic Orbit 189
Time Dependence of the Path 190
Exercises 191

Chapter 18. Scattering **195**

Definition of the Cross-Section 196
Calculation of the Cross-Section 198
Rutherford Cross-Section 200
Transformation to Laboratory System 203
Energy Balance 208
Exercises 210

Part V: Rigid Body **213**

Chapter 19. Kinematics **215**

Angular Velocity 216
Euler Angles 220

Chapter 20. Inertia Tensor **223**

Kinetic Energy 223
Angular Momentum 226
Principal Axis Transformation 228
Exercises 231

Chapter 21. Tensors **235**

Orthogonal Transformations 235
Definition 239
Calculation Rules 240
Tensor Equations 241
Examples of Tensors 242
Dyads 245
Tensor Fields 246
Exercises 248

Chapter 22. Euler's Equations **249**

Derivation of Euler's Equations 249
Free Rotation Around Principal Axes 252
Force-Free Symmetric Gyroscope 254
Exercises 258

Chapter 23. Heavy Gyroscope **259**

Lagrange Function 259
Heavy Symmetric Gyro 260
Earth as a Rotating Body 264
Exercises 270

Part VI: Small Oscillations **273**

Chapter 24. Forced Oscillations **275**

Equation of Motion 275
General Solution 277
Forced Oscillation 280
Exercises 284

Chapter 25. System with Many Degrees of Freedom **285**

Natural Frequencies 285
Normal Coordinates 290
Exercises 294

Chapter 26. Applications **295**

Two-dimensional Motion 295
Double Pendulum 297
Triatomic Molecule 299
Exercises 302

Part VII: Hamilton Formalism **307**

Chapter 27. Canonical Equations **309**

Hamilton Function 309
Conservation Laws 312
Phase Space 313
Exercises 316

Chapter 28. Canonical Transformations **319**

Hamilton's Principle 319
Generating Function $G(q, Q, t)$ 320
Poisson bracket 325
Example 326
Exercises 327

Chapter 29. Hamilton–Jacobi Equation **329**

Time-Independent Hamilton–Jacobi Equation 331
Example 332

Part VIII: Continuum Mechanics **335**

Chapter 30. String Vibration **337**

Generalizations 342
Vibrations of the Force-Free String 343
Exercises 347

Chapter 31. Beam Bending **349**

Static Bending in the Gravity Field 351

Chapter 32. Hydrodynamics **355**

Continuity Equation 357
Euler Equation 358
Equation of State 359
Applications 362
Sound Waves 367
Exercises 370

Chapter 33. Field Theories **373**

Lagrange Density 373
Examples 375
Noether Theorem 376

Part IX: Relativistic Mechanics **381**

Chapter 34. Relativity Principle **383**

Interval of Events . . . 386
Lorentz Transformation . . . 388
Special Lorentz Transformation . . . 389
Exercises . . . 393

Chapter 35. Length and Time Measurement **395**

Resting Rulers and Clocks . . . 395
Moving Ruler . . . 397
Moving Clock . . . 400
Simultaneity . . . 404
House Stick Experiment . . . 407
Exercises . . . 409

Chapter 36. Lorentz Group **413**

General Lorentz Transformation . . . 413
Successive Lorentz Transformation . . . 416
Velocity Addition Formula . . . 418

Chapter 37. Lorentz Tensor **421**

Calculation Rules . . . 423
Exercises . . . 428

Chapter 38. Equation of Motion **429**

Lorentz Force . . . 434
Relativistic Energy . . . 436
Relativistic Momentum . . . 439
Energy–Momentum Relation . . . 440
Exercises . . . 441

Chapter 39. Applications **443**

Kinetic Energy: Laboratory Versus Center-of-Mass System . . . 443

Hyperbolic Motion 447
Twin Paradox 449
Exercises 454

Chapter 40. Lagrange Function **455**

Exercises 461

Appendix A. Newtonian Force and Minkowski Force **463**

Different Statements in the Literature 463
Formal Derivation 464
Lorentz Force 466
Relativistic Mass 467
Symmetry Principles and the Logic of Theoretical Physics 468
Summary 469

Index 471

Introduction

The mechanics examines the laws according to which material bodies move. By this motion, one understands the change of the position as a function of the time. The motion takes place under the influence of forces that are assumed to be known.

In the 19th century, the opinion was widespread that physical processes are understood only when they can be explained mechanically (mechanistic world view). For example, heat can be attributed to the disordered motion of atoms. Today we know that this reduction is not possible for many known phenomena (electromagnetism, quantum effects).

A large part of this book deals with bodies idealized to mass points, or with systems of such mass points. Part I presents the mechanics of such systems on the basis of Newton's laws. Basic concepts such as mass point, trajectory, reference and coordinate frame, position and time, momentum, angular momentum, kinetic, and potential energy are introduced and discussed. The formulation of mechanics in the framework of the Lagrangian formalism (Part II) plays a central role. The relatively detailed Part III on variational principles leads to a general description of the connection between symmetries and conservation laws (Noethertheorem). The subsequent parts, IV–VI, investigate the most important application areas, like the motion in a central potential, the dynamics of a rigid body, and harmonic oscillations.

The alternative Hamiltonian formalism (Part VII) plays a minor role in practical applications. But it is needed for the introduction of quantum mechanics; furthermore it is the starting point for the discussion of the relations between mechanics and quantum mechanics. Besides the already mentioned applications mechanics includes the large fields of mechanics of elastic matter (e.g. string vibration, beam bending) and hydrodynamics. Each of these areas contains enough material for a lecture on its own. In Part VIII, only the simplest ideas and the basic equations are presented. Chapter 33 discusses the Lagrangian formalism for field theories beyond mechanics.

The last part, IX, gives a relatively detailed introduction to special relativity theory. In the resulting relativistic mechanics, among other things, the production of heavy particles and the twin paradoxon are discussed.

PART I

Elementary Newtonian Mechanics

Chapter 1

Trajectory

The trajectory of a mass point is a central concept in mechanics, comparable, for example, with the electromagnetic field in electrodynamics or the wave function in quantum mechanics. In this chapter, we introduce the trajectory curve and a number of related concepts (mass point, reference frame, space, and time) and explain them.

This chapter is limited to kinematics, i.e. to the mere description of the motion of a mass point. The following chapter deals with the dynamics, which means with the physical laws according to which this motion takes place.

We can describe the motion of a body only as motion relative to something else, for example, relative to ourselves or relative to other objects. If there would be nothing else than the considered body, we could not perceive or describe its motion. Preferably we will use certain objects as reference points for the description of the motion (and not ourselves), so that the motion is described independently of the particular observer. An obvious *frame of reference* is a laboratory room or a lecture hall. In such a reference frame, we may introduce spatial coordinates (like, for example, the distances x, y, z from the walls of a cuboid room). These coordinates are able to name all points in the room. In this reference, frame the time t can be defined by a suitable clock; we denote t as time coordinate. A reference frame with certain coordinates is also called a *coordinate frame* or coordinate system.

The laws of motion will depend on the choice of the reference frame. For example, billiards in a lecture hall is quite different from that on a merry-go-round. This means that it is subject to different laws of motion. For the formulation of these laws (Chapter 2) the reference frame must therefore be specified.

We now consider a *mass point*. A mass point is a body for whose motion only its position is relevant or of interest. In this sense, its motion is completely described by the specification of the position $\boldsymbol{r}$ at any time t, that is by the *trajectory* (also called trajectory curve)

$$\boldsymbol{r}(t) = x(t)\,\boldsymbol{e}_x + y(t)\,\boldsymbol{e}_y + z(t)\,\boldsymbol{e}_z \tag{1.1}$$

For example, the following objects could be considered as mass points: Particles (like electrons or protons) in a scattering process, atoms in a gas, billiard balls, planets in the solar system, galaxies in galaxy clusters. All these objects are indeed extended. The model conception of "mass point" is an idealization, which assumes that the trajectory curve can be treated without consideration of the other degrees of freedom; $\boldsymbol{r}$ refers to the center of mass of the body. Thus, for the calculation of the earth's orbit around the sun, the earth's rotation (and all other processes on the earth) can be ignored. In contrast to this, the rotation of a billiard ball can have an essential influence on the trajectory; so the application of the model mass point can also be faulty. In general, a necessary condition for the treatment as a point of mass will be that the size of the body is small compared to the other relevant dimensions of the system; thus the radius of the earth is small with respect to the semi-axes of its orbital ellipse. It is in this sense that the use of the term "point" in mass point is to be understood.

Already the familiar description (1.1) of a trajectory requires a number of non-trivial preconditions. In particular, a reference frame with Cartesian coordinates and a time coordinate must be defined; the concrete specification of x, y, z, and t requires the definition of the length and time measurement. The time measurement is discussed in the following. The measurement of lengths is assumed to be known. Cartesian coordinates imply that our three-dimensional space is flat;

this physical assumption about the structure of our space will be discussed subsequently.

Time Measurement

Time is defined by specifying a method for measuring the time coordinate t by a clock. A clock is an instrument that indicates (or counts) the number of periods of a periodic, continuous process. As a periodic process, one could, for example, choose the following:

- Pulse beat of the president of the United States.
- Pendulum clock.
- Earth rotation (day–night period).
- Atomic frequency.

The time unit may be defined as a multiple or a fraction of a given period. For example, one second may be defined as the 86 400th part of a day–night period or as the 9 191 631 770 multiple of the period of the transition between the two hyperfine structure levels of the ground state of cesium 133.

The laws for the trajectory $\boldsymbol{r}(t)$ and other physical laws will depend on the definition of t. We will prefer the time definition for which the laws of physics are most simple. Therefore, the time definition "pulse beat" is only useful for very rough considerations. Otherwise, the laws here become complicated because the pulse beat is subject to complex influences. In fact, for each time definition, there will be deviations from the (hypothetical) ideal time due to disturbances:

- Pulse beat: complex non-physical effects.
- Pendulum clock: feedback of drive, temperature.
- Earth rotation: friction due to tides.
- Atomic frequency: gravitational field of the earth.

Time measurement is therefore subject to continued refinements: The definition of time by the sequence of day and night implies that the angular velocity of the rotation of the earth is constant. From the physics based on it, one can however estimate that the earth does not really rotate uniformly; because the friction caused by tides leads to a torque and thus to a change of the angular momentum. This change is observable: The calculated position of the solar eclipse on the 14th January of the year 484 is shifted by 30° in the East–West direction with respect to the observed position. This means a correction of about two hours in 1 500 years. Consequently, we will no longer tie the time measurement *exactly* to day and night, because this would have the consequence that, for example, the speed of light and atomic frequencies would not be constant, but depend (slightly) on the effects of the tides. This would be impractical, since then the physical laws would become unnecessarily complicated.

The current time definition is done by atomic frequencies. The known influence of the gravitational field of the earth can be taken into account. As with each such definition, it is possible that there are still other, unknown disturbances. These disturbances could require a change of the time definition with increased measurement accuracy. The example with the day and night time shows that such effects can be found, even if one defines the time first by such disturbed processes. Physicists describe this situation with the expression "one can wash dishes with dirty water".

Euclidean Space

Another assumption in (1.1) is made by using a Cartesian coordinate frame: Such a frame exists only in *Euclidean* or *flat space*. The opposite to this is a curved space. A curved space can be defined by the statement that Cartesian coordinates are not possible; an example of this is the two-dimensional space of a spherical surface. In the Euclidean space one can, however, use curved coordinates (like spherical coordinates) instead of Cartesian coordinates; this does not change anything about the property "flat" of this space.

Gauß (1777–1855) measured the angle sum of the triangle Inselsberg–Brocken–Hohenhagen and obtained 180°. Thereby he verified experimentally that our three-dimensional space is Euclidean on the length scale of about 100 km. The primary intention of Gauß was probably not the verification of the Euclidicity of our space (as it is sometimes reported) but rather a land survey. But Gauß was well aware of the possibility of non-Euclidean space.

The measurement by Gauß presupposed that light rays propagate along a straight line. Light rays are in fact deflected in the gravitational field, namely by 1.75 arc-seconds for light passing the solar limb. These light rays may be described by straight lines in a more general space; then this space is curved. This curvature is very small (about 10^{-6} at the solar limb and 10^{-9} at the earth's surface). In a curved space, one can use *local* Cartesian coordinates; this can be seen from the example of the spherical surface.

In the following, we assume an Euclidean space.

Coordinates

We discuss the representation of the trajectory curve for different coordinates and the corresponding calculation of the velocity and acceleration. In (1.1), the position vector $\boldsymbol{r}$ was expressed by its Cartesian components x, y, and z; the basis vectors of the coordinate system S were denoted by $\boldsymbol{e}_x$, $\boldsymbol{e}_y$, and $\boldsymbol{e}_z$. If the basis vectors are given, then the vector $\boldsymbol{r}$ can be represented by its components:

$$\boldsymbol{r} = x\,\boldsymbol{e}_x + y\,\boldsymbol{e}_y + z\,\boldsymbol{e}_z := \begin{pmatrix} x \\ y \\ z \end{pmatrix} = R \tag{1.2}$$

We use the symbol ":=" for "represented by". The column vector we denote by R. The vector $\boldsymbol{r}$ is not equal to the column vector R. This can also be seen by the fact that in a rotated coordinate system S$'$ the same vector $\boldsymbol{r}$ is represented by a different column vector:

$$\boldsymbol{r} = x'\,\boldsymbol{e}'_x + y'\,\boldsymbol{e}'_y + z'\,\boldsymbol{e}'_z := \begin{pmatrix} x' \\ y' \\ z' \end{pmatrix} = R' \tag{1.3}$$

The column vectors R and R' are representations that refer to two different coordinate frames (with $\boldsymbol{e}_x$, $\boldsymbol{e}_y$, $\boldsymbol{e}_z$ or $\boldsymbol{e}'_x$, $\boldsymbol{e}'_y$, $\boldsymbol{e}'_z$, see also Figure 21.1). Often, however, one neglects the difference between equality and representation in the notation.

The scalar product $\boldsymbol{r} \cdot \boldsymbol{r}$ can be written as a matrix product:

$$\boldsymbol{r} \cdot \boldsymbol{r} = \boldsymbol{r}^2 = R^{\mathrm{T}} R = x^2 + y^2 + z^2 \tag{1.4}$$

Here, R^{T} ("R-transposed") is equal to the line vector (x, y, z).

Besides (1.1) we can also consider other representations of the trajectory curve $\boldsymbol{r}(t)$, for example,

$$\boldsymbol{r}(t) := \begin{cases} x(\lambda),\ y(\lambda),\ z(\lambda),\ t(\lambda) & \text{(parameter representation)} \\ \rho(t),\ \varphi(t),\ z(t) & \text{(cylindrical coordinates)} \\ r(t),\ \theta(t),\ \phi(t) & \text{(spherical coordinates)} \end{cases} \tag{1.5}$$

Velocity and Acceleration

We define the *velocity* of a point of mass by

$$\boldsymbol{v}(t) = \dot{\boldsymbol{r}}(t) = \frac{d\boldsymbol{r}}{dt} = \lim_{t' \to t} \frac{\boldsymbol{r}(t) - \boldsymbol{r}(t')}{t - t'} = \dot{x}(t)\, \boldsymbol{e}_x + \dot{y}(t)\, \boldsymbol{e}_y + \dot{z}(t)\, \boldsymbol{e}_z \tag{1.6}$$

and the *acceleration* by

$$\boldsymbol{a}(t) = \ddot{\boldsymbol{r}}(t) = \ddot{x}(t)\, \boldsymbol{e}_x + \ddot{y}(t)\, \boldsymbol{e}_y + \ddot{z}(t)\, \boldsymbol{e}_z \tag{1.7}$$

Note that the basis vectors $\boldsymbol{e}_x$, $\boldsymbol{e}_y$, and $\boldsymbol{e}_z$ are time independent. In contrast to this, the basis vectors for curvilinear coordinates depend on the position and thus for the motion of a particle effectively on time. We investigate these peculiarities by the example of polar coordinates for the two-dimensional motion of a particle.

The polar coordinates are given by

$$x = x(\rho, \varphi) = \rho \cos\varphi, \quad y = y(\rho, \varphi) = \rho \sin\varphi \tag{1.8}$$

We introduce the unit vectors $\boldsymbol{e}_\rho$ and $\boldsymbol{e}_\varphi$ which point in the direction of the change of $\boldsymbol{r}$ for the change $\rho \to \rho + d\rho$ or $\varphi \to \varphi + d\varphi$, see Figure 1.1. The orthogonal pair $(\boldsymbol{e}_\rho,\ \boldsymbol{e}_\varphi)$ is rotated by the angle φ relative to $(\boldsymbol{e}_x,\ \boldsymbol{e}_y)$. Therefore,

$$\begin{aligned} \boldsymbol{e}_\rho &= \cos\varphi\ \boldsymbol{e}_x + \sin\varphi\ \boldsymbol{e}_y \\ \boldsymbol{e}_\varphi &= -\sin\varphi\ \boldsymbol{e}_x + \cos\varphi\ \boldsymbol{e}_y \end{aligned} \tag{1.9}$$

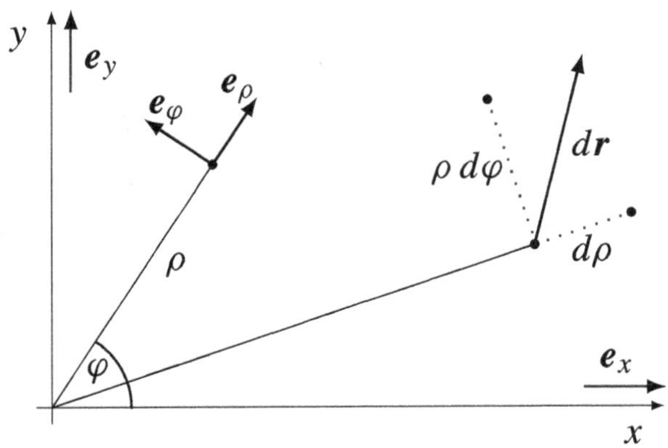

Figure 1.1 For polar coordinates, the direction of the basis vectors depends on the position. The line element reads $d\boldsymbol{r} = d\rho\, \boldsymbol{e}_\rho + \rho\, d\varphi\, \boldsymbol{e}_\varphi$.

For a trajectory described by $\rho(t)$ and $\varphi(t)$, we obtain the time derivative

$$\begin{aligned} \dot{\boldsymbol{e}}_\rho &= -\sin\varphi\, \dot\varphi\, \boldsymbol{e}_x + \cos\varphi\, \dot\varphi\, \boldsymbol{e}_y = \dot\varphi\, \boldsymbol{e}_\varphi \\ \dot{\boldsymbol{e}}_\varphi &= -\cos\varphi\, \dot\varphi\, \boldsymbol{e}_x - \sin\varphi\, \dot\varphi\, \boldsymbol{e}_y = -\dot\varphi\, \boldsymbol{e}_\rho \end{aligned} \tag{1.10}$$

The time derivative of the position vector

$$\boldsymbol{r}(t) = \rho(t)\, \boldsymbol{e}_\rho(t) \tag{1.11}$$

yields the velocity:

$$\boldsymbol{v} = \frac{d\boldsymbol{r}}{dt} = \dot\rho\, \boldsymbol{e}_\rho + \rho\, \dot{\boldsymbol{e}}_\rho = \dot\rho\, \boldsymbol{e}_\rho + \rho\, \dot\varphi\, \boldsymbol{e}_\varphi \tag{1.12}$$

Alternatively, we read from Figure 1.1 the infinitesimal line element $d\boldsymbol{r}$:

$$d\boldsymbol{r} = d\rho\, \boldsymbol{e}_\rho + \rho\, d\varphi\, \boldsymbol{e}_\varphi \tag{1.13}$$

Also, this yields (1.12). We differentiate (1.12) taking (1.10) into account. This yields the acceleration

$$\boldsymbol{a}(t) = \ddot{\boldsymbol{r}} = \frac{d\boldsymbol{v}}{dt} = \left(\ddot\rho - \rho\, \dot\varphi^2\right) \boldsymbol{e}_\rho + \left(\rho\, \ddot\varphi + 2\dot\rho\, \dot\varphi\right) \boldsymbol{e}_\varphi \tag{1.14}$$

Exercises

1.1 *Acceleration in spherical coordinates*

In spherical coordinates $(r,\ \theta,\ \phi)$ the trajectory of a mass point is of the form $\boldsymbol{r}(t) = r(t)\,\boldsymbol{e}_r(t)$. Determine the velocity and the acceleration in these spherical coordinates.

Chapter 2

Newton's Laws

Newton's laws can be considered as the basic laws of mechanics. Newton's first law specifies the preferred reference frames. Newton's second law is a differential equation for the trajectory of a mass point. It determines the dynamics, i.e. the time dependence of the motion.

All physical theories assume certain unproven, fundamental laws. These laws arise as generalization of (a finite number of) observations. They should be as simple as possible and should allow to describe or predict a large class of phenomena. These general laws are called natural laws. The description of phenomena by these laws is equivalent to their *explanation*. A law of nature implies an infinite number of predictions. The confirmation of predictions *verifies* the law but can not prove it (because the confirmation can be done only in a finite number of cases). However, a law of nature, once established, can be *falsified* by one single observation or experiment.

In mechanics, Newton's laws (sometimes also called Newton's axioms) can be understood as laws of nature. Later we will get to know alternative formulations of the basic laws, especially the Lagrangian formalism.

In his work *Philosophiae Naturalis Principia Mathematica*, Isaac Newton formulated three laws in 1687. Newton's first law (also called lex prima) concerns the reference frames:

First law: There are reference frames in which the force-free motion is described by $\dot{\boldsymbol{r}}(t) = \boldsymbol{v} = \text{const.}$ (2.1)

The preferred frame of reference thus specified is called *inertial frame* (or inertial frame of reference or inertial reference frame). An inertial frame with coordinates (like x, y, z, and t as in (1.1)) will be called an *inertial system*, abbreviated by IS.

Using the example of an observer on a carousel, one realizes that a statement like "$\boldsymbol{v} =$ const. for force-free motion" inevitably depends on the reference frame (just think of the attempt to play billiards on a carousel). Experimentally, one finds that the inertial frames are those reference frames that are at rest relative to the fixed-star sky, or that are moving with a constant velocity relative to the fixed-stars.

In all fields of physics, one formulates laws that presuppose a reference frame. In this context, the inertial frames play a distinguished role. For example, Maxwell's equations are valid only in inertial frames. It is possible to formulate the laws also in other reference frames. They have then in general a different and more complicated form; in the laws of mechanics, for example, additional inertial forces occur. The inertial frames are distinguished by the fact that here the physical laws are especially *simple*. In particular, because of (2.1), billiards is simpler in the lecture hall than on the carousel. The distinction of inertial frames and the complications in other reference frames will be discussed in more detail in Chapters 5 and 6.

Equation (2.1) states that the uniform motion (or rest) is a state in which the body remains. For example, a comet far away from other masses traverses the space with constant speed. In the area of everyday experience, the uniform motion can be observed only approximately, because there are practically always frictional forces.

Without forces, bodies move uniformly (first law). Forces, however, lead to a non-uniform motion. The term force is first taken from everyday language (muscle force, spring force, gravity); later it is specified by a measuring prescription. Newton's second law (also called lex secunda) describes the motion under the influence of a force:

$$\text{Second law:} \quad \frac{d\boldsymbol{p}}{dt} = \boldsymbol{F} \quad \text{in inertial frames} \tag{2.2}$$

Here, the momentum $\boldsymbol{p}$ is the product of the mass m and the velocity:

$$\boldsymbol{p} = m\boldsymbol{v} \tag{2.3}$$

Almost always, we are dealing with constant mass (except, for example, in the case of the equation of motion for a rocket), so that the second law can also be written as

$$m\,\ddot{\boldsymbol{r}} = \boldsymbol{F} \quad \text{(second law)} \tag{2.4}$$

Hereby, the mass is introduced as a property assigned to the body under consideration. Newton's second law includes the following definitions and statements:

1. Definition of mass.
2. Definition of force.
3. Physical statement about the trajectory.

We first explain in which way the second law defines mass and force as measurable quantities. We presuppose that length and time are already defined. This implies that the acceleration $\boldsymbol{a} = \ddot{\boldsymbol{r}}$ is a measurable quantity.

We consider a certain force, unknown in its magnitude (for example, a spring force), and two bodies 1 and 2. We measure the accelerations a_1 and a_2 caused by the unknown force. According to (2.4), the ratio m_1/m_2 is given by a_2/a_1; thus m_1/m_2 is defined as a measurable quantity. We now define arbitrarily the mass of a certain body as 1 mass unit; implicitly we assume that the property mass is an invariable property of the body. The name of the mass unit is as arbitrary as the choice of the reference body; it is well known that one chooses the kilogram (kg) as the mass unit. By this and by the measurement prescription for m_1/m_2 the mass of each body is defined.

After mass and acceleration are measurable, equation (2.4) specifies the force as a measurable quantity. The unit of force is

$$1\,\text{kg}\,\frac{\text{m}}{\text{s}^2} = 1\,\text{N} = 1\,\text{Newton} \tag{2.5}$$

We sum up: The measurement of masses and forces is based on the second law.

According to the second law, mass is a measure of the resistance of a body to the change of its velocity. The greater m is, the smaller is — for a given force — the change of velocity. The quantity m is therefore called *inertial mass*.

Another concept is *gravitational mass*, which is proportional to the strength of the gravitational force on a body; this gravitational mass is determined experimentally by a force measurement. Introduced in this form, the gravitational mass is a property of a body like, for example, its charge; the charge is determined by the strength of the force on a body in an electromagnetic field. Like the charge, the gravitational mass could be a property of a body that is independent of the inertial mass. But experimentally it turns out that the ratio of inertial to gravitational mass is always the same (with a relative accuracy up to 10^{-12}). Therefore, one does not introduce a new measurable quantity "gravitational mass"; rather, one just uses m for both quantities in the equations of motion. In the Newtonian mechanics, the equality of inertial and gravitational mass appears to be accidental. In contrast to this, general relativity takes it as its central starting point (Chapter 6).

The second law is not only a defining equation for the mass and the force. Rather it is also a physical law about dynamics; it contains physical statements. For example, for a constant force, the nontrivial statement $x \propto t^2$ holds for the one-dimensional force-free motion. This statement can be checked by measurement; thus the second law can be verified or falsified. The statements following the second law are indeed confirmed by many experiments. This is, however, only true as long as the occurring velocities are small compared to the velocity of light c.

For velocities comparable with c the predictions of the second law (like $x \propto t^2$ for constant force) are wrong. Thus, the law is in principle falsified and therefore to be rejected. Since the second law makes correct predictions in wide areas, one does not go that fare. Rather one supplements the second law with the restriction that it is to be used only for velocities that are small relative to the velocity of light.

Newton's third law (also called lex tertia) reads: The force with which the environment acts on a mass point, always corresponds to an equal opposite force with which the mass point acts on its

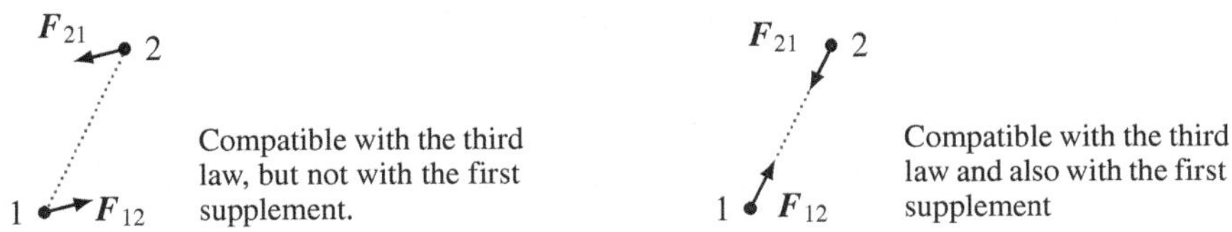

Figure 2.1 The forces between two mass points must, according to the third law, be opposite equal to each other. In addition, we restrict ourselves to forces that are parallel (or antiparallel) to the relative vector.

environment, or

$$\text{Third law:} \quad \boldsymbol{F}_{\text{actio}} = -\boldsymbol{F}_{\text{reactio}} \tag{2.6}$$

For the forces, which two mass points mutually exert on each other (Figure 2.1) this implies

$$\boldsymbol{F}_{12} = -\boldsymbol{F}_{21} \quad \text{(third law)} \tag{2.7}$$

The statement (2.6) holds in a more general sense in all parts of physics: Every effect, which the environment exerts on a body, causes a corresponding counter-effect.

For some derivations, we need additional assumptions about the occurring forces. We formulate these additional assumptions as the first and second supplements. The first supplement states that the forces that two mass points exert on each other (Figure 2.1) act in the direction of the line connecting them, i.e.

$$(\boldsymbol{r}_1 - \boldsymbol{r}_2) \times \boldsymbol{F}_{12} = 0 \quad \text{(first supplement)} \tag{2.8}$$

The second supplement reads: If several forces $\boldsymbol{F}_i$ act on one mass point, the total force $\boldsymbol{F}$ is the sum of the individual forces

$$\boldsymbol{F} = \sum_i \boldsymbol{F}_i \quad \text{(second supplement)} \tag{2.9}$$

This is also called the superposition principle of the forces.

These two supplements are, for example, fulfilled for Newton's gravitational forces between massive particles and for the Coulomb forces between charged particles. They represent a restriction on the allowed form of the forces. They are, however, less fundamental than Newton's laws themselves. A counterexample to the first supplement is the magnetic forces between moving charges; for this, we refer to a later discussion following (4.15). In a medium, nonlinear electromagnetic field effects may occur; then the second supplement would not apply to the Coulomb forces between charged particles in this medium.

Applications

We shall confine ourselves to forces in mechanics that depend only on the position and velocity of the particle, and on the time,

$$\boldsymbol{F} = \boldsymbol{F}(\boldsymbol{r}(t),\, \dot{\boldsymbol{r}}(t),\, t) \tag{2.10}$$

We thus exclude, for example, that $\boldsymbol{F}$ depends on the acceleration or on higher derivatives, or on the motion of the particle at earlier times. Velocity-dependent forces are, for example, the friction force and the Lorentz force.

When solving problems with the help of Newton's laws, the following steps typically occur:

1. Establishment of the force law.
2. Mathematical solution of the differential equation:

$$m\,\ddot{\boldsymbol{r}}(t) = \boldsymbol{F}(\boldsymbol{r}(t),\, \dot{\boldsymbol{r}}(t),\, t) \tag{2.11}$$

3. Determination of the integration constants of the solution which depend on the initial conditions for the position and the velocity at a certain time (for example, values for $\boldsymbol{r}(0)$ and $\dot{\boldsymbol{r}}(0)$).
4. Discussion of the solution: The solution could be displayed graphically, or conserved quantities (such as energy or angular momentum) could be looked for.

Compared to the problem defined in this way, many generalizations are conceivable. In particular, instead of the coordinates x, y, and z more general coordinates q_i (with $i = 1, 2, \ldots, f$) may be adequate. The dynamics (time-dependent behavior) of such systems is described by the functions $q_i(t)$. For these functions, the equations of motion (2.11) have to be generalized.

One-dimensional Motion

A specialization of (2.11) may be the restriction to two or one dimension or a simple expression for the force. We first consider the specialization to one dimension,

$$m\,\ddot{x}(t) = F(x(t),\, \dot{x}(t),\, t) \tag{2.12}$$

The reader should be familiar with the solution of the following special equations of motion:

No forces	$m\ddot{x} = 0$	(2.13)
Homogeneous gravity field:	$m\ddot{x} = -mg$	(2.14)
Gravity field and friction:	$m\ddot{x} = -mg - \gamma\dot{x}$	(2.15)
Free oscillation with friction:	$m\ddot{x} = -kx - 2m\lambda\dot{x}$	(2.16)
Enforced oscillation:	$\ddot{x} + \omega_0^2\, x + 2\lambda\dot{x} = f\,\cos(\omega t)$	(2.17)

The last case will be discussed in detail in Chapter 24.

We study the solution for a force that depends on the position only:

$$m\,\ddot{x}(t) = F(x(t)) \tag{2.18}$$

If we multiply both sides by $\dot{x}(t)$ the resulting equation can be written as

$$\frac{m}{2}\,\frac{d}{dt}\,\dot{x}(t)^2 = -\frac{d}{dt}\,U(x(t)) \tag{2.19}$$

with the *potential*

$$U(x) = -\int dx\; F(x) + \text{const.} \tag{2.20}$$

and $F(x) = -U'(x)$. The function $U(x)$ is also called the potential energy. The constant in (2.20) is of no importance for (2.19). The integration of (2.19) gives

$$\frac{m}{2}\,\dot{x}(t)^2 = E - U(x(t)) \tag{2.21}$$

The integration constant E is the sum of the potential energy U and of the kinetic energy $m\dot{x}^2/2$; these quantities will be discussed in more detail in the following chapter. We solve (2.21) for $\dot{x} = dx/dt$ and obtain

$$dt = \frac{dx}{\sqrt{2[E - U(x)]/m}} \tag{2.22}$$

The integration of this equation yields

$$t - t_0 = \int_{x_0}^{x} \frac{dx'}{\sqrt{2[E - U(x')]/m}} \tag{2.23}$$

The initial conditions for $x(t_0)$ and $\dot{x}(t_0)$ determine the two integration constants E and x_0. The solution of the integral (2.23) yields $t = t(x)$ and thus implicitly the wanted function $x = x(t)$.

In special cases, such as the oscillator with $U(x) = kx^2/2$, the integral (2.23) can be solved analytically. In any case, the solution can be discussed qualitatively on the basis of the graph of $U(x)$ in Figure 2.2: The vertical distance between $U(x)$ and the horizontal E gives $m\dot{x}^2/2$. If one still selects a direction of motion ($\dot{x} > 0$ or $\dot{x} < 0$), then the change of $\dot{x}^2$ can be read from the change of the vertical distance. Approaching an intersection point of $U(x)$ with the horizontal E implies $\dot{x} \to 0$. These points are called turning points

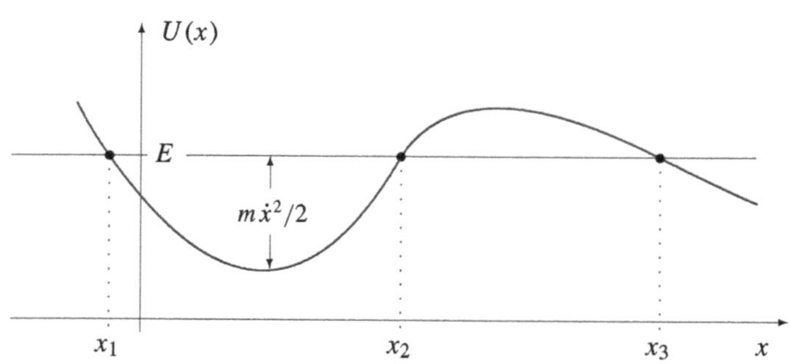

Figure 2.2 The one-dimensional motion in a potential $U(x)$ can be determined using the energy conservation, $E = m\dot{x}^2/2 + U(x) = \text{const}$. The distance between $U(x)$ and the horizontal at E equals the kinetic energy $m\dot{x}^2/2$. This allows a graphical discussion of possible motions.

because the direction of motion is reversed there. If the motion runs between two turning points ($x_1 \le x \le x_2$ in Figure 2.2) an oscillation with period T arises:

$$T = 2\int_{x_1}^{x_2} \frac{dx}{\sqrt{2[E - U(x)]/m}} \tag{2.24}$$

In contrast to this, the region $x_2 < x < x_3$ in Figure 2.2 is inaccessible.

We denote the position of the minimum of the potential by x_0 (here $dU/dx = 0$). For $E = U(x_0)$, this means that $x = x_0$ is a static solution. At the minimum of the potential, this solution will be stable. The corresponding static solution at the maximum of the potential would be unstable (because any small disturbance leads to a runaway of the particle).

The features occurring here (energy conservation leads to a first-order differential equation, graphical discussion of the solution) will be encountered later in more general cases.

Exercises

2.1 *Crashing Satellite*

An earth satellite (mass m) moves under the influence of the gravitational force and a frictional force:

$$\boldsymbol{F} = \boldsymbol{F}_{\text{grav}} + \boldsymbol{F}_{\text{diss}} = -m\,\frac{\alpha}{r^2}\,\boldsymbol{e}_r - m\,\gamma(r)\,\boldsymbol{v}$$

Here, r is the distance to the center of the earth, $\alpha = GM$ with the earth's mass M and $\gamma(r) > 0$. Set up the equations of motion in spherical coordinates r, θ, ϕ (using the result of Exercise 1.1). Determine how $\gamma(r)$, β and ϵ have to be chosen so that

$$r(t) = r_0\left(1 - \beta t\right)^{2/3}, \quad \theta(t) = -\frac{1}{\epsilon}\,\ln\left(1 - \beta t\right)^{2/3}, \quad \phi(t) = \text{const.} \tag{2.25}$$

solves the equations of motion. What is the shape of the trajectory curve? Determine the magnitude $|\boldsymbol{v}|$ of the velocity as a function of r.

2.2 *Raindrops in the gravitational field*

A spherical water droplet (radius R, volume V, mass m) falls in the atmosphere saturated with water vapor vertically downwards. It underlies the gravity and a friction force,

$$F = F_{\text{grav}} + F_{\text{diss}} = mg - \lambda R^2 v \quad (\lambda > 0)$$

The water droplet starts with velocity $v(0) = 0$. By condensation, the volume of the water drop grows in proportion to its surface; the radius and mass of the drop are thus time-dependent. Set up the equation of motion, and integrate it by using R instead of time t as the independent variable.

2.3 *Period of oscillation of an anharmonic oscillator*

Let a body of mass m move in the potential

$$U(x) = \frac{f}{2}x^2 + \alpha x^4$$

Calculate the period T of the oscillation for the slightly anharmonic case (for $\alpha E \ll f^2$, where E is the energy).

Instructions: Use the substitution $\sin^2\varphi = U(x)/E$ and express x and dx as a function of φ up to first order in α.

2.4 *Influence of the time definition on the equation of motion*

In an inertial frame the time T is measured by an inaccurate clock. For T there will be a certain relation $T = T(t)$ to the (true) inertial frame time t.

With this inaccurate clock, one measure the force-free one-dimensional motion of a body $d^2x/dT^2 = a_0 = F/m$ in contrast to Newton's laws. This demonstrates the dependence of physical laws on the definition of time.

Determine the apparent force F. For a concrete clock with a weakening spring, let $T(t) = \lambda^{-1}\ln(1 + \lambda t)$. What results then for the apparent force F?

Chapter 3

Conservation Laws

Conservation laws play an important role in physics; they facilitate or even make possible the solution of problems. Starting from Newton's second law, we investigate the conservation laws for momentum, angular momentum and energy. Basic concepts like angular momentum, work, kinetic and potential energy, conservative and dissipative forces are introduced.

Momentum Conservation

If no force acts on a particle, its *momentum* is conserved:

$$\boldsymbol{F} = 0 \quad \overset{(2.2)}{\longrightarrow} \quad \frac{d\boldsymbol{p}}{dt} = 0 \quad \longrightarrow \quad \boldsymbol{p} = \text{const.} \qquad \text{(conservation of momentum)} \tag{3.1}$$

Conservation of Angular Momentum

We multiply Newton's equation of motion (2.4) for the position $\boldsymbol{r}(t)$ of a mass point vectorially by $\boldsymbol{r}(t)$:

$$m\,\boldsymbol{r}(t) \times \ddot{\boldsymbol{r}}(t) = \boldsymbol{r}(t) \times \boldsymbol{F} \tag{3.2}$$

We define the *angular momentum* $\boldsymbol{\ell}$ of a particle by

$$\boldsymbol{\ell} = \boldsymbol{r}(t) \times \boldsymbol{p}(t) = \boldsymbol{r} \times (m\dot{\boldsymbol{r}}) \tag{3.3}$$

The right-hand side of (3.2) is the *torque* $\boldsymbol{M}$ acting on the particle:

$$\boldsymbol{M} = \boldsymbol{r} \times \boldsymbol{F} \tag{3.4}$$

Both quantities, the angular momentum and the torque, depend on the origin of the chosen inertial system. In case of a displacement of the inertial frame, $\boldsymbol{\ell}$ and $\boldsymbol{M}$ change (in contrast to this, momentum $\boldsymbol{p}$ and force $\boldsymbol{F}$ are unchanged).

Since $\dot{\boldsymbol{r}} \times \dot{\boldsymbol{r}} = 0$, $d\boldsymbol{\ell}/dt$ is equal to the left-hand side of (3.2). Thus, (3.2) becomes the equation of motion for the angular momentum:

$$\frac{d\boldsymbol{\ell}}{dt} = \boldsymbol{M} \tag{3.5}$$

Thus, the change in angular momentum over time is equal to the torque. This gives rise to the law of conservation of angular momentum: If the torque vanishes, then the angular momentum is conserved:

$$\boldsymbol{M} = 0 \quad \longrightarrow \quad \frac{d\boldsymbol{\ell}}{dt} = 0 \quad \longrightarrow \quad \boldsymbol{\ell} = \text{const.} \qquad \text{(conservation of angular momentum)} \tag{3.6}$$

Central force

For $\boldsymbol{F} \neq 0$, $\boldsymbol{M} = \boldsymbol{r} \times \boldsymbol{F}$ is only equal to zero if $\boldsymbol{F}$ is parallel to $\boldsymbol{r}$. The force must act in the direction to the center of the reference frame (or opposite to it). For a particle moving under the influence of such a *central force*, the angular momentum is conserved:

$$\boldsymbol{F} \parallel \pm\boldsymbol{r} \quad \longrightarrow \quad \boldsymbol{M} = 0 \quad \longrightarrow \quad \boldsymbol{\ell} = \text{const.} \qquad \text{(central force)} \tag{3.7}$$

Since $\boldsymbol{\ell} = \text{const.}$, we may put the z-axis of the inertial frame parallel to $\boldsymbol{\ell}$, i.e.,

$$\boldsymbol{\ell} = \ell\, \boldsymbol{e}_z = m\,(\boldsymbol{r} \times \dot{\boldsymbol{r}}) \tag{3.8}$$

Thus, $\boldsymbol{r}$ and $\dot{\boldsymbol{r}}$ lie in the x–y plane. For polar coordinates, we obtain with (1.11) and (1.12)

$$\frac{\ell}{m} = \rho^2\,\dot{\varphi} = 2\,\frac{dA}{dt} = \text{const.} \tag{3.9}$$

This is the so-called *law of equal areas* or Kepler's second law of planetary motion (Figure 3.1): During each time interval dt, the driving beam $\overline{\text{SP}}$ sweeps out the same area dA. For larger distances $\overline{\text{SP}}$, the angle change is slower; for shorter $\overline{\text{SP}}$, the angle variation is faster.

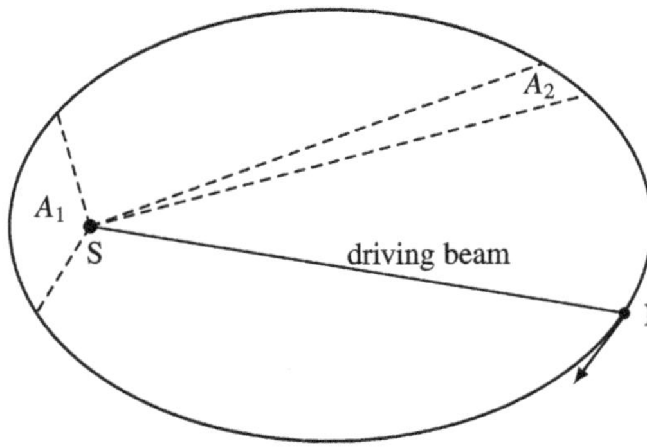

Figure 3.1 A planet P orbits a star S under the influence of the central gravitational force. Since $\ell = m\rho^2\,\dot{\varphi} =$ const., the smaller the $\dot{\varphi}$, the greater the ρ. At equal time intervals, the driving beam $\overline{\mathrm{SP}}$ sweeps out equal areas, for example, $A_1 = A_2$. This fact is called the equal area law or Kepler's second law of planetary motion (Kepler's laws are discussed in more detail in Chapter 17).

Energy Conservation

Work

Let a particle move under the influence of an external force from $\boldsymbol{r}$ to $\boldsymbol{r} + d\boldsymbol{r}$. Then we denote the scalar product

$$dW = \boldsymbol{F} \cdot d\boldsymbol{r} \tag{3.10}$$

as the *work* dW done by the force on the particle. Along a finite path C from $\boldsymbol{r}_1$ to $\boldsymbol{r}_2$, one obtains then the work

$$W = \int_C dW = \int_{\boldsymbol{r}_1, C}^{\boldsymbol{r}_2} d\boldsymbol{r} \cdot \boldsymbol{F} \tag{3.11}$$

Here, $d\boldsymbol{r}$ is the line element along the way C. The work W depends on the starting and ending points, but in general also on the path C between these points. The work W is equal to the energy transferred from the force field $\boldsymbol{F}(\boldsymbol{r})$ to the particle. As an example, one may imagine an electron accelerated in the electric field of a plate capacitor.

The work done per time is called *power* or power output P,

$$P = \frac{dW}{dt} = \frac{\boldsymbol{F} \cdot d\boldsymbol{r}}{dt} = \boldsymbol{F} \cdot \dot{\boldsymbol{r}} \tag{3.12}$$

The work is expressed in the energy units

$$\text{joule} = \text{J} = \text{N}\,\text{m} = \frac{\text{kg}\,\text{m}^2}{\text{s}^2} \tag{3.13}$$

The power output is measured in watt (W = J/s).

Kinetic and potential energy

We consider the scalar product of the Newtonian equation with $\dot{\boldsymbol{r}}$:

$$m\,\ddot{\boldsymbol{r}} \cdot \dot{\boldsymbol{r}} = \boldsymbol{F} \cdot \dot{\boldsymbol{r}} \tag{3.14}$$

We can express this in the form

$$\frac{d}{dt}\,\frac{m\,\dot{\boldsymbol{r}}^2}{2} = \boldsymbol{F} \cdot \dot{\boldsymbol{r}} = P \tag{3.15}$$

P is the energy per time transferred to the system (it may be positive or negative). The transferred energy changes the *kinetic energy*

$$T = \frac{m}{2}\,\dot{\boldsymbol{r}}^2 \tag{3.16}$$

of the particle, i.e., the energy associated with the motion. This form of the kinetic energy follows, as shown here, from Newton's second law. Note that the kinetic energy depends on the considered inertial system (because for a different inertial frame, the velocity will in general be different).

We now divide the force into a *conservative* and a *dissipative* part:

$$\boldsymbol{F} = \boldsymbol{F}_{\text{cons}} + \boldsymbol{F}_{\text{diss}} \tag{3.17}$$

This division is defined by the condition that $\boldsymbol{F}_{\text{cons}}$ contains all shares which may be written in the form

$$\boldsymbol{F}_{\text{cons}} \cdot \dot{\boldsymbol{r}} = -\frac{dU(\boldsymbol{r})}{dt} \tag{3.18}$$

We say that the force $\boldsymbol{F}_{\text{cons}}$ has a *potential* $U(\boldsymbol{r})$. As an example, consider (2.15), $m\ddot{x} = -mg - \gamma\dot{x}$. For this, $F_{\text{cons}} = -mg$ can be written as $-dU(x)/dx$ with the potential $U = mgx$. For $F_{\text{diss}} = -\gamma\dot{x}$, on the other hand, this is not possible because $F_{\text{diss}}\,\dot{x}$ is quadratic in $\dot{x}$, while $dU(x)/dt$ is necessarily linear in $\dot{x}$.

We substitute (3.17) and (3.18) into (3.15):

$$\frac{d}{dt}\left(\frac{m\,\dot{\boldsymbol{r}}^2}{2}+U(\boldsymbol{r})\right)=\boldsymbol{F}_{\text{diss}}\cdot\dot{\boldsymbol{r}} \tag{3.19}$$

This gives rise to the conservation law of energy:

$$\text{Forces conservative} \quad \longrightarrow \quad \frac{m\,\dot{\boldsymbol{r}}^2}{2}+U(\boldsymbol{r})=E=\text{const.} \tag{3.20}$$

Thus, in a conservative force field, the sum of kinetic energy and U is conserved. The function $U(\boldsymbol{r})$ introduced in (3.18) is an energy and is given the name *potential energy* or also *potential.* The conservation law (3.20) also justifies the designation *conservative* for the force (3.18), meaning "energy-conserving".

Dissipative forces lead to a transformation of mechanical energy $(T+U)$ of the considered particle into other forms of energy. This can mean dissipation in the narrower sense, i.e., conversion into heat energy; this happens, for example, through friction forces. Generally, time-dependent forces will cause an exchange of energy with the environment.

Potential

We investigate the determination of the potential from a given conservative force. For this purpose, we perform the time derivative of the potential $U(\boldsymbol{r})=U(x,y,z)$ in (3.18):

$$\boldsymbol{F}_{\text{cons}}\cdot\dot{\boldsymbol{r}}=-\frac{\partial U}{\partial x}\frac{dx}{dt}-\frac{\partial U}{\partial y}\frac{dy}{dt}-\frac{\partial U}{\partial z}\frac{dz}{dt}=-\dot{\boldsymbol{r}}\cdot\operatorname{grad}U \tag{3.21}$$

From this, it follows for $\boldsymbol{F}_{\text{cons}}$ the form

$$\boldsymbol{F}_{\text{cons}}=-\operatorname{grad}U(\boldsymbol{r})\,+\,\dot{\boldsymbol{r}}\times\boldsymbol{B}(\boldsymbol{r},t) \tag{3.22}$$

where $\boldsymbol{B}$ is an arbitrary vector field. The standard example for the second term is the force $(q/c)\,\dot{\boldsymbol{r}}\times\boldsymbol{B}$ on a charged particle in the magnetic field $\boldsymbol{B}(\boldsymbol{r},t)$. This force is conservative because it conserves energy; it satisfies the condition (3.18) with $U=\text{const.}$

In the following, we restrict ourselves to the case

$$\boldsymbol{F} = -\operatorname{grad} U(\boldsymbol{r}) \qquad \text{(conservative force)} \tag{3.23}$$

and do not write the index "cons" anymore. Necessary and sufficient condition for a given force $\boldsymbol{F}(\boldsymbol{r})$ to be written in this form is the disappearance of the curl[1] of $\boldsymbol{F}$,

$$\operatorname{curl} \boldsymbol{F}(\boldsymbol{r}) = 0 \quad \longleftrightarrow \quad \boldsymbol{F}(\boldsymbol{r}) = -\operatorname{grad} U(\boldsymbol{r}) \tag{3.24}$$

For a detailed exposition of the vector operations and the justification of (3.24), we refer to the first three chapters of the textbook *Elektrodynamik* [2]. For the conclusion direction $\leftarrow$, one calculates the rotation of the force $\boldsymbol{F} = -\operatorname{grad} U$,

$$\boldsymbol{e}_x \cdot \operatorname{curl} \boldsymbol{F}(\boldsymbol{r}) = \frac{\partial F_z}{\partial y} - \frac{\partial F_y}{\partial z} = -\frac{\partial^2 U}{\partial y\, \partial z} + \frac{\partial^2 U}{\partial z\, \partial y} = 0 \tag{3.25}$$

For the other conclusion direction $\rightarrow$, we consider two independent paths C_a and C_b, both leading from $\boldsymbol{r}_0$ to $\boldsymbol{r}$. Now, the integral difference $\int_{C_a} - \int_{C_b}$ over $\boldsymbol{F} \cdot d\boldsymbol{r}'$ is equal to the closed integral $\oint \boldsymbol{F} \cdot d\boldsymbol{r}'$ which, according to Stokes's theorem, is converted into the integral $\int d\boldsymbol{A} \cdot \operatorname{curl} \boldsymbol{F}$ over the enclosed area. Since this integral vanishes according to (3.24), it follows

$$W = \int_{\boldsymbol{r}_0,\, C_a}^{\boldsymbol{r}} d\boldsymbol{r}' \cdot \boldsymbol{F} = \int_{\boldsymbol{r}_0,\, C_b}^{\boldsymbol{r}} d\boldsymbol{r}' \cdot \boldsymbol{F} \tag{3.26}$$

Due to this path independence, the work W is a function of $\boldsymbol{r}$ only. The potential $U(\boldsymbol{r})$ can now be identified as the work $-W$ needed to move the particle against the force $\boldsymbol{F}$ to the point $\boldsymbol{r}$:

$$U(\boldsymbol{r}) - U(\boldsymbol{r}_0) = -W = -\int_{\boldsymbol{r}_0}^{\boldsymbol{r}} d\boldsymbol{r}' \cdot \boldsymbol{F}(\boldsymbol{r}') \tag{3.27}$$

[1]For Cartesian coordinates, we may define $\operatorname{curl} \boldsymbol{F} = \boldsymbol{\nabla} \times \boldsymbol{F}$ as the vector product with the so-called nabla operator $\boldsymbol{\nabla} = \boldsymbol{e}_x(\partial/\partial x) + \boldsymbol{e}_y(\partial/\partial y) + \boldsymbol{e}_z(\partial/\partial z)$. In European countries, one will often find the expression "rotation" instead of "curl" (where $\operatorname{rot} \boldsymbol{F} = \operatorname{curl} \boldsymbol{F}$ holds). The gradient introduced in (3.23) may also be expressed using the nabla operator, $\operatorname{grad} U(\boldsymbol{r}) = \boldsymbol{\nabla}\, U(\boldsymbol{r})$.

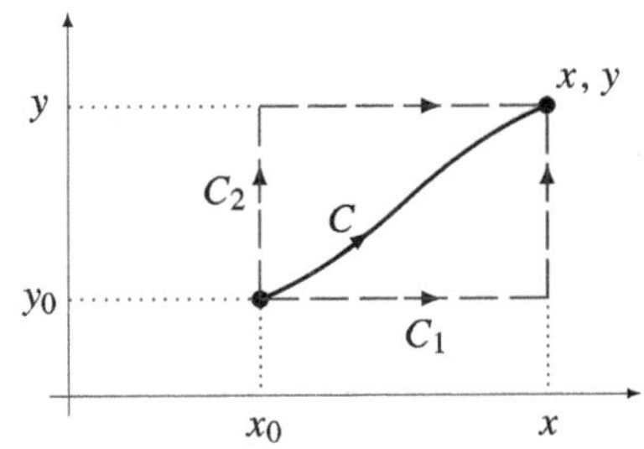

Figure 3.2 From a force of the form $\boldsymbol{F} = -\operatorname{grad} U(\boldsymbol{r})$, one obtains via the integral $-\int d\boldsymbol{r} \cdot \boldsymbol{F}$ the potential difference $U(x,y,z) - U(x_0,y_0,z_0)$. The result is independent of the path C. In the two-dimensional figure, two simple possible paths C_1 and C_2 are plotted.

For the potential determined in this way, $\operatorname{grad} U(\boldsymbol{r}) = -\boldsymbol{F}$ holds. The reference point $\boldsymbol{r}_0$ is arbitrary; the potential is determined only up to constant (here $U(\boldsymbol{r}_0)$).

For the calculation of the potential from a given force, we can now choose any path (Figure 3.2). For the path

$$C_1 : \; x_0\,,y_0\,,z_0 \;\stackrel{dy=dz=0}{\longrightarrow}\; x\,,y_0\,,z_0 \;\stackrel{dx=dz=0}{\longrightarrow}\; x,\, y,\, z_0 \;\stackrel{dx=dy=0}{\longrightarrow}\; x\, y\, z \tag{3.28}$$

we get

$$U(\boldsymbol{r}) - U(\boldsymbol{r}_0) = -\int_{x_0}^{x} dx'\; F_x(x', y_0, z_0) - \int_{y_0}^{y} dy'\; F_y(x, y', z_0) \\ - \int_{z_0}^{z} dz'\; F_z(x, y, z') \tag{3.29}$$

Hereby, the potential $U(\boldsymbol{r})$ is determined up to a constant. A constant in the potential is, however, irrelevant.

In this chapter, we have given the conservation laws in their simplest form; we will encounter them again in more general formulations (Chapters 9, 11, and 15). Newton's law for the trajectory $\boldsymbol{r}(t)$ is a second-order differential equation. In contrast to this, the conservation laws are first-order differential equations (easier to solve). Therefore, the conservation laws are also called *first integrals of motion*.

Exercises

3.1 *Forced oscillation*

Determine the general solution of the driven, damped Oscillator

$$\ddot{x} + \omega_0^2\, x + 2\lambda\dot{x} = f\,\cos(\omega t) \tag{3.30}$$

with $0 < \lambda < \omega_0$. Consider the solution specifically for large times, and calculate the time-averaged power loss P due to friction. For the case $\epsilon = \omega - \omega_0 \ll \omega_0$ and $\lambda \ll \omega_0$, determine P as a function of f, ϵ and λ.

3.2 *Work does not depend on path*

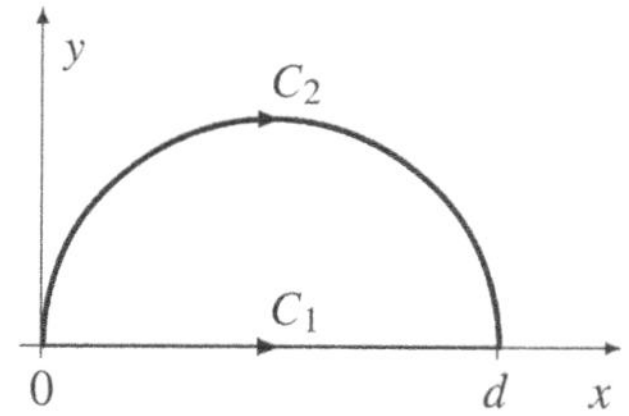

The path integral

$$W = \int_C dW = \int_{\boldsymbol{r}_1, C}^{\boldsymbol{r}_2} d\boldsymbol{r}\cdot\boldsymbol{F}(\boldsymbol{r})$$

is to be calculated for the spring force $\boldsymbol{F} = -k\boldsymbol{r}$ and for different paths.

Consider the two paths, straight line C_1 and a semicircle C_2 (radius $d/2$), both with the starting point $\boldsymbol{r}_1 := (x, y, z) = (0, 0, 0)$ and the end point $\boldsymbol{r}_2 := (d, 0, 0)$. Calculate the work A that must be done in order to move a particle from $\boldsymbol{r}_1$ to $\boldsymbol{r}_2$.

3.3 *Free fall with friction*

For a sphere moving in a viscous fluid in the gravitational field, the equation of motion reads

$$m\ddot{z} = -mg - \gamma\dot{z} \tag{3.31}$$

Solve this equation for a sphere that is initially at rest at $z = 0$. Use this solution to verify the energy-balance equation (3.19).

3.4 *Conveyor Belt — Energy Balance*

On a horizontal conveyor belt that moves with the constant velocity v_0, matter falls out of a hopper with the rate $R = dm/dt =$ mass/time. The matter comes to rest on the conveyor belt and is transported further with v_0. With which force F must the conveyor belt be driven in order to cause the change of momentum of the matter to be taken up? Compare the power P of the conveyor belt with the rate dT/dt with which kinetic energy T is transferred to the matter. Are the forces that occur conservative?

Chapter 4

System of Mass Points

We generalize the previous investigations to a system of N mass points. We determine the time derivative of the momentum, angular momentum, and energy of such a system.

Let the trajectory of the νth mass point be $\boldsymbol{r}_\nu(t)$, its mass m_ν and the force acting on it $\boldsymbol{F}_\nu$. For each of the N mass points, Newton's second law holds,

$$m_\nu\, \ddot{\boldsymbol{r}}_\nu(t) = \boldsymbol{F}_\nu \quad (\nu = 1, \ldots, N) \tag{4.1}$$

For the forces, we distinguish between *internal* and *external* forces. The internal forces are those that the mass points of the system exert on each other; for charged particles, these could be Coulomb forces. We restrict ourselves to two-body forces. This means that the forces exerted by m_μ on m_ν depend on the coordinates of m_ν and m_μ only (and not from the coordinates of other particles). The *external* forces $\boldsymbol{F}_\nu^{\rm ext}$ are those forces, which act from outside the system, for example, the gravitational forces on the individual mass points or the forces caused by an external electromagnetic field. With this division (4.1) becomes

$$m_\nu\, \ddot{\boldsymbol{r}}_\nu = \boldsymbol{F}_\nu = \boldsymbol{F}_\nu^{\rm ext} + \sum_{\mu=1,\, \mu\neq\nu}^{N} \boldsymbol{F}_{\nu\mu} \tag{4.2}$$

Momentum

The *center of mass* of the mass points is given by

$$\boldsymbol{R} = \frac{1}{M} \sum_{\nu=1}^{N} m_\nu \, \boldsymbol{r}_\nu \tag{4.3}$$

where

$$M = \sum_{\nu=1}^{N} m_\nu \tag{4.4}$$

is the total mass of the system. The equation of motion for the center of mass is obtained by summing (4.2) over ν:

$$\sum_{\nu=1}^{N} m_\nu \, \ddot{\boldsymbol{r}}_\nu = M\ddot{\boldsymbol{R}} = \sum_{\nu=1}^{N} \; \sum_{\mu=1,\,\mu\neq\nu}^{N} \boldsymbol{F}_{\nu\mu} + \sum_{\nu=1}^{N} \boldsymbol{F}_\nu^{\text{ext}} = \sum_{\nu=1}^{N} \boldsymbol{F}_\nu^{\text{ext}} \tag{4.5}$$

Due to Newton's third law (2.7), the internal forces drop out. Formally, this can be seen by renaming the indices, exchanging the order of summation, and using the third law:

$$\sum_{\nu,\mu,\nu\neq\mu} \boldsymbol{F}_{\nu\mu} \overset{(\nu\leftrightarrow\mu)}{=} \sum_{\mu,\nu,\mu\neq\nu} \boldsymbol{F}_{\mu\nu} = \sum_{\nu,\mu,\nu\neq\mu} \boldsymbol{F}_{\mu\nu} \overset{(3.\text{law})}{=} - \sum_{\nu,\mu,\nu\neq\mu} \boldsymbol{F}_{\nu\mu} \tag{4.6}$$

Since this sum is equal to the negative sum, it must vanish; this was used in (4.5). From (4.5) follows the *center-of-mass theorem*: The center of mass of a system moves as if the mass is unified and as if the sum of the external forces acts on it:

$$M\ddot{\boldsymbol{R}} = \sum_{\nu=1}^{N} \boldsymbol{F}_\nu^{\text{ext}} = \boldsymbol{F} \tag{4.7}$$

Thus, the internal forces have no effect on the motion of the center of mass. This statement can be formulated somewhat casually like this: You can't pull yourself out of the swamp by your own hair.

The center-of-mass theorem justifies in retrospect the idealization of real bodies as mass points: As far as the inner movements of a body (e.g. rotations, vibrations) are not of interest, we can treat a body as a mass point, for which Newton's second law is valid in the form (4.7).

A *closed system* is a system, which has no interactions with anything outside the system. For a closed system, there are no external forces, so $\boldsymbol{F}_\nu^{\text{ext}} = 0$. Thus, the center-of-mass momentum $\boldsymbol{P}$ for a closed system reads

$$\text{Closed system:} \quad \boldsymbol{P} = M\dot{\boldsymbol{R}} = \text{const.} \tag{4.8}$$

For the description of the motion of the mass points within the system, one will then preferably use the inertial frame in which the center of mass rests.

Angular Momentum

Analogous to the procedure in Chapter 3, we derive the equation for the time change of the angular momentum. We consider the vector product of (4.1) with $\boldsymbol{r}_\nu$ and sum over ν:

$$\sum_{\nu=1}^{N} \boldsymbol{r}_\nu \times m_\nu \ddot{\boldsymbol{r}}_\nu = \sum_{\nu=1}^{N} \boldsymbol{r}_\nu \times \boldsymbol{F}_\nu \tag{4.9}$$

This becomes

$$\frac{d}{dt} \sum_{\nu=1}^{N} m_\nu \left(\boldsymbol{r}_\nu \times \dot{\boldsymbol{r}}_\nu\right) = \sum_{\nu=1}^{N} \boldsymbol{r}_\nu \times \boldsymbol{F}_\nu^{\text{ext}} + \sum_{\nu=1}^{N} \boldsymbol{r}_\nu \times \sum_{\mu,\,\mu\neq\nu}^{N} \boldsymbol{F}_{\nu\mu} \tag{4.10}$$

On the left side we identify

$$\boldsymbol{L} = \sum_{\nu=1}^{N} m_\nu \left(\boldsymbol{r}_\nu \times \dot{\boldsymbol{r}}_\nu\right) = \sum_{\nu=1}^{N} \boldsymbol{\ell}_\nu \tag{4.11}$$

as the total angular momentum of the system. It depends, as well as the single angular momentum $\boldsymbol{\ell}_\nu$, on the origin of the inertial frame used. We consider the last term in (4.10):

$$\sum_{\nu,\,\mu,\,\mu\neq\nu}^{N} \boldsymbol{r}_\nu \times \boldsymbol{F}_{\nu\mu} \overset{(\nu\leftrightarrow\mu)}{=} \sum_{\nu,\,\mu,\,\mu\neq\nu}^{N} \boldsymbol{r}_\mu \times \boldsymbol{F}_{\mu\nu} \overset{\text{(third law)}}{=} -\sum_{\nu,\,\mu,\,\mu\neq\nu}^{N} \boldsymbol{r}_\mu \times \boldsymbol{F}_{\nu\mu} \tag{4.12}$$

With the help of the first supplement we get

$$\sum_{\nu,\mu,\nu\neq\mu}^{N} \boldsymbol{r}_\nu \times \boldsymbol{F}_{\nu\mu} \overset{(4.12)}{=} \frac{1}{2} \sum_{\nu,\mu,\nu\neq\mu}^{N} (\boldsymbol{r}_\nu - \boldsymbol{r}_\mu) \times \boldsymbol{F}_{\nu\mu} \overset{(2.8)}{=} 0 \qquad (4.13)$$

Thus, the internal forces do not yield a resultant torque; they cannot change the total angular momentum of the system.

With (4.13), (4.10) becomes

$$\frac{d\boldsymbol{L}}{dt} = \sum_{\nu=1}^{N} \boldsymbol{r}_\nu \times \boldsymbol{F}_\nu^{\text{ext}} = \boldsymbol{M} \qquad (4.14)$$

Thus, the time derivative of the total angular momentum is equal to the total torque of the external forces. This equation will be applied to the rigid body in Part V.

For vanishing total torque the total angular momentum is conserved. In particular,

$$\text{Closed system:} \quad \boldsymbol{L} = \text{const.} \qquad (4.15)$$

In deriving this statement, we have used the first supplement (2.8). The condition (2.8) does not apply, for example, to the magnetic forces between moving charged particles. In the closed system, however, the total angular momentum is preserved even in this case. The conservation law follows in fact from a rather general condition (isotropy of space). In a system of moving charged particles, the electromagnetic field necessarily contains angular momentum, too. The conservation of angular momentum is valid then in the form $\boldsymbol{L} = \boldsymbol{L}_{\text{mass}} + \boldsymbol{L}_{\text{field}} = \text{const.}$, where $\boldsymbol{L}_{\text{mass}}$ is the angular momentum considered in (4.11), and $\boldsymbol{L}_{\text{field}}$ is the angular momentum of the electromagnetic field.

Energy

Analogous to the procedure in Chapter 3, we derive the equation for the time change of the energy. We consider the scalar product of (4.2) with $\dot{\boldsymbol{r}}_\nu$ and sum over ν:

$$\sum_{\nu=1}^{N} m_\nu\, \ddot{\boldsymbol{r}}_\nu \cdot \dot{\boldsymbol{r}}_\nu = \sum_{\nu=1}^{N} \boldsymbol{F}_\nu \cdot \dot{\boldsymbol{r}}_\nu \tag{4.16}$$

The left side is the time derivative of the kinetic energy

$$T = \sum_{\nu=1}^{N} \frac{m_\nu}{2}\, \dot{\boldsymbol{r}}_\nu^{\,2} \tag{4.17}$$

We again divide the forces into conservative and dissipative forces,

$$\boldsymbol{F}_\nu = \boldsymbol{F}_{\nu,\,\text{cons}} + \boldsymbol{F}_{\nu,\,\text{diss}} \tag{4.18}$$

Here, $\boldsymbol{F}_{\nu,\,\text{cons}}$ is the part for which a potential U exists, so that

$$\sum_{\nu=1}^{N} \boldsymbol{F}_{\nu,\,\text{cons}} \cdot \dot{\boldsymbol{r}}_\nu = -\,\frac{dU(\boldsymbol{r}_1, \boldsymbol{r}_2, \ldots, \boldsymbol{r}_N)}{dt} = -\sum_{\nu=1}^{N} \frac{\partial U}{\partial \boldsymbol{r}_\nu} \cdot \dot{\boldsymbol{r}}_\nu \tag{4.19}$$

The partial derivative with respect to the vector $\boldsymbol{r}_\nu := (x_\nu, y_\nu, z_\nu)$ means

$$\frac{\partial U(\boldsymbol{r}_1, \ldots, \boldsymbol{r}_N)}{\partial \boldsymbol{r}_\nu} = \frac{\partial U}{\partial x_\nu}\, \boldsymbol{e}_x + \frac{\partial U}{\partial y_\nu}\, \boldsymbol{e}_y + \frac{\partial U}{\partial z_\nu}\, \boldsymbol{e}_z \tag{4.20}$$

For a function $U(\boldsymbol{r})$ that depends only on one vector $\boldsymbol{r}$ this becomes $\operatorname{grad} U$. We restrict ourselves to the case that (4.19) can be written as

$$\boldsymbol{F}_{\nu,\,\text{cons}} = -\,\frac{\partial U(\boldsymbol{r}_1, \boldsymbol{r}_2, \ldots, \boldsymbol{r}_N)}{\partial \boldsymbol{r}_\nu} \tag{4.21}$$

Then (4.16) becomes

$$\frac{d}{dt}\,(T + U) = \sum_{\nu=1}^{N} \boldsymbol{F}_{\nu,\,\text{diss}} \cdot \dot{\boldsymbol{r}}_\nu \tag{4.22}$$

In the absence of dissipative forces, the energy theorem holds:

$$\text{Conservative forces:} \quad T + U = E = \text{ const.} \tag{4.23}$$

Similarly as in (4.8) and (4.15), one might expect the statement, that the energy is conserved in the closed system. But this is not valid for $T + U$ if there are dissipative internal forces. The energy theorem in the form

$$\text{Closed system:} \quad E = T + U + \text{other forms of energy} = \text{const.} \tag{4.24}$$

requires that *all* energy contributions are taken into account (like, for example, the heat energy that is generated by the friction forces).

Determination of Potentials

We discuss the determination of the potential U in (4.21) from given conservative forces; in the following, we omit the Index "cons", i.e. $\boldsymbol{F}_{\nu,\,\text{cons}} = \boldsymbol{F}_\nu$. For the assumed division into internal and external forces, (4.21) becomes

$$\begin{aligned} \boldsymbol{F}_\nu &= \boldsymbol{F}_\nu^{\text{ext}}(\boldsymbol{r}_\nu) + \sum_{\mu,\,\mu\neq\nu} \boldsymbol{F}_{\nu\mu}(\boldsymbol{r}_\nu, \boldsymbol{r}_\mu) \\ &= -\frac{\partial U^{\text{ext}}(\boldsymbol{r}_1, \boldsymbol{r}_2, \ldots, \boldsymbol{r}_N)}{\partial \boldsymbol{r}_\nu} - \frac{\partial U^{\text{int}}(\boldsymbol{r}_1, \boldsymbol{r}_2, \ldots, \boldsymbol{r}_N)}{\partial \boldsymbol{r}_\nu} \end{aligned} \tag{4.25}$$

The potential U^{ext} yields the external forces, and the potential U^{int} the inner ones. The forces between the particles are two-body forces that depend only on the positions of the two bodies, and not on the position of other (third) bodies. Then the potentials can be written in the following form:

$$U^{\text{ext}} = \sum_{\nu=1}^{N} U_\nu(\boldsymbol{r}_\nu) \quad \text{and} \quad U^{\text{int}} = \sum_{\nu=2}^{N} \sum_{\mu=1}^{\nu-1} U_{\nu\mu}(\boldsymbol{r}_\nu, \boldsymbol{r}_\mu) \tag{4.26}$$

From

$$\boldsymbol{F}_\nu^{\text{ext}}(\boldsymbol{r}) = -\operatorname{grad} U_\nu(\boldsymbol{r}) \tag{4.27}$$

we obtain the desired relation $\boldsymbol{F}_\nu^{\text{ext}}(\boldsymbol{r}_\nu) = -\partial U^{\text{ext}}/\partial \boldsymbol{r}_\nu$. The relation (4.27) is of the form (3.23); the potential $U_\nu(\boldsymbol{r})$ can therefore be derived form the force $\boldsymbol{F}_\nu^{\text{ext}}$ (as in (3.19)).

For the internal forces $\boldsymbol{F}_{\nu\mu}$ the third law requires

$$\boldsymbol{F}_{\nu\mu}(\boldsymbol{r}_\nu, \boldsymbol{r}_\mu) = -\boldsymbol{F}_{\mu\nu}(\boldsymbol{r}_\nu, \boldsymbol{r}_\mu) \tag{4.28}$$

or

$$-\frac{\partial U_{\nu\mu}(\boldsymbol{r}_\nu, \boldsymbol{r}_\mu)}{\partial \boldsymbol{r}_\nu} = \frac{\partial U_{\nu\mu}(\boldsymbol{r}_\nu, \boldsymbol{r}_\mu)}{\partial \boldsymbol{r}_\mu} \tag{4.29}$$

Therefore, the potential $U_{\nu\mu}$ can depend on the difference $\boldsymbol{r}_\nu - \boldsymbol{r}_\mu$ only. Furthermore, it can depend only on scalar quantities, i.e. on $\boldsymbol{a} \cdot (\boldsymbol{r}_\nu - \boldsymbol{r}_\mu)$ (with a given vector $\boldsymbol{a}$) or from $(\boldsymbol{r}_\nu - \boldsymbol{r}_\mu)^2$. Dependence of $\boldsymbol{a} \cdot (\boldsymbol{r}_\nu - \boldsymbol{r}_\mu)$ would give forces $F_{\nu\mu}$ which are parallel to the vector $\pm\boldsymbol{a}$. Such forces are excluded by the first supplement (2.8). Thus,

$$U_{\nu\mu} = U_{\nu\mu}(|\boldsymbol{r}_\nu - \boldsymbol{r}_\mu|) \tag{4.30}$$

For this, $\boldsymbol{F}_{\nu\mu}$ is parallel to $\pm(\boldsymbol{r}_\nu - \boldsymbol{r}_\mu)$ as in Figure 2.1 on the right. We now consider a particular index pair and use $\boldsymbol{r} = \boldsymbol{r}_\nu - \boldsymbol{r}_\mu$. Then

$$\boldsymbol{F}_{\nu\mu}(\boldsymbol{r}) = -\operatorname{grad} U_{\nu\mu}(|\boldsymbol{r}|) = -\operatorname{grad} U_{\nu\mu}(r) \tag{4.31}$$

and

$$\boldsymbol{F}_{\nu\mu}(\boldsymbol{r}) = F_{\nu\mu}(r)\, \boldsymbol{e}_r \tag{4.32}$$

From

$$F_{\nu\mu}(r) = -\frac{dU_{\nu\mu}(r)}{dr} \tag{4.33}$$

the potential $U_{\nu\mu}(r)$ can be determined by simple integration. The total potential energy is then

$$U(\boldsymbol{r}_1, \boldsymbol{r}_2, \ldots, \boldsymbol{r}_N) = \sum_{\nu=1}^{N} U_\nu(\boldsymbol{r}_\nu) + \sum_{\nu=2}^{N} \sum_{\mu=1}^{\nu-1} U_{\nu\mu}(|\boldsymbol{r}_\nu - \boldsymbol{r}_\mu|) \tag{4.34}$$

Since the contribution $U_{\nu\mu}$ includes both $\boldsymbol{F}_{\nu\mu}$ and $\boldsymbol{F}_{\mu\nu}$, each pair of indexes $(\nu\mu)$ occurs only once; there is no term with $\nu = \mu$. Since $U_{\nu\mu} = U_{\mu\nu}$, one could also use an unrestricted double sum in (4.34) if one adds a factor $1/2$ and sets $U_{\nu\nu} = 0$.

As an example, we consider particles with the charges q_ν in the homogeneous gravitational field at the surface of the earth (acceleration due to gravity $\boldsymbol{g} = -g\,\boldsymbol{e}_z$):

$$U(\boldsymbol{r}_1, \ldots, \boldsymbol{r}_N) = \sum_{\nu=1}^{N} m_\nu\, g\, z_\nu + \sum_{\nu=2}^{N} \sum_{\mu=1}^{\nu-1} \frac{q_\nu\, q_\mu}{|\boldsymbol{r}_\nu - \boldsymbol{r}_\mu|} \tag{4.35}$$

Alternative to the gravitational field, one could consider an electrostatic field with the potential $\Phi(\boldsymbol{r})$:

$$U(\boldsymbol{r}_1, \ldots, \boldsymbol{r}_N) = \sum_{\nu=1}^{N} q_\nu\, \Phi(\boldsymbol{r}_\nu) + \sum_{\nu=2}^{N} \sum_{\mu=1}^{\nu-1} \frac{q_\nu\, q_\mu}{|\boldsymbol{r}_\nu - \boldsymbol{r}_\mu|} \tag{4.36}$$

This does not take into account magnetic forces which can be a good approximation since magnetic forces are of size $v_\nu v_\mu / c^2$ relative to the Coulomb forces (v_ν and v_μ denote the velocities of the particles, and c is the velocity of light).

Exercises

4.1 *Potential for Coulomb force*

We consider particles with charges q_ν. The force that the particle μ exerts on the particle ν is

$$\boldsymbol{F}_{\nu\mu} = \frac{q_\nu\, q_\mu\, (\boldsymbol{r}_\nu - \boldsymbol{r}_\mu)}{|\boldsymbol{r}_\nu - \boldsymbol{r}_\mu|^3}$$

Show $\operatorname{curl}_\nu \boldsymbol{F}_{\nu\mu} = 0$. Determine the corresponding potential contribution $U_{\nu\mu}$.

Chapter 5

Inertial Frames

Newton's first law introduces inertial frames as the preferred reference frames; the validity of the central second law is explicitly restricted to inertial reference frames. We discuss the special role of inertial frames and the equivalence of different inertial frames. The Galilei transformations lead from one inertial system to another inertial system. In the following chapter, we investigate the modifications that arise for non-inertial frames.

Reference frames can be realized by the laboratory of the experimenter. For example, we may consider a laboratory room on earth, a satellite laboratory, or a carousel. To designate points of space in the reference frame, one introduces coordinates. Also, a suitable time coordinate t is defined, for example, by a specific clock. A reference frame with certain spatial and time coordinates is called a coordinate system. Inertial frames with such coordinates are called *inertial systems* in the following, and abbreviated as IS.

Newton's laws are valid in the reference frames that are at rest with respect to the fixed-star heaven or which move relative to it with constant velocity. These reference frames are called inertial frames. For these frames, Galilei stated the *relativity principle*: *All inertial frames are equivalent*. This means that physical processes in all inertial frames occur in the same way. Formally, this means that the basic laws have the same form in all inertial frames. In non-inertial frames, the fundamental laws have a different form; the equations of motion for mass points are modified by inertial forces (Chapter 6).

For the connection between the inertial frames and the fixed-star sky, we consider the following experiment: In the evening, one looks at the stars with loosely relaxed arms. Then one turns around one's axis, thus performing a pirouette. This leads to the following experimental findings:

1. There is a preferred reference frame, in which the arms hang loosely downwards and one does not get dizzy. This reference frame is thus experimentally distinguished as compared to a rotating frame.
2. In the preferred reference frame, the stars are at rest.

From the second observation, the question arises whether the stars are *accidentally* resting in the preferred frame, or whether the stars might determine this preferred frame (i.e. the inertial frame). The assertion that the mass distribution in the universe determines the inertial frames is called *Mach principle*. This relation can be investigated in the framework of the general theory of relativity.

Galilean Transformation

Newton's laws are relativistic laws in the sense of Galilei; they have the same form in all inertial frames. From this, we may determine the transformations between different inertial systems.

We consider an inertial frame IS with Cartesian coordinates x_1, x_2, x_3 and the time coordinate t. A point in this IS we denote as *event*:

$$\text{Event:} \quad (x_1,\, x_2,\, x_3,\, t) \tag{5.1}$$

Specifically, an event can mean that at a certain time and at a certain position something happens, for example, the collision of two point-like particles. Coordinate systems have the purpose to give names to various events; the name consists of the specific coordinate values.

In another IS′, the same event has other coordinate values x'_1, x'_2, x'_3, and t'. Since it is the same event, the old and the new coordinates must be related to one another. This relation is the Galilean transformation we are looking for.

In Newtonian mechanics, the time indicated by a clock is assumed to be independent of its motion. This is confirmed by observation — but only for velocities that are small compared to the speed of light. Then one can use the same (absolute) time in any reference frame, so that in particular $t' = t$ holds for IS and IS$'$. Allowing different time zero points, this becomes

$$t' = t - t_0 \tag{5.2}$$

A constant displacement by t_0 is without influence on Newton's laws (because of $dt' = dt$).

A trajectory $x_1(t)$, $x_2(t)$, $x_3(t)$ is a sequence of events; at every time t the considered mass point is at a certain position (x_1, x_2, x_3). By the transition from IS to IS$'$, the event (x_i, t) becomes (x_i', t') and correspondingly the trajectory $x_i(t)$ gets $x_i'(t')$; here, i stands for 1, 2, and 3. We start the discussion with the first law,

$$m\,\ddot{x}_i(t) = 0 \quad \text{in IS} \tag{5.3}$$

We consider now a reference frame S$'$ (Figure 5.1) whose origin lies at $\boldsymbol{d}(t) = \sum d_i(t)\,\boldsymbol{e}_i$. Let the axes of IS and S$'$ be parallel; both systems have the same basis vectors $\boldsymbol{e}_i$. Whether S$'$ is an inertial frame depends on $\boldsymbol{d}(t)$; we leave this open for now. We denote the

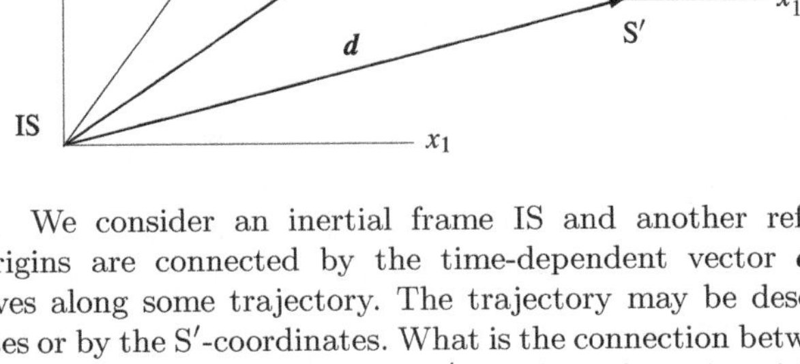

Figure 5.1 We consider an inertial frame IS and another reference frame S$'$ whose origins are connected by the time-dependent vector $\boldsymbol{d}(t)$. A mass point P moves along some trajectory. The trajectory may be described by the IS-coordinates or by the S$'$-coordinates. What is the connection between these two coordinate sets? What is the condition for S$'$ in order to be an inertial system, too?

position vector of the mass point P by $\boldsymbol{r} = \sum x_i \, \boldsymbol{e}_i$ in IS and by $\boldsymbol{r}' = \sum x_i' \, \boldsymbol{e}_i$ in S′. Then

$$\boldsymbol{r}(t) = \boldsymbol{r}'(t) + \boldsymbol{d}(t) \quad \text{or} \quad x_i(t) = x_i'(t) + d_i(t) \tag{5.4}$$

We substitute this into the valid equation (5.3):

$$m\,\ddot{x}_i'(t) = -m\,\ddot{d}_i(t) \quad \text{in S}' \tag{5.5}$$

Now, S′ is an IS′ if (5.5) again takes the form of the first Newton law, i.e.

$$m\,\ddot{x}_i'(t') = 0 \quad \text{in IS}' \tag{5.6}$$

Since $dt = dt'$, we could replace t' with t. For (5.5) and (5.6) to be of the same form, we must have

$$\ddot{d}_i(t) = 0, \quad \text{so } d_i(t) = v_i\,t + a_i \tag{5.7}$$

This means that IS′ moves with respect to IS with *constant* velocity $\boldsymbol{v} = \sum v_i \, \boldsymbol{e}_i$; furthermore IS′ can be shifted by a constant vector $\boldsymbol{a} = \sum a_i \, \boldsymbol{e}_i$. From (5.2), (5.4), and (5.7), it follows the transformation between IS and IS′:

$$x_i' = x_i - v_i t - a_i, \quad t' = t - t_0 \tag{5.8}$$

This is a *Galilean transformation.* The application of such a transformation to (5.3) leads to the same form (5.6); the law of motion is *form invariant* or *covariant* under Galilean transformations. Galileo's *principle of relativity* postulates this covariance for all fundamental laws.

We assume now that $\boldsymbol{d}(t)$ is restricted by (5.7), i.e. S′ in Figure 5.1 is indeed an inertial frame (replace S′ by IS′). In Figure 5.1, the axes of S′ or IS′ could be rotated with respect to IS. A time-dependent rotation between the coordinate axes of IS and S′ would however lead to additional terms (Chapter 6); that means S′ would no longer be an inertial reference frame. Therefore, only a relative rotation of the coordinate axes by *constant* angles is admitted. Such a time-independent rotation is obtained by an orthogonal transformation $x_i' = \sum_j \alpha_{ij}\, x_j$ (with $\alpha\alpha^{\mathrm{T}} = 1$). Multiplying $m\ddot{x}_j = 0$ from the left by α_{ij} and summing over j, we get $m\ddot{x}_i' = 0$. The form of the first law is preserved; a rotated (constant angles) inertial frame is again an inertial frame.

The general Galilean transformation is obtained from (5.8) if we allow in addition this kind of rotation:

General Galilean transformation:

$$x'_i = \sum_{j=1}^{3} \alpha_{ij}\, x_j - v_i\, t - a_i, \quad t' = t - t_0 \tag{5.9}$$

This general Galilean transformation consists of the following transformations:

1. Rotation described by α_{ij}.
2. Spatial displacement by $v_i\, t$.
3. Spatial displacement by a_i.
4. Temporal displacement by t_0.

The Galilean transformation thus depends on 10 parameters: The constant rotation is determined by three angles (e.g. by the Euler angles, see Part V) plus the seven quantities v_i, a_i and t_0. The v_i are parameters of the transformation (not related to the velocity of the particle).

Two successive Galilean transformations from IS to IS′ and from IS′ to IS″ lead again to an inertial frame IS″, they define a new Galilei transformation (from IS to IS″). To every transformation (from IS to IS′) there exits an inverse transformation (from IS′ to IS). The trivial identity transformation from IS to IS is also a Galilean transformation. The Galilean transformations thus form a group, which is called *Galilei group*.

In the discussion of the Galilean transformation, the emphasis usually lies on the relative motion between IS and IS′. Therefore, one often considers the special Galilei transformation that depends on one parameter only:

Special Galilean transformation:

$$x' = x - vt, \quad y' = y, \quad z' = z, \quad t' = t \tag{5.10}$$

We consider now the Galilei transformation for the velocities. From (5.9), it follows that

$$dx'_i = \sum_{j=1}^{3} \alpha_{ij}\, dx_j - v_i\, dt, \quad dt' = dt \tag{5.11}$$

From this, we get

$$\dot{x}'_i(t') = \frac{dx'_i}{dt'} = \frac{dx'_i}{dt} = \sum_{j=1}^{3} \alpha_{ij}\, \dot{x}_j(t) - v_i \tag{5.12}$$

By further differentiation, we obtain

$$\ddot{x}'_j(t') = \sum_{i=1}^{3} \alpha_{ij}\, \ddot{x}_i(t) \tag{5.13}$$

Hereby, one can check again that the first law is covariant under the general Galilean transformation:

$$m\, \ddot{x}_i(t) = 0 \quad \overset{(5.13)}{\longleftrightarrow} \quad m\, \ddot{x}'_j(t') = 0 \tag{5.14}$$

Second Law

We consider the behavior of the second law under a Galilean transformation. With the notation $\boldsymbol{F} = \sum_i F_i \boldsymbol{e}_i$, we assume (as already before) that the force is a vector. This means that the components F_i transform during a rotation like the x_i, thus $F'_j = \sum_j \alpha_{ji} F_i$. Therefore, we multiply both sides of the second law $m\ddot{x}_i = F_i$ by α_{ji} and sum over i. Thus, we obtain

$$m\, \ddot{x}_i(t) = F_i(x, \dot{x}, t) \quad \text{transforms to} \quad m\, \ddot{x}'_j(t') = F'_j(x', \dot{x}', t') \tag{5.15}$$

At the same time, we perform the Galilean transformation in the arguments of the force; we restrict ourselves to the arguments given in (2.11). The undashed arguments in $F_i(x, \dot{x}, t)$ have to be replaced by the dashed ones according to (5.9). Here, x stands for (x_1, x_2, x_3) and $\dot{x}$ for $(\dot{x}_1, \dot{x}_2, \dot{x}_3)$.

The dash at F'_j now implies two points: On the one hand, the components of this force refer to the rotated coordinate axes. On the other hand $F_i(\ldots)$ and $F'_i(\ldots)$ are in general *different* functions of their arguments (because of the transformation of these arguments). As a simple example, we subject the function $f(x) = x$ to the special Galilean transformation (5.10). Then $f(x) = x = x' + v\,t = x' + v\,t' = g(x', t')$. Obviously, $f(x)$ is a different function of the arguments than $g(x', t')$, even if $f = g$ holds. This difference is indicated in (5.15) by the (inconspicuous) dash; we do not introduce a new letter.

After the Galilean transformation, the Newtonian laws have the same form (left and right part of (5.15)); they are form invariant or *covariant*. Under certain circumstances, the forces F_i and F_i' depend on their arguments in the same way. Then Newton's laws are *invariant* under Galilean transformations. We explain this distinction by means of examples.

First, we perform a Galilean transformation for the common friction force:

$$\boldsymbol{F}(\dot{\boldsymbol{r}}) = -\gamma\,\dot{\boldsymbol{r}} = -\gamma\,(\dot{\boldsymbol{r}}' + \boldsymbol{v}) = \boldsymbol{F}'(\dot{\boldsymbol{r}}') \tag{5.16}$$

The new force is a different function of the argument because $\boldsymbol{F}'(\dot{\boldsymbol{r}}') \neq -\gamma\,\dot{\boldsymbol{r}}'$. Since $\boldsymbol{F}$ and $\boldsymbol{F}'$ are different functions, the law of motion is not invariant under Galilean transformations. In this case, the two inertial frames are not equivalent. There is rather a preferred inertial frame, namely the one in which the medium causing the friction is at rest (the IS where $\boldsymbol{F} = -\gamma\,\dot{\boldsymbol{r}}$).

Other examples of non-invariant forces are forces that depend on the position or/and the time; shifts in position or time will alter these forces. An external force may also have a preferred direction (for example, the gravitational field at the earth's surface). In this case, the invariance under rotations is obviously violated.

Now, we come to an example with an invariant force. A planet (mass m_1) under the influence of the sun (mass m_2) obeys the equation of motion

$$m_1\,\ddot{\boldsymbol{r}}_1 = -G\,m_1\,m_2\,\frac{\boldsymbol{r}_1 - \boldsymbol{r}_2}{|\boldsymbol{r}_1 - \boldsymbol{r}_2|^3} \tag{5.17}$$

A Galilean transformation (where (5.9) applies to $\boldsymbol{r}_1$ as well as to $\boldsymbol{r}_2$) yields

$$m_1\,\ddot{\boldsymbol{r}}_1' = -G\,m_1\,m_2\,\frac{\boldsymbol{r}_1' - \boldsymbol{r}_2'}{|\boldsymbol{r}_1' - \boldsymbol{r}_2'|^3} \tag{5.18}$$

The spatial displacements cancel out the differences on the right side. The identical forms (5.17) and (5.18) imply that we get the same kind of solutions in IS and IS′.

Note that in all considered cases, we have $\boldsymbol{F} = \boldsymbol{F}'$. This follows simply from the fact that we apply a Galilei transformation to the equation $m\ddot{\boldsymbol{r}} = \boldsymbol{F}$, and that for this transformation $\ddot{\boldsymbol{r}} = \ddot{\boldsymbol{r}}'$ holds. Since the forces depend on the same argument types, namely $\boldsymbol{F}(\boldsymbol{r}, \dot{\boldsymbol{r}}, t)$ and $\boldsymbol{F}'(\boldsymbol{r}', \dot{\boldsymbol{r}}', t')$, the equations of motion are always covariant (of the same form). However, they are invariant only if $\boldsymbol{F}$ and $\boldsymbol{F}'$ depend in the same way on their arguments as in (5.17) and (5.18)

Passive and Active Transformation

The Galilean transformation can connect (i) two inertial frames or (ii) two physical systems with each other:

(i) *Passive transformation*: We observe *one* physical system from *two* different inertial frames, IS and IS′. The processes in the system are described by the observers in IS and IS′ by laws of the same form (*covariance*, synonymous to form invariance). Thus, both observers apply Newton's laws; thereby, the Galilei transformation links the coordinates that a specific event has in IS and IS′.

(ii) *Active transformation*: Within *one* reference frame IS we consider *two* physical systems which are connected by a Galilean transformation. Also, here the processes in both systems are described by laws of the same form (covariance). The processes run in the two systems are, however, in general, different because the forces change due to the transformation.

Especially for *closed* systems the laws of motion are not only covariant, but they are also *invariant* under Galilean transformations (next section). In this case, the processes run in both systems in the same way.

The insertion of a Galilei transformation into Newton's laws is a formal step defined by the transformation equations (5.9). This step can have two quite different meanings: On the one hand, two observers (each resting in IS or IS′) can observe the same physical system. Then it is clear that they are describing the same physical process, they only use different coordinates (the respective standard coordinates of IS and IS′). On the other hand, an IS-observer can observe two different physical systems namely the one given at first, and then another one which results from an (active) Galilei transformation (e.g. by an actual boost, rotation, or displacement of the system). In general, the physical processes in these two systems will be different.

We have introduced the Galilei transformations under the first point of view (passive transformation). In the following, we discuss the second point of view (active transformation) in some detail. We mark an active transformation by an asterisk as index:

$$x_i \to x_i^* = \sum_{k=1}^{3} \alpha_{ik}\, x_k - v_i\, t - a_i, \qquad t \to t^* = t - t_0 \tag{5.19}$$

Here, x_i, t and x_i^*, t^* are coordinates in the *same* IS. They may denote the position of a mass point and of a second-shifted one. We now put the transformation (5.10) into Newton's second law. Then we obtain, similarly as in (5.15), the laws of motion for the two mass points:

$$m\, \ddot{x}_i(t) = F_i(x, \dot{x}, t) \quad \text{transformed to} \quad m\, \ddot{x}_i^*(t^*) = F_i^*(x^*, \dot{x}^*, t^*) \tag{5.20}$$

Only if the forces $F_i(x, \dot{x}, t)$ and $F_i^*(x^*, \dot{x}^*, t^*)$ depend in the same way on their arguments, the motion proceeds in the same way. As the discussion in the last section shows, this is a non-trivial condition. It is, for example, not satisfied for frictional forces, and in general also not for external forces. On the other hand, the condition holds for the gravitational forces, (5.17) and (5.18).

Invariance of Newton's Laws

A system with no interaction with the environment is called a *closed system*. For a closed system of mass points (Chapter 4), the external forces vanish. The internal forces are assumed to be conservative. According to (4.30), they can be represented by potentials:

$$\boldsymbol{F}_\nu = -\sum_{\mu,\,\mu\neq\nu}^{N} \frac{\partial U_{\nu\mu}(|\boldsymbol{r}_\nu - \boldsymbol{r}_\mu|)}{\partial \boldsymbol{r}_\nu} \qquad (\nu = 1, 2, \ldots, N) \tag{5.21}$$

The distance $|\boldsymbol{r}_\nu - \boldsymbol{r}_\mu|$ is invariant under the displacements and rotations of a Galilean transformation. Since

$$U(|\boldsymbol{r}^*_\nu - \boldsymbol{r}^*_\mu|) = U(|\boldsymbol{r}_\nu - \boldsymbol{r}_\mu|) \tag{5.22}$$

the forces $\boldsymbol{F}^*_\nu$ and $\boldsymbol{F}_\nu$ depend in the same way on their arguments. Therefore, the equations of motion are invariant, and the motion is the same in both systems.

For closed systems, the *invariance* of the equations of motion under Galilean transformations means that the motion in the displaced (boosted, rotated) system proceeds in the same way. So, for example, our solar system at a distance of 10^8 light-years would run in the same way (as stated, we neglect external forces like the gravitational field of our galaxy (Milky Way)). The invariance of the laws of motion also means that our solar system tomorrow or in 10^8 years will work the same way, or in a rotated state, or if it moves with an arbitrary constant velocity relative to its present state. The actual motion depends on the initial conditions. The temporal translational invariance does not mean that the actual motion of our solar system in 10^8 years will have any resemblance at all with the present situation. It is only said that all processes in 10^8 years with the same initial conditions would proceed as they do now.

As a counterexample, let us consider a horizontal billiard plate and a slightly tilted billiard plate. These two systems are connected by an active Galilean transformation (rotation by a constant angle). Since there is an external force (gravity), these are not closed systems. For both systems Newton's laws can be used (because of the covariance). The actual motion (the billiard game) runs very differently, because the outer force field (gravity field) is not rotationally invariant; the forces in the two systems are different.

The independence of position, orientation, time, and relative velocity discussed here are consequences of the invariance of Newton's laws (for closed systems) under general Galilean transformations. It is to be reminded that these laws were not derived but merely postulated. Actual observations are (except for high velocities) in agreement with the laws and the independence of position, orientation, time, and relative velocity.

For the investigation of the Galilean invariance and its consequences (conservation laws), the Lagrange formalism (to be introduced in Part II) is much more suitable than the Newtonian equations of motion. The present discussion of the Galilean invariance and its consequences (conservation laws) will be continued in Chapter 11.

Validity of the Galilean Transformation

The front of an electromagnetic wave (or a wave packet) moves with the speed of light $\dot{x} = c$ in the x-direction in an inertial frame IS. We consider this wave front in another IS$'$. According to the Galilei transformation, the wave front has different velocity in IS$'$:

$$\frac{dx}{dt} = c \quad \xrightarrow[\text{Galilean transformation}]{x' = x - v\,t',\ t' = t} \quad \frac{dx'}{dt'} = c - v \tag{5.23}$$

In contrast to this, Maxwell's equations always yield the velocity c for a wave front (because c is a parameter of Maxwell's equations). Therefore, the Maxwell equations are not covariant under the Galilei transformation.

These findings result in the following two possibilities:

1. Maxwell's equations are not valid in all inertial frames. They are non-relativistic and therefore not fundamental equations in the sense of Galileo's principle of relativity.

   This is a quite plausible point of view: For other waves (for example, sound waves) there is a preferred inertial frame IS, namely the IS, in which the wave-bearing medium (such as water or air) is at rest. The speed of sound is just the propagation velocity in this preferred IS. Maxwell himself assumed the existence of a medium in which light propagates; this postulated medium is called ether. The Maxwell equations should then be valid only in the IS which is at rest relative to the ether.

2. Maxwell's equations hold in all inertial frames. Then the Galilei transformation cannot be the correct transformation between different inertial frames.

   If the ether would exist as a light-bearing medium, then the different velocities of light could be measured depending on the inertial frame. Surprisingly, however, the same speed of light c is measured in different inertial frames. Therefore, the concept of the ether had to be abandoned. Einstein formulated the relativity principle "All inertial frames are equivalent" anew: The fundamental laws *inclusive* of Maxwell's equations are of the same form in all inertial frames. Then the Galilei transformations can no longer be the (exactly) correct transformations between inertial systems. The correct transformation is rather the Lorentz transformations (Part IX). This implies then also that Newton's laws are not exactly correct. The Galilei transformations and Newton's laws remain, however, valid in the limiting case of small velocities ($v \ll c$).

Chapter 6

Accelerated Reference Frames

The inertial systems are distinguished by the fact that in them Newton's laws are valid. Non-inertial systems are reference systems that are accelerated relative to an inertial frame. Newton's laws are not valid in accelerated reference systems. We determine the additional terms in the equations of motion for a linearly accelerated and for a rotating reference frame.

Linearly Accelerated Reference Frame

We refer again to Figure 5.1 with the two reference systems, IS and S$'$. We assume equal clock rate ($t' = t$) in IS and S$'$. Let the origin of S$'$ be subject to constant acceleration ($\boldsymbol{a} = \text{const.}$) relative to IS,

$$\boldsymbol{d}(t) = \frac{\boldsymbol{a}\, t^2}{2} \tag{6.1}$$

The transformation (5.4) then reads

$$\boldsymbol{r}(t) = \boldsymbol{r}'(t) + \frac{\boldsymbol{a} t^2}{2} \tag{6.2}$$

We consider the force-free motion of a particle. Then Newton's first law (valid in IS) implies

$$m\,\ddot{\boldsymbol{r}}(t) = 0 \quad \text{in IS} \quad \stackrel{(6.2)}{\longrightarrow} \quad m\,\ddot{\boldsymbol{r}}'(t') = -m\boldsymbol{a} \quad \text{in S}' \tag{6.3}$$

The additional term on the right-hand side means that Newton's first law does not hold in S$'$. This additional term corresponds to a constant force field $\boldsymbol{F} = -m\boldsymbol{a}$ if we compare the equation of motion in S′ with the form of the second law. Since these forces have their origin in the inertia term $m\ddot{\boldsymbol{r}}$, such forces are called *inertial forces*. In the accelerated system, these inertial forces are real forces; a passenger feels such forces when his vehicle is accelerated.

It is noteworthy that inertial forces occur in the same manner as gravitational forces. For a mass point in the homogeneous gravitational field at the earth's surface the equation of motion reads

$$m\ddot{\boldsymbol{r}} = m\boldsymbol{g} \tag{6.4}$$

The transformation (6.2) would lead to

$$m\ddot{\boldsymbol{r}}' = m(\boldsymbol{g} - \boldsymbol{a}) \quad \text{in S}' \tag{6.5}$$

If we choose a *free-falling elevator* as S′, i.e. $\boldsymbol{a} = \boldsymbol{g}$, then we find in S′ the same equations of motion as in an inertial frame IS without gravity. By the transition into a free-falling reference frame S′ the gravitational forces are eliminated. The precondition for this is that inertial and gravitational mass are equal: The inertial mass $m_{\rm i}$ is the mass defined by Newton's laws. The *gravitational* mass $m_{\rm g}$ is the proportionality constant that determines the force on a body in a gravitational field. Therefore, (6.4) is initially

$$m_{\rm i}\,\ddot{\boldsymbol{r}} = m_{\rm g}\,\boldsymbol{g} \tag{6.6}$$

A transformation to the free-falling elevator system ($\boldsymbol{a} = \boldsymbol{g}$) then leads to

$$m_{\rm i}\,\ddot{\boldsymbol{r}}' = (m_{\rm g} - m_{\rm i})\,\boldsymbol{g} \stackrel{!}{=} 0 \quad \text{in S}' \tag{6.7}$$

The possibility of eliminating the gravitational forces from the equation of motion is based on the equality of inertial and gravitational mass. The gravitational mass could be, like, for example, the charge, a property that is independent of the inertial mass. Experimentally one finds that both masses are always proportional to each other; with a suitable system of physical units, they are then equal. This was tacitly assumed in (6.4).

Einstein's equivalence principle postulates the equality of the inertial and gravitational mass, and consequently the equivalence of inertial and gravitational forces. From this, Einstein developed a relativistic theory of gravitation, the *General Theory of Relativity*. The basic consideration is as follows: In a local, free-falling reference frame, *all* processes (not only the mechanical ones) run as in the absence of gravity. Such a frame can be realized by a laboratory in a satellite orbiting the earth (the motion of a satellite in orbit is identical to a free fall). In a laboratory in this satellite, the known laws apply *without* gravitation, especially the laws of the Special Theory of Relativity (Part IX). The relativistic laws *with* gravity are then obtained by the transformation from the satellite laboratory to the considered reference frame, for instance, the laboratory on earth. The transformation between these two frames is a transformation between systems accelerated relative to each other. Starting from a law without gravity (like $m\ddot{\boldsymbol{r}}' = 0$), the transformation leads to the corresponding law with gravity (like $m\ddot{\boldsymbol{r}} = m\boldsymbol{g}$, (6.4)). In contrast to the equations considered here, the accelerations in the real gravitational field are position-dependent; different satellite laboratories have different accelerations relative to a point on the earth.

Rotating Reference System

Rotating reference frames are of special interest, because the earth itself represents such a reference frame, and also because it is used in the treatment of gyroscopic motions (Part V). In Figure 6.1, a rotating system S′ with the coordinates x', y', z' is sketched. This system shall rotate relative to an inertial frame IS (coordinates x, y, z) with the angular velocity

$$\boldsymbol{\omega} = \frac{d\boldsymbol{\varphi}}{dt} \tag{6.8}$$

The vectors $\boldsymbol{\omega}$ and $d\boldsymbol{\varphi}$ point in the direction of the rotation axis (z-axis); $|d\boldsymbol{\varphi}|$ is the angle by which S′ is rotated during dt. We assume that the clocks in IS and S′ show the same time, i.e. $t' = t$.

We consider a vector $\boldsymbol{G}$ which is time-independent in S′; in S′ the vector has a constant length and forms constant angles with the coordinate axes. From Figure 6.1 we see that for the change of this vector due to the rotation (index "rot") of S′ one

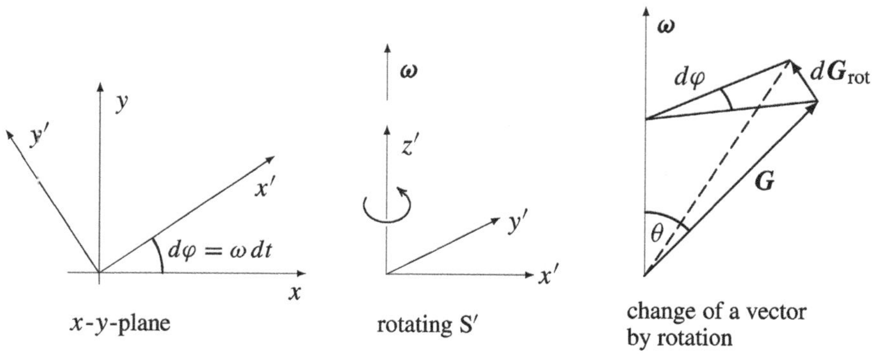

Figure 6.1 We consider a reference frame S′ which rotates relative to an inertial frame IS with the angular velocity ω around the z-axis. The right part shows the change of a vector due to a rotation with $d\varphi$.

obtains: $|d\boldsymbol{G}_{\rm rot}| = |\boldsymbol{G}|\,|d\boldsymbol{\varphi}|\sin\theta$, $d\boldsymbol{G}_{\rm rot} \perp \boldsymbol{\omega}$, and $d\boldsymbol{G}_{\rm rot} \perp \boldsymbol{G}$. From this follows

$$d\boldsymbol{G}_{\rm rot} = d\boldsymbol{\varphi} \times \boldsymbol{G} = \boldsymbol{\omega}\, dt \times \boldsymbol{G} \tag{6.9}$$

We now consider an arbitrary (in general time-dependent) vector $\boldsymbol{G}(t)$. In S′, let it change during dt by $d\boldsymbol{G}_{\rm S'}$. Then the change of $\boldsymbol{G}$ in IS equals

$$d\boldsymbol{G}_{\rm IS} = d\boldsymbol{G}_{\rm S'} + d\boldsymbol{G}_{\rm rot} \tag{6.10}$$

The time derivative of this yields

$$\left(\frac{d\boldsymbol{G}}{dt}\right)_{\rm IS} = \left(\frac{d\boldsymbol{G}}{dt}\right)_{\rm S'} + \boldsymbol{\omega} \times \boldsymbol{G} \tag{6.11}$$

Position and velocity

The position vector of a mass point can be expressed using the basis vectors $\boldsymbol{e}_i$ of IS or the basis vectors $\boldsymbol{e}'_i(t)$ of the rotating S′,

$$\boldsymbol{r} = \sum_{i=1}^{3} x_i(t)\, \boldsymbol{e}_i = \sum_{i=1}^{3} x'_i(t)\; \boldsymbol{e}'_i(t) = \boldsymbol{r}' \tag{6.12}$$

Since IS and S′ have the same origin, $\boldsymbol{r} = \boldsymbol{r}'$ holds. The time dependency refers to the inertial system. The basis vectors $\boldsymbol{e}'_i(t)$ of S′ depend on t, whereas the basis vectors $\boldsymbol{e}_i$ of IS do not.

We now calculate the velocity:

$$\dot{\boldsymbol{r}} = \frac{d\boldsymbol{r}}{dt} = \sum_{i=1}^{3} \frac{dx_i'}{dt}\, \boldsymbol{e}_i' + \sum_{i=1}^{3} x_i'\, \frac{d\boldsymbol{e}_i'}{dt} = \dot{\boldsymbol{r}}' + \boldsymbol{\omega} \times \boldsymbol{r}' \tag{6.13}$$

The first term on the r.h.s.

$$\dot{\boldsymbol{r}}' = \sum_{i=1}^{3} \frac{dx_i'}{dt}\, \boldsymbol{e}_i' \tag{6.14}$$

is the change of the position vector relative to S′, so equal to $(d\boldsymbol{r}/dt)_{\mathrm{S}'}$. In the second term on the r.h.s of (6.13), we use $(d\boldsymbol{e}_i'/dt) = (d\boldsymbol{e}_i'/dt)_{\mathrm{rot}} = \boldsymbol{\omega} \times \boldsymbol{e}_i'$. These two terms in (6.13) correspond to those in (6.11).

Equation of motion

We determine the equation of motion of a force-free mass point in S′. We restrict ourselves to a rotation with constant angular velocity

$$\boldsymbol{\omega} = \text{const.} \tag{6.15}$$

We apply (6.11) to the velocity vector $\dot{\boldsymbol{r}} = \dot{\boldsymbol{r}}' + \boldsymbol{\omega} \times \boldsymbol{r}'$,

$$\left(\frac{d\dot{\boldsymbol{r}}}{dt}\right)_{\mathrm{IS}} = \left(\frac{d(\dot{\boldsymbol{r}}' + \boldsymbol{\omega} \times \boldsymbol{r}')}{dt}\right)_{\mathrm{S}'} + \boldsymbol{\omega} \times \left(\dot{\boldsymbol{r}}' + \boldsymbol{\omega} \times \boldsymbol{r}'\right) \tag{6.16}$$

Analogous to (6.14), the acceleration in S′ reads:

$$\ddot{\boldsymbol{r}}' = \sum_{i=1}^{3} \frac{d^2 x_i'}{dt^2}\, \boldsymbol{e}_i' = \sum_{i=1}^{3} \frac{d^2 x_i'}{dt'^2}\, \boldsymbol{e}_i' \tag{6.17}$$

Thus, (6.16) becomes

$$\ddot{\boldsymbol{r}} = \ddot{\boldsymbol{r}}' + 2\left(\boldsymbol{\omega} \times \dot{\boldsymbol{r}}'\right) + \boldsymbol{\omega} \times \left(\boldsymbol{\omega} \times \boldsymbol{r}'\right) \tag{6.18}$$

For a force-free particle, the first law holds in the inertial frame IS:

$$m\ddot{\boldsymbol{r}} = \left(\frac{d^2\boldsymbol{r}}{dt^2}\right)_{\mathrm{IS}} = 0 \quad \text{in IS} \tag{6.19}$$

From the last two equations we obtain

$$\text{Force-free particle in the rotating frame:} \quad m\,\ddot{\boldsymbol{r}}' = -2m\left(\boldsymbol{\omega}\times\dot{\boldsymbol{r}}'\right) - m\,\boldsymbol{\omega}\times\left(\boldsymbol{\omega}\times\boldsymbol{r}'\right) \tag{6.20}$$

The inertial forces on the right side are described as *Coriolis force* and *centrifugal force*. The Coriolis force is proportional to ω and to the velocity of the mass point. It is perpendicular to this velocity, i.e. to the direction of motion; on a carousel it is, therefore, difficult to go straight. The centrifugal force is proportional to ω^2 and to the distance of the mass point from the axis of rotation. It points away from the axis of rotation; on a rapidly rotating carousel, you have to hold on to avoid sliding away to the outside.

If we insert the transformation (6.18) into Newton's second law, then the Coriolis force and the centrifugal force are added to the forces on the right side. From the valid equation (second law) in the inertial frame IS, one obtains the valid equation in the accelerated system. The new equation is not called second law; in the second law there are no acceleration forces on the right-hand side.

Lab on earth

A laboratory on the earth's surface is an accelerated reference frame S′, Figure 6.2. Let the center of mass of the earth be the origin of an (approximate) inertial frame (Thereby we neglect the acceleration of the earth due to its orbit around the sun). The vector from the origin of this IS to the origin of S′ is denoted by $\boldsymbol{d}(t)$. Due to the rotation of the earth, the origin of S′ moves on a circle perpendicular to $\boldsymbol{\omega}$ with radius $R\sin\theta$. The centrifugal force then points in the direction $\boldsymbol{e}_x = \boldsymbol{e}_{x'}\cos\theta + \boldsymbol{e}_{z'}\sin\theta$ (perpendicular to the axis of rotation $\boldsymbol{\omega}$, see Figure 6.2), and the acceleration thus points in the direction $-\boldsymbol{e}_x$. The magnitude of the acceleration is $\omega^2 R\sin\theta$. Thus, we get

$$\ddot{\boldsymbol{d}}(t) = -\,\omega^2 R\sin\theta\left(\boldsymbol{e}_x'\cos\theta + \boldsymbol{e}_z'\sin\theta\right) \tag{6.21}$$

In addition to this circular motion, S′ rotates about an axis parallel to $\boldsymbol{\omega}$ through its origin, exactly by one full rotation if the earth rotates once around itself. The angular velocity of S′ is therefore equal to

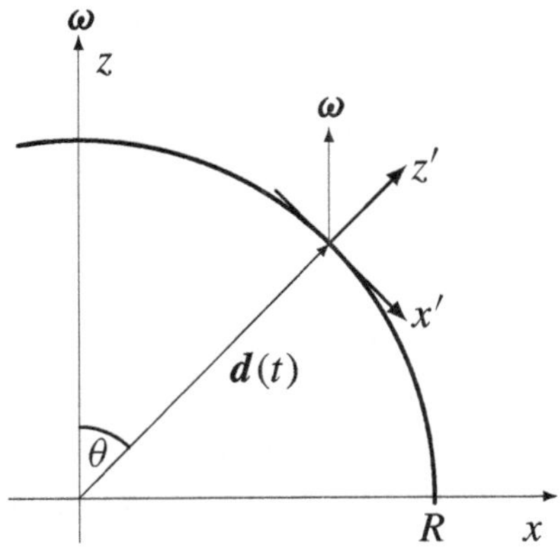

Figure 6.2 The earth (center at $x = y = z = 0$) rotates with the angular velocity $\boldsymbol{\omega}$ around the z-axis. The circle with radius R indicates the earth's surface. A laboratory system on the earth's surface is an accelerated reference system S′ with coordinates x', y', z'; the x'-axis points south. The origin of S′ moves on a circle with radius $R \sin\theta$ around the z-axis. Furthermore, S′ rotates with the angular velocity $\boldsymbol{\omega}$.

that of the earth, $\boldsymbol{\omega} = \omega\, \boldsymbol{e}_z$. The combination of all inertial forces yields

$$m\,\ddot{\boldsymbol{r}}' = \boldsymbol{F}' + m\boldsymbol{g}_{\rm S} - m\ddot{\boldsymbol{d}} - 2m\,(\boldsymbol{\omega}\times\dot{\boldsymbol{r}}') - m\,\boldsymbol{\omega}\times(\boldsymbol{\omega}\times\boldsymbol{r}') \qquad (6.22)$$

The non-inertial forces consist of gravity $m\boldsymbol{g}_{\rm S}$ and other forces $\boldsymbol{F}'$. The motion of S′ was divided into a circular motion of the origin plus a rotation. Thus, the centrifugal force was divided into terms $-m\ddot{\boldsymbol{d}}$ and $-m\boldsymbol{\omega}\times(\boldsymbol{\omega}\times\boldsymbol{r}')$. For $\theta = \pi/2$ the first term gives the familiar result $m\omega^2 R\,\boldsymbol{e}'_z$.

The shape of a rotating ball of liquid is formed under the effect of gravitation $m\boldsymbol{g}_{\rm S}$ and the centrifugal force $-m\ddot{\boldsymbol{d}}$ just so that the total force $m\boldsymbol{g}_{\rm S} - m\ddot{\boldsymbol{d}}$ is perpendicular to the surface. (Otherwise, there would be residual forces acting in the direction of a displacement at the surface.) The flattening of the earth corresponds approximately to this setting. The resulting force is therefore perpendicular to the earth's surface:

$$\boldsymbol{g}_{\rm S} - \ddot{\boldsymbol{d}} \approx -g\,\boldsymbol{e}'_z \qquad (6.23)$$

Here, g is the effective (geographic latitude-dependent) acceleration due to gravity, and $\boldsymbol{e}'_z$ is the direction of the perpendicular to the oblate earth's surface. The oblateness of the earth is rather small; for a numerical value see (23.17). For the other forces in (6.22), the corrections due to deviations from the spherical shape can therefore be neglected.

The centrifugal force due to the rotation of the earth leads to the contributions $-m\ddot{\boldsymbol{d}}$ and $-m\,\boldsymbol{\omega}\times(\boldsymbol{\omega}\times\boldsymbol{r}')$. The first contribution is absorbed into the effective acceleration due to gravity, (6.23). The second contribution can mostly be neglected against the Coriolis force in (6.22) because of the relatively small angular velocity $\omega = 2\pi/\text{day}$. The formal condition for this is $\omega \ll v/\ell$ where $\ell \sim |\boldsymbol{r}'|$ and $v \sim |\dot{\boldsymbol{r}}'|$.

With the given approximations, (6.22) becomes

$$m\,\ddot{\boldsymbol{r}}' = \boldsymbol{F}' - mg\boldsymbol{e}'_z - 2m\,(\boldsymbol{\omega}\times\dot{\boldsymbol{r}}'). \tag{6.24}$$

where $\boldsymbol{\omega} = \omega\,\boldsymbol{e}_z = \omega\,(\cos\theta\,\boldsymbol{e}'_z - \sin\theta\,\boldsymbol{e}'_x)$. For horizontal motion near the ground ($z' \approx 0$), like for instance for winds, we obtain then

$$\begin{aligned} m\,\ddot{x}' &= F'_x + 2m\omega\,\dot{y}'\cos\theta \\ m\,\ddot{y}' &= F'_y - 2m\omega\,\dot{x}'\cos\theta \end{aligned} \tag{6.25}$$

For $\theta < \pi/2$, a velocity $\dot{x}' > 0$ leads to a force in $-y'$ direction, and a velocity $\dot{y}' > 0$ leads to a force in x'-direction. Therefore, a motion in the northern hemisphere ($\theta < \pi/2$) is deflected to the right. For $\theta > \pi/2$ (southern hemisphere) the cosine is negative and the motion is deflected to the left. An example is the direction of rotation of the winds at a low-pressure zone (counterclockwise in the northern hemisphere) or a high-pressure zone (clockwise in the northern hemisphere).

The inertial forces in (6.25) are just the Coriolis forces for a rotation around the z'-axis with the effective frequency

$$\boldsymbol{\omega}_{\text{eff}} = \omega\,\cos\theta\,\boldsymbol{e}'_z \tag{6.26}$$

Therefore, the motion is like a merry-go-round that rotates with the frequency $\boldsymbol{\omega}_{\text{eff}}$. This frequency depends on the geographic latitude $\pi/2 - \theta$.

The small, horizontal deflections of a (Foucault) pendulum (mass M, length L) can be described by (6.25). Due to (6.26), the plane of oscillation of the Foucault pendulum rotates with $-\omega_{\text{eff}}$ relative to the ground.

The example of the Foucault pendulum shows that the description of the motion in the accelerated system is more complicated than in an inertial frame. In the inertial system, the plane of oscillation of a Foucault pendulum at the north pole is simply constant, and the earth's surface rotates relative to this once in 24 hours. In S′, on the other hand, the equations of motion (6.25) have to be solved in order to calculate the rotation of the pendulum plane (once in 24 hours).

Final remarks

The fact that Newton's laws do not hold in accelerated reference frames does not exclude such frames. It is a matter of expediency whether we treat a problem in an inertial frame or in another reference frame. The advantage of the inertial frame lies in the simplicity of the equations of motion. A rotating reference frame comes into question in particular if it corresponds to the actual observing system (e.g., the laboratory on a rotating earth) or if other quantities are simpler in it than in the inertial frame (like the inertia tensor of a rotating body, Part V).

Exercises

6.1 *Coriolis force in free fall*

On a square in Central Europe (with latitude $\varphi_0 = 50°$) there is a tower of height $H = 200\,\text{m}$. The flat square represents the x'–y'-plane, the tower the z'-axis of S′. Due to the earth rotation, S′ is a rotating frame (in which the ω^2-terms are negligible). Calculate in S′ where a body falling freely from the tower (initial velocity zero) will hit the ground (how far away from the z'-axis?). Verify the result by treating the free fall in an inertial frame.

PART II
Lagrangian Formalism

Chapter 7

Lagrange's Equations of First Kind

The Lagrangian formalism (this part) is an elegant and simple method to establish equations of motion. In this chapter, the Newtonian laws are generalized to the Lagrangian equations of the first kind. This generalization is necessary to treat problems with constraints.

Constraints

For many problems, Newton's laws are not directly applicable. As an example, consider the gravity pendulum sketched in Figure 7.1. The pendulum is restricted to a plane (plane pendulum). Moreover, the mass point is held on a circular path by a thread (or a rod) of length l. The restriction of the trajectory $\boldsymbol{r}(t) := (x, y, z)$ may be described by the following *constraints*:

$$z = 0, \quad x^2 + y^2 - l^2 = 0 \tag{7.1}$$

The thread exerts a force $\boldsymbol{Z}$ on the mass point, which is called *constraint force*. Thus, Newton's second law becomes

$$m\ddot{\boldsymbol{r}} = \boldsymbol{F} + \boldsymbol{Z} \tag{7.2}$$

The problem now is that the constraint (7.1) is given, but we do not know the constraint force $\boldsymbol{Z}$ in (7.2). The constraint force $\mathbf{Z}$ can usually not be calculated directly, because it depends on the actual motion.

In this chapter, we present a procedure by which the constraint forces can be determined. This leads to the Lagrange equations of

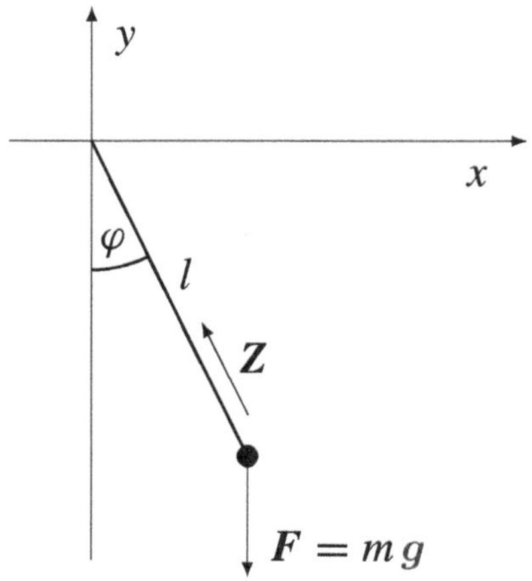

Figure 7.1 The plane gravity pendulum: The mass m is acted upon by gravity $\boldsymbol{F}$ and an unknown force $\boldsymbol{Z}$ exerted by the thread.

the first kind. Often one is not interested in the constraint forces themselves. Then these forces can be eliminated by introducing suitable coordinates; this results in the Lagrange equations of second kind (Chapter 9). For the plane pendulum, the angle φ is such a generalized coordinate; the Lagrange equations of the second kind then reduce to an equation of motion for $\varphi(t)$.

We formulate the problem introduced by (7.1) and (7.2) in a more general way. For this purpose we write the constraints in the form

$$g_1(\boldsymbol{r}, t) = 0, \quad g_2(\boldsymbol{r}, t) = 0 \quad \text{(holonomic constraint)} \tag{7.3}$$

For a particle, there may be one or two such constraints; three independent constraints would fix all three coordinates x, y, and z, i.e. they would not allow any movement at all. Geometrically, a condition like in (7.3) represents a surface; one can imagine $g(x, y, z) = 0$ to be resolved to $z = z_g(x, y)$. A simple example is the motion on a horizontal table, Figure 7.2. For this, the constraint reduces the motion of the plane

$$g(\boldsymbol{r}, t) = z = 0 \tag{7.4}$$

One of the conditions (7.3) restricts the motion to a surface. Two conditions restrict the motion to the intersection of two surfaces, i.e.

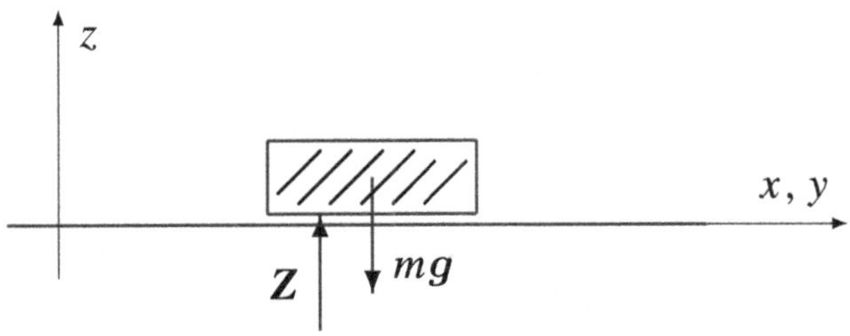

Figure 7.2 A simple example of a constraint is the sliding on a horizontal table. Since the body under consideration does not perform any motion in the z-direction, the resulting forces in this direction must vanish. Therefore, the constraint force is $\boldsymbol{Z} = -m\boldsymbol{g}$.

to a curve. In (7.1), $g_2 = z = 0$ represents a plane (the picture plane in Figure 7.1), and $g_1 = x^2 + y^2 - l^2$ represents a circular cylinder. The intersection of these two surfaces is the circle in the picture plane, on which the mass can move.

For several particles (7.3) becomes

$$g_\alpha(\boldsymbol{r}_1, \boldsymbol{r}_2, \ldots, \boldsymbol{r}_N, t) = 0, \quad (\alpha = 1, 2, \ldots, R) \tag{7.5}$$

The possible number R of conditions is bounded by $R \leq 3N - 1$.

Constraints of the type (7.3)–(7.5) are called *holonomous*, all other constraints are called *non-holonomic*. A non-holonomic condition is, for example, the condition $r \leq R$, which constrains the motion to the interior of a sphere with radius R. Another example of a non-holonomic constraint is a condition involving the velocities of the mass points.

We consider only holonomic conditions in the following. Here, we allow in (7.3) that the constraint depends explicitly on the time t. For example, the thread length of the pendulum in Figure 7.1 could be time-dependent; for this purpose imagine that at $x = y = 0$ the thread is guided through a small eyelet (a small closed loop). Pulling on the thread from the other side of the eyelet leads to a time-dependent condition

$$g(\boldsymbol{r}, t) = x^2 + y^2 - l(t)^2 = 0 \tag{7.6}$$

The direction of the constraining force is not changed; a thread can only exert forces in the direction of tension. Time-dependent conditions are also called rheonomic constraints in contrast to scleronomic (time-independent) constraints.

Constraint Forces

Using the example of the plane pendulum, it is easy to realize that the constraint force depends on the actual motion: On the one hand, it has to compensate for the component of gravity in the direction of the thread, and on the other hand it has to counteract the centrifugal force ($ml\dot{\varphi}^2$). In very simple cases (as in Figure 7.2), the constraint force can also be independent of the motion. In all cases, however, the constraint condition determines the *direction* of the constraint force. By taking advantage of this, we arrive at a solution for (7.2) and related problems.

If a particle is restricted by a holonomic constraint to a surface this does not imply any restriction or influence on the motion within the surface. This means that the constraint force has no component tangential to the surface, it must be orthogonal to the surface:

$$g(\boldsymbol{r}, t) = 0 \quad \longrightarrow \quad \boldsymbol{Z} \parallel \operatorname{grad} g(\boldsymbol{r}, t) \tag{7.7}$$

This allows for the following approach to the constraint force

$$\boldsymbol{Z}(\boldsymbol{r}, t) = \lambda(t) \operatorname{grad} g(\boldsymbol{r}, t) \tag{7.8}$$

Here, $\lambda(t)$ is an unknown function; it is not known because of possible time dependence of $g(\boldsymbol{r}, t)$ and because of the dependence on the actual motion. Let us substitute (7.8) into (7.2), we obtain the *Lagrange equations of first kind*:

$$m\ddot{\boldsymbol{r}} = \boldsymbol{F} + \lambda \operatorname{grad} g(\boldsymbol{r}, t), \quad g(\boldsymbol{r}, t) = 0 \tag{7.9}$$

These are four equations (three second-order differential equations and one algebraic equation) for four unknown functions $x(t)$, $y(t)$, $z(t)$, and $\lambda(t)$; the force $\boldsymbol{F}$ is considered to be given. Together with the initial conditions, these equations are sufficient to determine all wanted functions. In Chapter 8, a systematic solution procedure is presented.

We now consider two conditions (7.3) for a particle. We set for each constraint a force of the form (7.8) and add the two forces:

$$\left.\begin{array}{l} g_1(\boldsymbol{r},t)=0 \\ g_2(\boldsymbol{r},t)=0 \end{array}\right\} \rightarrow \boldsymbol{Z}(\boldsymbol{r},t)=\lambda_1(t)\,\mathrm{grad}\,g_1(\boldsymbol{r},t)+\lambda_2(t)\,\mathrm{grad}\,g_2(\boldsymbol{r},t) \tag{7.10}$$

Here, $\lambda_1(t)$ and $\lambda_2(t)$ are two unknown functions. The reasoning for this form of the constraint is the following: The two constraints define a curve; the motion of the particle is constrained to this curve. Then the combined constraint forces do not act in the direction of the curve, i.e. not in the direction of the tangent vector at the point under consideration. Now, $\mathrm{grad}\,g_1$ and $\mathrm{grad}\,g_2$ form two independent basis vectors which are perpendicular to the curve; because $\mathrm{grad}\,g_\alpha$ is perpendicular to the surface $g_\alpha = 0$ in which the curve lies. Thus, (7.10) is a general approach for an arbitrary force that is perpendicular to the curve. With (7.10), (7.2) yields the Lagrange equations of first kind:

$$m\,\ddot{\boldsymbol{r}}=\boldsymbol{F}+\sum_{\alpha=1}^{2}\lambda_\alpha(t)\,\mathrm{grad}\,g_\alpha(\boldsymbol{r},t),\qquad g_\alpha(\boldsymbol{r},t)=0 \tag{7.11}$$

These are five equations (three differential equations of second order and two algebraic equations) for five unknown functions $x(t)$, $y(t)$, $z(t)$, $\lambda_1(t)$, and $\lambda_2(t)$.

General Case

We write the Lagrange equations (7.11) for Cartesian coordinates, which we number $(x_1, x_2, x_3) = (x, y, z)$:

$$\begin{aligned} m\,\ddot{x}_n &= F_n+\sum_{\alpha=1}^{2}\lambda_\alpha(t)\,\frac{\partial g_\alpha(x_1,x_2,x_3,t)}{\partial x_n}\qquad (n=1,2,3) \\ &g_\alpha(x_1,x_2,x_3,t)=0\quad(\alpha=1,2) \end{aligned} \tag{7.12}$$

We generalize this to the case of $\nu = 1, \ldots, N$ particles, which are subject to the constraints (7.5). The Cartesian coordinates of the

N particles are denoted by

$$x_n = x_{3\nu+j-3} = \boldsymbol{r}_\nu \cdot \boldsymbol{e}_j \quad (\nu = 1, \ldots, N, \quad j = 1, 2, 3, \quad n = 1, \ldots, 3N) \tag{7.13}$$

Thus, for example, the position vector $\boldsymbol{r}_3$ of the third mass point is represented by the components x_7, x_8, and x_9. The mass of the νth particle occurs as a coefficient of $\ddot{x}_n$ with $n = 3\nu - 2$, $3\nu - 1$, and 3ν; this mass we therefore denote by $m_{3\nu-2} = m_{3\nu-1} = m_{3\nu}$. Thus, the generalization of (7.12) is

Lagrange's equations first kind:

$$m_n \ddot{x}_n = F_n + \sum_{\alpha=1}^{R} \lambda_\alpha(t) \, \frac{\partial g_\alpha(x_1, \ldots, x_{3N}, t)}{\partial x_n} \quad (n = 1, 2, \ldots, 3N)$$

$$g_\alpha(x_1, \ldots, x_{3N}, t) = 0 \quad (\alpha = 1, 2, \ldots, R) \tag{7.14}$$

These are $3N + R$ equations ($3N$ second-order differential equations and R algebraic equations) for $3N + R$ unknown functions $x_n(t)$ and $\lambda_\alpha(t)$. The forces F_n are taken as given as in Newton's second law. The form (7.14) assumes Cartesian coordinates. For other coordinates, one has to use the corresponding coordinate transformation.

We discuss the meaning of the generalization (7.14) for two particles and one constraint $g(\boldsymbol{r}_1, \boldsymbol{r}_2, t) = 0$. From (7.14) follow the constraint forces acting on particles 1 and 2 $\boldsymbol{Z}_1$ and $\boldsymbol{Z}_2$:

$$\boldsymbol{Z}_1 = \lambda(t) \, \frac{\partial g(\boldsymbol{r}_1, \boldsymbol{r}_2, t)}{\partial \boldsymbol{r}_1}, \qquad \boldsymbol{Z}_2 = \lambda(t) \, \frac{\partial g(\boldsymbol{r}_1, \boldsymbol{r}_2, t)}{\partial \boldsymbol{r}_2} \tag{7.15}$$

The partial derivatives are defined as in (4.20). Since it is one constraint, only a single function $\lambda(t)$ occurs. This is illustrated by the following examples:

1. The constraint depends on one coordinate only, for example, $g(\boldsymbol{r}_1, \boldsymbol{r}_2, t) = g(\boldsymbol{r}_1, t) = 0$. Then $\boldsymbol{Z}_1 = \lambda(t) \; \mathrm{grad}_1 \, g(\boldsymbol{r}_1, t)$ (as in (7.8)) and $\boldsymbol{Z}_2 = 0$.
2. The constraint acts directly between the two particles. For example, by

$$g(\boldsymbol{r}_1, \boldsymbol{r}_2, t) = g(|\boldsymbol{r}_1 - \boldsymbol{r}_2|) = |\boldsymbol{r}_1 - \boldsymbol{r}_2| - l = 0 \tag{7.16}$$

the distance between the particles could be fixed. Consider Figure 2.1 with a massless rod between the two mass points. In this case, the equality of the λ's is necessary because of the third law,

$$\boldsymbol{Z}_1 = -\boldsymbol{Z}_2 \ \text{ requires } \ \lambda_1 = \lambda_2 \ \text{ in } \ \boldsymbol{Z}_i = \lambda_i \, \text{grad}_i \, g \qquad (7.17)$$

Due to the dependence of $|\boldsymbol{r}_1 - \boldsymbol{r}_2|$, the gradients are of opposite equal size.

The two examples can be compared with the division into external (first case) and internal (second case) forces, which we introduced in Chapter 4. A given constraint, however, can also be a combination of these cases, because the two constraints $g_1 = 0$ and $g_2 = 0$ can always be replaced by two independent linear combinations $a g_1 + b g_2 = 0$. This does not change the set of equations (7.14).

The form of the constraint forces has been made plausible here by means of examples. The resulting Lagrangian equations of the 1st kind represent a non-trivial generalization of Newton's laws. Like the laws themselves, they are fundamental laws, which cannot be proven; they can only be verified or falsified.

d'Alembert principle

Our approach here and in Chapter 9 differs from that in comparable presentations; for the reader's orientation, this is briefly discussed. Often, "virtual displacements" $\delta\boldsymbol{r}$ (or more generally $\delta\boldsymbol{r}_i$) are introduced; these are infinitesimal changes of the position vector $\boldsymbol{r}$, which are compatible with the constraints at a given time. Instead of the statement $\boldsymbol{Z} \parallel \text{grad}\, g$ comes the "principle of virtual work" $\boldsymbol{Z} \cdot \delta\boldsymbol{r} = 0$, or more generally $\sum \boldsymbol{Z}_i \cdot \delta\boldsymbol{r}_i = 0$. The projection of the equations of motion onto the virtual displacements eliminates the constraint forces and thus leads to the *d'Alembert principle* $\sum(m_i \ddot{\boldsymbol{r}}_i - \boldsymbol{F}_i) \cdot \delta\boldsymbol{r}_i = 0$. This principle can be used to derive the Lagrange equations of the first and second kind.

These principles (principle of virtual work, d'Alembert principle) have their particular meaning in the historical development of mechanics. The generalization of Newton's laws to the Lagrangian equations can, however, be done more directly and more simply (as here and in Chapter 9).

Conservation Laws

We discuss the question of the conservation of momentum, angular momentum, and energy in the case of constraints. The possible statements are handicapped by the fact that the constraint forces are usually unknown at first. Moreover, constraints often violate the symmetries underlying the conservation laws (Chapter 11).

For a mass, point apply

$$\frac{d}{dt}\,(m\dot{\boldsymbol{r}}) = \boldsymbol{F} + \boldsymbol{Z} \tag{7.18}$$

and

$$\frac{d\boldsymbol{\ell}}{dt} = \frac{d}{dt}\,(m\,\boldsymbol{r}\times\dot{\boldsymbol{r}}) = \boldsymbol{r}\times(\boldsymbol{F}+\boldsymbol{Z}) \tag{7.19}$$

If the right-hand side vanishes, these equations imply the conservation of momentum and angular momentum, respectively. For example, for the motion on the horizontal table, the momentum is conserved because of $\boldsymbol{Z} = -\boldsymbol{F}$. As in Chapter 4, the discussion can be generalized to a system of N mass points.

To discuss the conservation of energy, we multiply the equations of motion in (7.14) by $\dot{x}_n$ and sum over n. We use

$$\sum_{n=1}^{3N} m_n\,\ddot{x}_n\,\dot{x}_n = \frac{d}{dt}\sum_{n=1}^{3N}\frac{m_n}{2}\,\dot{x}_n^{\,2} = \frac{d}{dt}\,T \tag{7.20}$$

and assume conservative forces F_n:

$$\sum_{n=1}^{3N} F_n\,\dot{x}_n = -\sum_{n=1}^{3N}\frac{\partial U}{\partial x_n}\,\dot{x}_n = -\frac{d}{dt}\,U(x_1,\ldots,x_{3N}) \tag{7.21}$$

Thus, we obtain from (7.14)

$$\frac{d}{dt}\,(T+U) = \sum_{\alpha=1}^{R}\sum_{n=1}^{3N}\lambda_\alpha\,\frac{\partial g_\alpha}{\partial x_n}\,\dot{x}_n \tag{7.22}$$

The solution $x_n(t)$ must satisfy the constraints $g_\alpha(x_1,\ldots,x_{3N},t) = 0$. Therefore, the total time derivative of the constraint

$g_\alpha = 0$ vanishes:

$$\sum_{n=1}^{3N} \frac{\partial g_\alpha}{\partial x_n}\, \dot{x}_n + \frac{\partial g_\alpha}{\partial t} = 0 \tag{7.23}$$

The combination of the last two equations gives

$$\frac{d}{dt}\,(T + U) = -\sum_{\alpha=1}^{R} \lambda_\alpha \,\frac{\partial g_\alpha}{\partial t} \tag{7.24}$$

Thus, the energy conservation holds in the form:

$$\left.\begin{array}{l}\text{forces conservative and constraint}\\ \text{conditions time-independent}\end{array}\right\} \longrightarrow\; T + U = \text{const.} \tag{7.25}$$

For the gravity pendulum in Figure 7.1, this conservation law holds because the gravity is conservative and because (7.1) does not explicitly depend on time. If, on the other hand, the length of the thread depends on time (7.6), this would change the energy of the system by an outside actor. This actor (pulling on the thread from outside) could excite oscillations of the pendulum.

Chapter 8

Applications I

We present the general procedure for the solution of the Lagrange equations of the first kind. This method is applied to the following examples: The inclined plane, the mass point on the rotating rod, and Atwood's falling machine.

General Procedure

The Lagrangian equations of first kind are:

$$m_n \ddot{x}_n = F_n + \sum_{\alpha=1}^{R} \lambda_\alpha \frac{\partial g_\alpha(x_1, \ldots, x_{3N}, t)}{\partial x_n} \quad (n = 1, 2, \ldots, 3N)$$
$$g_\alpha(x_1, \ldots, x_{3N}, t) = 0 \quad (\alpha = 1, 2, \ldots, R) \tag{8.1}$$

Its application to concrete problems can be divided into the following steps:

1. Formulation of constraints.
2. Formulation of the Lagrange equations.
3. Elimination of the λ_α.
4. Solution of the equations of motion.
5. Determination of the integration constants.
6. Determination of the constraint forces.

This is followed by a possible discussion of the solution (graphical representation of the solution, physical meaning of the constraint forces, conserved quantities).

Points 1 and 2 were examined in the previous chapter in examples and in general. In the third step, the λ_α are expressed as functions of the (still unknown) x_n and $\dot{x}_n$, and inserted into the equations of motion. The resulting equations of motion are then to be solved (points 4 and 5). The constraint forces (point 6) can be calculated from the λ_α, the g_α, and the actual solution $x_n(t)$.

The procedure for eliminating λ_α is as follows: One performs the second total time derivative of the constraints,

$$\frac{d^2 g_\alpha}{dt^2} = 0 \quad \longrightarrow \quad \sum_{n=1}^{3N} \frac{\partial g_\alpha}{\partial x_n}\, \ddot{x}_n = G_\alpha(x, \dot{x}, t) \quad (\alpha = 1, \ldots, R) \quad (8.2)$$

The essential point here is that the second derivative $\ddot{x}_n$ can only occur linearly. All other terms are abbreviated by $G_\alpha(x, \dot{x}, t)$. These terms depend only on

$$x = (x_1, \ldots, x_{3N}), \quad \dot{x} = (\dot{x}_1, \ldots, \dot{x}_{3N}) \quad (8.3)$$

and on t; we will use this shorthand notation for the arguments in $G_\alpha(x, \dot{x}, t)$ frequently in the following. For the $\ddot{x}_n$ in (8.2) we now insert the equations of motion, i.e., the first line of (8.1):

$$\sum_{n=1}^{3N} \frac{\partial g_\alpha(x,t)}{\partial x_n}\, \frac{1}{m_n} \left(F_n(x, \dot{x}, t) + \sum_{\beta=1}^{R} \lambda_\beta\, \frac{\partial g_\beta}{\partial x_n} \right) = G_\alpha(x, \dot{x}, t) \quad (8.4)$$

Here, we have indicated in the notation that also the given forces F_n can depend on x, $\dot{x}$, and t. As one can see, (8.4) is a *linear inhomogeneous system of equations* for the unknown quantities λ_α; the coefficients can depend on x, $\dot{x}$ and t (but not on $\ddot{x}$). The number R of equations is equal to that of the unknowns so that we can obtain from them the λ_α in the form

$$\lambda_\alpha = \lambda_\alpha(x, \dot{x}, t) \quad (\alpha = 1, \ldots, R) \quad (8.5)$$

The actual solution $\lambda_\alpha(t)$ results later from this by inserting the solution $x_n(t)$. The λ_α from (8.5) determine, the constraint forces:

$$Z_n = \sum_{\alpha=1}^{R} \lambda_\alpha(x, \dot{x}, t)\, \frac{\partial g_\alpha(x,t)}{\partial x_n} \quad (8.6)$$

In the equations of motion

$$m\,\ddot{x}_n = F_n(x,\dot{x},t) + Z_n(x,\dot{x},t) \tag{8.7}$$

are now on the right-hand side *known* functions of x, $\dot{x}$, and t. This brings the problem to a form that we already considered in the framework of Newton's mechanics (Part I); it is not a fundamentally new problem. The integration constants are determined by the initial conditions *and* the constraints. By substituting this solution into (8.6) the constraint forces $Z_n(t) = Z_n(x(t),\dot{x}(t),t)$ can be calculated (item 6).

We apply the given solution procedure to three examples.

Inclined Plane

We consider the frictionless sliding of a body on an inclined plane, Figure 8.1. This simple problem can be solved directly without the use of the formalism set up here. In this example, however, we can follow easily the systematic way in the solution of the Lagrange equations of the first kind.

Formulation of constraints:

$$g_1(\boldsymbol{r},t) = x\sin\alpha - z\cos\alpha = 0, \quad g_2(\boldsymbol{r},t) = y = 0 \tag{8.8}$$

Lagrange equations of first kind: We write the equations

$$m\,\ddot{\boldsymbol{r}} = -m\,g\,\boldsymbol{e}_z + \lambda_1\,\mathrm{grad}\,g_1 + \lambda_2\,\mathrm{grad}\,g_2 \tag{8.9}$$

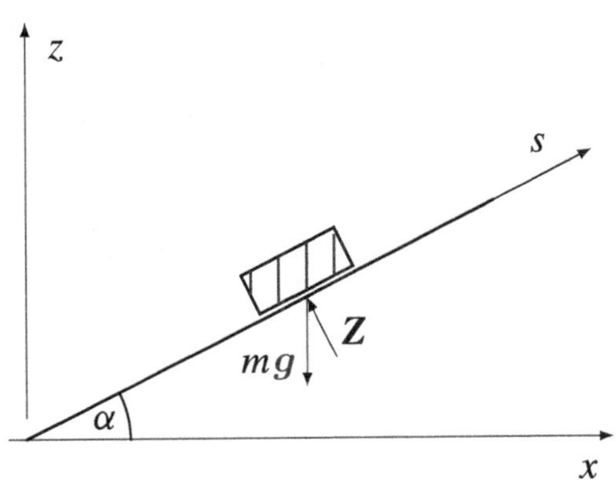

Figure 8.1 A body slides frictionless in the gravity field on an inclined plane.

conveniently in components:

$$m\,\ddot{x} = \lambda_1 \sin\alpha, \quad m\,\ddot{y} = \lambda_2, \quad m\,\ddot{z} = -\lambda_1 \cos\alpha - mg \tag{8.10}$$

Elimination of λ_1 and λ_2: Differentiating (8.8) twice yields

$$\ddot{x}\sin\alpha - \ddot{z}\cos\alpha = 0 \quad \text{and} \quad \ddot{y} = 0 \tag{8.11}$$

According to the given procedure, the second derivatives from (8.10) are to be inserted:

$$\lambda_1 \sin^2\alpha + \lambda_1 \cos^2\alpha + mg\cos\alpha = 0, \quad \lambda_2 = 0 \tag{8.12}$$

This gives

$$\lambda_1 = -mg\cos\alpha, \quad \lambda_2 = 0 \tag{8.13}$$

Here, λ_1 and λ_2 are constants; in general, this method returns the λ_α only as functions of x, y, z, $\dot{x}$, $\dot{y}$, $\dot{z}$, and t. We insert the λ_α into the equations of motion (8.10),

$$m\,\ddot{x} = -mg\sin\alpha\cos\alpha, \quad m\,\ddot{y} = 0, \quad m\,\ddot{z} = -mg\sin^2\alpha \tag{8.14}$$

Solution of the equation of motion: The general solution of (8.14) is

$$\begin{aligned} x(t) &= -\frac{1}{2}\,g\,t^2 \sin\alpha\cos\alpha + a_1\,t + a_2 \\ y(t) &= b_1\,t + b_2 \\ z(t) &= -\frac{1}{2}\,g\,t^2 \sin^2\alpha + c_1\,t + c_2 \end{aligned} \tag{8.15}$$

Determination of integration constants: The integration constants a_1, a_2, b_1, b_2, c_1, and c_2 are to be determined in such a way that the constraint conditions and the initial conditions are satisfied. We first consider the constraint conditions (8.8),

$$(a_1 \sin\alpha - c_1\cos\alpha)\,t + (a_2\sin\alpha - c_2\cos\alpha) = 0, \quad b_1\,t + b_2 = 0 \tag{8.16}$$

Since this must be satisfied for arbitrary values of t, the following holds:

$$\begin{aligned} a_1\sin\alpha - c_1\cos\alpha = 0, &\quad b_1 = 0 \\ a_2\sin\alpha - c_2\cos\alpha = 0, &\quad b_2 = 0 \end{aligned} \tag{8.17}$$

This leaves only two of the original six constants in (8.15) open. By

$$
\begin{aligned}
a_1 &= v_0 \cos\alpha, \qquad & c_1 &= v_0 \sin\alpha \\
a_2 &= s_0 \cos\alpha & c_2 &= s_0 \sin\alpha
\end{aligned}
\tag{8.18}
$$

(8.17) is solved; for the two remaining constants we have introduced s_0 and v_0. The general solution compatible with the constraints is thus

$$
\left.\begin{aligned}
x(t) &= s(t)\cos\alpha \\
y(t) &= 0 \\
z(t) &= s(t)\sin\alpha
\end{aligned}\right\}
\quad \text{with} \quad s(t) = -\frac{1}{2}\, g\, t^2 \sin\alpha + v_0\, t + s_0
\tag{8.19}
$$

The two remaining integration constants can be fixed by the initial conditions.

Determination of the constraint force: In

$$
\boldsymbol{Z} = \lambda_1 \operatorname{grad} g_1 + \lambda_2 \operatorname{grad} g_2
\tag{8.20}
$$

we set the λ_α from (8.13), the g_α from (8.8) and the solution $\boldsymbol{r}(t)$. This gives

$$
\boldsymbol{Z} = -mg \sin\alpha \cos\alpha\, \boldsymbol{e}_x + mg \cos^2\alpha\, \boldsymbol{e}_z, \qquad |\boldsymbol{Z}| = mg\cos\alpha
\tag{8.21}
$$

This constraint force just compensates for the component of gravity perpendicular to the inclined plane.

Discussion of the solution: The particle moves along the in Figure 8.1 drawn s-axis with $s(t) = -g\,t^2 \sin\alpha/2 + v_0\, t + s_0$. This corresponds to the equation of motion

$$
m\,\ddot{s}(t) = -mg\sin\alpha
\tag{8.22}
$$

which can be understood as follows: One decomposes the force $\boldsymbol{F} = m\boldsymbol{g}$ into a component perpendicular to the inclined plane (result $mg\cos\alpha$) and a component parallel to it (result $mg\sin\alpha$). The perpendicular component is compensated by the constraint force $\lambda_1 \operatorname{grad} g_1$. The parallel component leads to accelerated motion according to (8.22).

The force $\boldsymbol{F} = m\boldsymbol{g}$ is conservative; it results from the potential $U = mgz$. Moreover, the constraint conditions (8.8) are time-independent. Therefore, the conservation of energy holds:

$$T + U = \frac{m}{2}\,\dot{\boldsymbol{r}}^2 + mgz = \frac{m}{2}\,\dot{s}^2 + mgs\sin\alpha = \text{const.} \tag{8.23}$$

This first-order differential equation could be used to solve for $s(t)$. Therefore, it can replace the equation of motion completely.

Constraint to a plane

We consider a motion that is restricted to a plane, like $z = 0$ in (7.1) or Figure 7.2. In this case, the constraint force just compensates the component of the other forces perpendicular to the plane. The motion in this direction (i.e. in z-direction for the constraint $z = 0$) can then be left out of consideration from the beginning. We show this formally for the general case (8.1). The constraint

$$g_1(x_1, \ldots, x_{3N}, t) = x_n - a = 0 \tag{8.24}$$

is arbitrarily called the first one (index 1). Then the equation of motion for x_n reads

$$m_n\,\ddot{x}_n = F_n + \lambda_1 + \sum_{\alpha=2}^{R} \lambda_\alpha\,\frac{\partial g_\alpha}{\partial x_n} \tag{8.25}$$

Differentiating twice the condition (8.24) gives $\ddot{x}_n = 0$. The equations of motion (8.25) are to be inserted into this:

$$0 = F_n + \lambda_1 + \sum_{\alpha=2}^{R} \lambda_\alpha\,\frac{\partial g_\alpha}{\partial x_n} \tag{8.26}$$

According to the general procedure, this λ_α must be inserted into (8.25). This gives

$$m_n\,\ddot{x}_n = 0 \quad \to \quad x_n = A + B\,t \tag{8.27}$$

The constants of the solution are to be determined in conformity with the constraint, i.e. $A = a$ and $B = 0$.

Since λ_1 does not appear in the other equations of motion, the procedure can be shortened by omitting the coordinate x_n from the beginning: Given a constraint of the type $x_n = \text{const.}$ one considers only the $3N-1$ equations of motion for the $x_{n'\neq n}(t)$ and the $R-1$ other constraints. This applies correspondingly if there are several constraints of the kind $x_n = \text{const.}$

Mass Point on the Rotating Rod

We consider a mass point that can move along a rod (Figure 8.2); imagine a sphere with a hole in which the rod fits. The rod rotates with a constant angular velocity ω. Apart from the resulting constraint forces, no other forces should act on the particle. In particular, the mass point slides without friction on the rod.

As discussed in the last section, we can ignore the equation for the z-motion from the beginning. We therefore only consider the motion for $x(t)$ and $y(t)$ with the one constraint

$$g = \arctan \frac{y}{x} - \omega t = \varphi - \omega\, t = 0 \tag{8.28}$$

The given arrangement suggests using polar coordinates, $x = \rho \cos\varphi$ and $y = \rho \sin\varphi$. We write the Lagrange equations of first kind in the form (7.9)

$$m\,\ddot{\boldsymbol{r}} = \lambda \,\text{grad}\, g \tag{8.29}$$

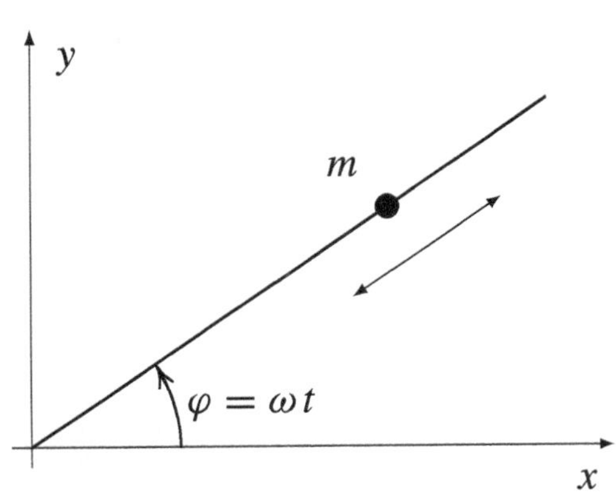

Figure 8.2 A mass point is moving frictionless along a rod, which rotates with given angular velocity ω.

The position vector lies in the $x-y$-plane, $\boldsymbol{r} = x(t)\,\boldsymbol{e}_x + y(t)\,\boldsymbol{e}_y = \rho(t)\,\boldsymbol{e}_\rho$. With the gradient operator in polar coordinates

$$\text{grad} = \boldsymbol{e}_\rho\,\frac{\partial}{\partial\rho} + \boldsymbol{e}_\varphi\,\frac{1}{\rho}\,\frac{\partial}{\partial\varphi} \tag{8.30}$$

we evaluate the right-hand side of (8.29). For the position vector $\boldsymbol{r}$, the acceleration $\ddot{\boldsymbol{r}}$ is given by (1.14). Thus, the two components of (8.29) are

$$m\,\ddot\rho - m\,\rho\,\dot\varphi^2 = 0, \quad m\,\rho\,\ddot\varphi + 2\,m\,\dot\rho\,\dot\varphi = \frac{\lambda}{\rho} \tag{8.31}$$

We differentiate the constraint (8.28) twice with respect to time,

$$\frac{d^2 g}{dt^2} = \ddot\varphi = 0 \tag{8.32}$$

Into this, we substitute $\ddot\varphi$ from the equations of motion and obtain

$$\lambda = 2\,m\,\rho\,\dot\rho\,\dot\varphi = \lambda(\rho,\dot\rho,\dot\varphi) \tag{8.33}$$

This is not yet the solution $\lambda(t)$, but only the functional dependence of λ on the coordinates and the velocities. We put this λ into the equations of motion (8.31),

$$m\,\ddot\rho - m\,\rho\,\dot\varphi^2 = 0, \quad m\,\rho\,\ddot\varphi = 0 \tag{8.34}$$

We note that $\rho(t) = 0$ is a trivial (unstable) solution of these equations of motion. For $\rho \neq 0$ the second equation becomes $\ddot\varphi = 0$. Its integration gives $\varphi = At + B$. The integration constants are to be fixed in such a way that the constraints are fulfilled, i.e.

$$\varphi(t) = \omega\,t \tag{8.35}$$

This follows also directly from the constraint. The remaining differential equation is

$$\ddot\rho(t) - \omega^2\rho(t) = 0 \tag{8.36}$$

This result can be interpreted as one-dimensional motion in ρ-direction under the influence of the centrifugal force $F_\rho = m\,\omega^2\rho$.

The equation of motion (8.36) is a linear homogeneous differential equation with constant coefficients. The solution is therefore of the form $\rho(t) \propto \exp(\kappa t)$. This leads to $\kappa^2 - \omega^2 = 0$ and to the general solution

$$\rho(t) = A \exp(\omega t) + B \exp(-\omega t) \tag{8.37}$$

The constants A and B are determined by the initial conditions. In general, the mass is accelerated with exponentially increasing velocity towards the outside. This is only not true in the special case where the initial conditions are chosen such that $\rho(t) = B \exp(-\omega t)$.

We calculate the constraint force:

$$\boldsymbol{Z} = \lambda \operatorname{grad} g = 2m\,\dot{\rho}\,\omega\,\boldsymbol{e}_\varphi = 2m\,\omega^2\,\boldsymbol{e}_\varphi\,[A \exp(\omega t) - B \exp(-\omega t)] \tag{8.38}$$

The opposite force $-\boldsymbol{Z}$ acts on the rod. For $A \neq 0$, the force $|\boldsymbol{Z}|$ grows exponentially. A real rod will bend and break.

The energy $E = T = m\dot{\boldsymbol{r}}^2/2$ is not conserved, because the constraint depends explicitly on the time.

Atwood's Falling Machine

As a final example, consider a massless pulley (a wheel on an axle or shaft, radius R) supporting a rope that connects two masses, Figure 8.3. A homogeneous gravitational field acts an the masses. For the six Cartesian coordinates x_1, y_1, z_1, x_2, y_2 and z_2 there are

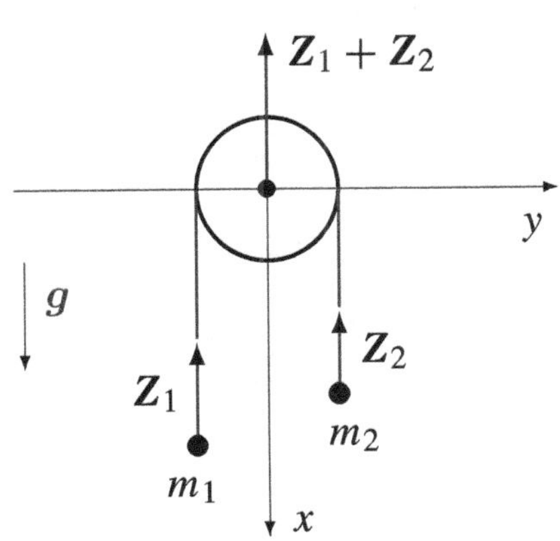

Figure 8.3 Atwood's falling machine: Two masses in the gravitational field are connected by a rope.

five constraints:

$$y_1 = -R, \quad z_1 = 0, \quad y_2 = R, \quad z_2 = 0 \tag{8.39}$$

$$g(x_1, x_2) = x_1 + x_2 - l = 0 \tag{8.40}$$

Here, $l = L - \pi R$ is given by the rope length L. The first four conditions result in vanishing constraint forces and trivial solutions of the equations of motion. We, therefore, consider only the equations of motion for x_1 and x_2 with the constraint condition (8.40)

$$m_1 \ddot{x}_1 = m_1 g + \lambda, \quad m_2 \ddot{x}_2 = m_2 g + \lambda \tag{8.41}$$

Differentiating the constraint (8.40) twice with respect to time gives $\ddot{x}_1 + \ddot{x}_2 = 0$. Herein, we insert the equations of motion (8.41)

$$g + \frac{\lambda}{m_1} + g + \frac{\lambda}{m_2} = 0, \quad \lambda = -2g\,\frac{m_1 m_2}{m_1 + m_2} \tag{8.42}$$

The constraint forces on the two masses are equal, $\boldsymbol{Z}_1 = \boldsymbol{Z}_2 = \lambda\, \boldsymbol{e}_x$. The axis of the wheel must then support the force $\boldsymbol{Z}_1 + \boldsymbol{Z}_2$. For $m_1 = m_2 = m$, $\boldsymbol{Z}_1 + \boldsymbol{Z}_2 = -2m\boldsymbol{g}$ is equal to the weight of the two masses. For $m_1 \neq m_2$, on the other hand, the force on the pulley is smaller than $(m_1 + m_2)\,\boldsymbol{g}$, because a part of the weight forces serves for the acceleration of the masses. We substitute λ from (8.42) into (8.41):

$$(m_1 + m_2)\, \ddot{x}_1 = (m_1 - m_2)\, g \tag{8.43}$$

Here, the sum of the masses enters as the inertial mass, the difference as the gravitational mass. The solution of this equation is

$$x_1(t) = \frac{m_1 - m_2}{m_1 + m_2}\,\frac{g t^2}{2} + c_1 t + c_2 \tag{8.44}$$

The energy $E = T + U = m_1 \dot{x}_1^2/2 + m_2 \dot{x}_2^2/2 - m_1 g x_1 - m_2 g x_2$ is constant, because the forces are conservative and because the constraint (8.40) does not explicitly depend on the time.

Exercises

8.1 *Mass point on the curve in the gravity field*

In the vertical z–x-plane a mass point glides frictionless on the curve $z = f(x)$. Gravity acts on the mass point yielding $\boldsymbol{F} = -mg\boldsymbol{e}_z$. Set up the Lagrangian equations of first kind.

8.2 *Mass point on the spherical surface*

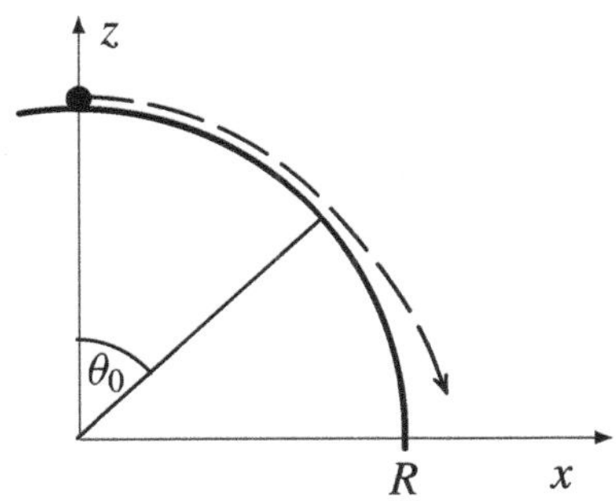

A point of mass lies in the gravitational field on the uppermost point of a sphere. It starts to slide down there. At what point does it lift off the sphere? Use energy conservation.

8.3 *Dumbbell on concentric circles*

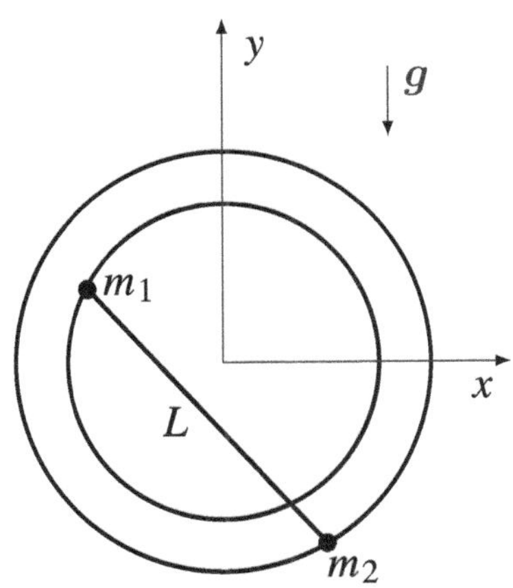

Two mass points ($m_1 = m_2 = m$) can move frictionless on two concentric circles (radii r and R). The two mass points are connected by a massless rod of length L; let $R - r < L < R + r$. The acceleration due to gravity is $\boldsymbol{g} = -g\,\boldsymbol{e}_y$.

Set up the Lagrange equations of first kind. Determine the equilibrium position of the masses (i) from the Lagrange equations and alternatively (ii) from the condition $U_{\text{pot}} = mg(y_1 + y_2) =$ minimal.

8.4 *Accelerated inclined plane*

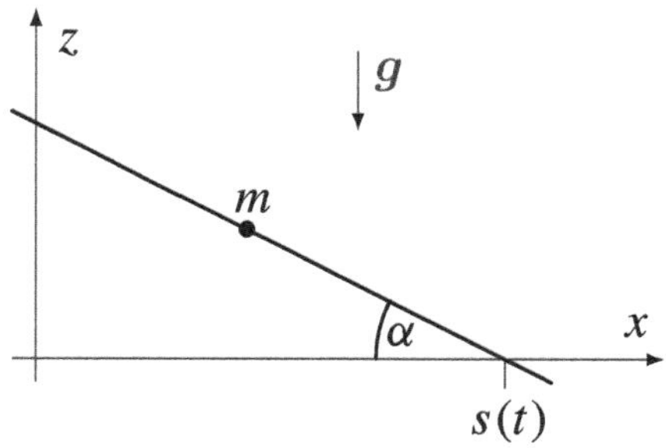

A point of mass slides without friction on an inclined plane which is accelerated in x-direction, $s(t) = a\,t^2/2$. The inclination α of the plane is constant. Set up the constraint condition and the Lagrange equations of first kind. Solve the equations of motion and determine the constraint forces.

Chapter 9

Lagrange's Equations Second Kind

In many cases, one is not interested in the calculation of the constraint forces. Then it is much more convenient to choose a formulation in which the constraint forces are eliminated from the equations of motion. This leads to the Lagrangian equations of the second kind.

Generalized Coordinates

We start from the Lagrange equations of first kind (7.14) for N mass points,

$$m_n \, \ddot{x}_n = F_n + \sum_{\alpha=1}^{R} \lambda_\alpha \, \frac{\partial g_\alpha(x_1, \ldots, x_{3N}, t)}{\partial x_n} \qquad (n = 1, 2, \ldots, 3N)$$

$$g_\alpha(x_1, \ldots, x_{3N}, t) = 0 \quad (\alpha = 1, 2, \ldots, R) \tag{9.1}$$

For R constraints, only

$$f = 3N - R \quad \text{(number of degrees of freedom)} \tag{9.2}$$

of the $3N$ Cartesian coordinates are independent of each other. We call this the number of *degrees of freedom* of the system because by f numbers the positions of the N mass points can be specified. The decisive step consists now in the choice of f suitable *generalized coordinates*

$$q_1, \, q_2, \, \ldots, \, q_f \quad \text{(generalized coordinates)} \tag{9.3}$$

The choice of the new coordinates must be such that the q_i defines the position of all mass points,

$$x_n = x_n(q_1, q_2, \ldots, q_f, t) \quad (n = 1, 2, \ldots, 3N) \tag{9.4}$$

and that the constraints are satisfied for *any values of the* q_i,

$$g_\alpha(x_1(q_1, \ldots, q_f, t), \ldots, x_{3N}(q_1, \ldots, q_f, t), t) \equiv 0 \quad \text{for all } q_i \tag{9.5}$$

We consider some examples:

- For the plane gravitational pendulum with variable length $l(t)$ we choose the angle φ as the generalized coordinate (Figure 7.1). The generalized coordinate φ specifies all three Cartesian coordinates:

$$\begin{aligned} x &= x(\varphi, t) = l(t) \sin\varphi \\ y &= y(\varphi, t) = -l(t) \cos\varphi \\ z &= z(\varphi, t) = 0 \end{aligned} \tag{9.6}$$

  The constraint conditions (7.1) are automatically satisfied by this. In particular,

$$\begin{aligned} g(x(\varphi, t),\, y(\varphi, t),\, z(\varphi, t), t) &= x(\varphi, t)^2 + y(\varphi, t)^2 - l(t)^2 \\ &= l^2 \cos^2\varphi + l^2 \sin^2\varphi - l^2 \equiv 0 \end{aligned} \tag{9.7}$$

  The constraints are satisfied for any value of φ; they do not impose any constraint on the φ motion.
- A point of mass moves on a spherical surface (radius R). The trajectory is subject to the constraint

$$g(x, y, z) = x^2 + y^2 + z^2 - R^2 = 0 \tag{9.8}$$

  As generalized coordinates we choose the angles θ and ϕ:

$$\begin{aligned} x &= x(\theta, \phi) = R \sin\theta \cos\phi \\ y &= y(\theta, \phi) = R \sin\theta \sin\phi \\ z &= z(\theta, \phi) = R \cos\theta \end{aligned} \tag{9.9}$$

  Here, R is a parameter and not a coordinate. Again the constraint (9.8) is satisfied for arbitrary values of θ and ϕ.

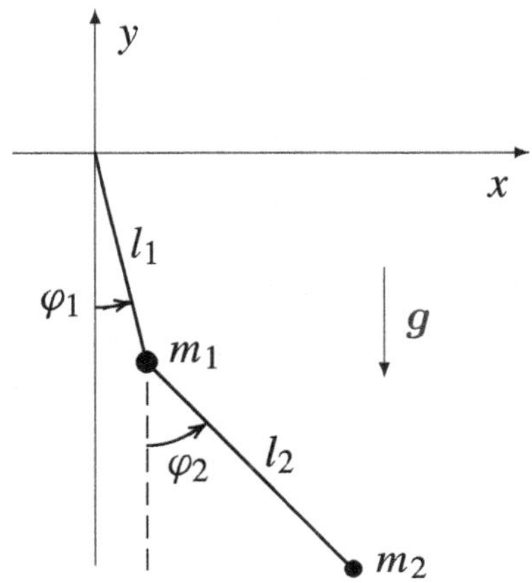

Figure 9.1 For the gravity double pendulum the angles φ_1 and φ_2 are introduced as generalized coordinates.

- For the plane double pendulum, Figure 9.1, we choose as generalized coordinates the angles φ_1 and φ_2:

$$\begin{aligned} x_1 &= l_1 \sin\varphi_1 & x_2 &= l_1 \sin\varphi_1 + l_2 \sin\varphi_2 \\ y_1 &= -l_1 \cos\varphi_1 & y_2 &= -l_1 \cos\varphi_1 - l_2 \cos\varphi_2 \\ z_1 &= 0 & z_2 &= 0 \end{aligned} \tag{9.10}$$

  It is clear that with this the constraints (not explicitly written down here) are fulfilled.

Elimination of Constraint Forces

The generalized coordinates were chosen such that the constraints are automatically fulfilled. Therefore, the $q_i(t)$-movement is not subject to any constraint forces; this is also shown by the following formal derivation.

The g_α in (9.5) does not depend on the q_i; because the conditions $g_\alpha = 0$ are fulfilled for any q_i-values. As an example, consider the φ-dependence of g in (9.7). The independence of q_i means that the total derivative of g_α in (9.5) with respect to the q_i vanishes:

$$\frac{dg_\alpha}{dq_k} = 0, \quad \text{so} \quad \sum_{n=1}^{3N} \frac{\partial g_\alpha}{\partial x_n}\frac{\partial x_n}{\partial q_k} = 0 \quad (k = 1, 2, \ldots, f) \tag{9.11}$$

Hereby, we eliminate the constraint forces: we multiply (9.1) by $\partial x_n/\partial q_k$ and sum over n. This gives

$$\sum_{n=1}^{3N} m_n\, \ddot{x}_n\, \frac{\partial x_n}{\partial q_k} = \sum_{n=1}^{3N} F_n\, \frac{\partial x_n}{\partial q_k} \quad (k = 1, 2, \ldots, f) \tag{9.12}$$

In these equations x_n stands for $x_n(q_1, q_2, \ldots, q_f, t)$; this holds also for the arguments of $F_n = F_n(x, \dot{x}, t)$. These equations, therefore, represent f equations of motion for the f functions $q_k(t)$. Constraint forces do not appear anymore, and there are no more constraints to be taken into account.

Lagrange Function

We put (9.12) into a form that is particularly well suited for practical applications. The resulting equations are the Lagrange equations of the second kind.

For the arguments of functions, we use the shortcut notation

$$\begin{aligned} x &= (x_1, \ldots, x_{3N}), & \dot{x} &= (\dot{x}_1, \ldots, \dot{x}_{3N}) \\ q &= (q_1, \ldots, q_f), & \dot{q} &= (\dot{q}_1, \ldots, \dot{q}_f) \end{aligned} \tag{9.13}$$

The $\dot{q}_i$ are denoted as *generalized* velocities. We differentiate the transformation equations (9.4)

$$x_n = x_n(q, t) = x_n(q_1, q_2, \ldots, q_f, t) \quad (n = 1, 2, \ldots, 3N) \tag{9.14}$$

with respect to the time,

$$\dot{x}_n = \frac{d}{dt}\, x_n(q, t) = \sum_{k=1}^{f} \frac{\partial x_n(q, t)}{\partial q_k}\, \dot{q}_k + \frac{\partial x_n(q, t)}{\partial t} = \dot{x}_n(q, \dot{q}, t) \tag{9.15}$$

Hereby, $\dot{x}_n$ is defined as a function of q, $\dot{q}$, and t. For this function, we obtain from (9.15)

$$\frac{\partial \dot{x}_n(q, \dot{q}, t)}{\partial \dot{q}_k} = \frac{\partial x_n(q, t)}{\partial q_k} \tag{9.16}$$

From Newton's laws, we obtain the form of the kinetic energy in Cartesian coordinates:

$$T = T(\dot{x}) = \sum_{n=1}^{3N} \frac{m_n}{2}\, \dot{x}_n^2 \tag{9.17}$$

Here, we substitute $\dot{x}_n(q, \dot{q}, t)$ from (9.15):

$$T = T(q, \dot{q}, t) = \sum_{i,k=1}^{f} m_{ik}(q,t)\, \dot{q}_i\, \dot{q}_k + \sum_{k=1}^{f} b_k(q,t)\, \dot{q}_k + c(q,t) \quad (9.18)$$

Since $\dot{x}_n$ is linear in the $\dot{q}_k$, the kinetic energy is maximally quadratic in the $\dot{q}_k$. The resulting coefficients are denoted by $m_{ik}(q,t)$, $b_k(q,t)$, and $c(q,t)$; here we set $m_{ik} = m_{ki}$. By (9.18) the kinetic energy is given as a function of q and $\dot{q}$.

The quantities T in (9.17) and (9.18) are *different functions* of the arguments; from the mathematical point of view, one should therefore use different symbols (say T and T^*). Since T denotes the same physical quantity (the kinetic energy), one uses the same letter (the physical point of view). The notation, $T(\dot{x})$ or $T(q, \dot{q}, t)$, then shows which function is meant.

If the $x_n(q,t)$ does *not* depend explicitly on time, then (9.15) becomes $\dot{x}_n = \sum_k (\partial x_n / \partial q_k)\, \dot{q}_k$. Thus, (9.18) reduces to

$$T = T(q, \dot{q}) = \sum_{i,k=1}^{f} m_{ik}(q)\, \dot{q}_i\, \dot{q}_k \quad (\text{if } \partial x_n/\partial t = 0) \qquad (9.19)$$

The following derivations do not assume this special form. From $T = \sum_n m_n\, \dot{x}_n^2/2$ follow:

$$\frac{\partial T(q,\dot{q},t)}{\partial q_k} = \sum_{n=1}^{3N} m_n\, \dot{x}_n\, \frac{\partial \dot{x}_n}{\partial q_k} \qquad (9.20)$$

and

$$\frac{\partial T(q,\dot{q},t)}{\partial \dot{q}_k} = \sum_{n=1}^{3N} m_n\, \dot{x}_n\, \frac{\partial \dot{x}_n}{\partial \dot{q}_k} \overset{(9.16)}{=} \sum_{n=1}^{3N} m_n\, \dot{x}_n\, \frac{\partial x_n}{\partial q_k} \qquad (9.21)$$

We differentiate (9.21) with respect to the time

$$\frac{d}{dt}\frac{\partial T}{\partial \dot{q}_k} = \sum_{n=1}^{3N} m_n\, \ddot{x}_n\, \frac{\partial x_n}{\partial q_k} + \sum_{n=1}^{3N} m_n\, \dot{x}_n\, \frac{\partial \dot{x}_n}{\partial q_k} \qquad (9.22)$$

In the last term we have interchanged d/dt and $\partial/\partial q_k$ which only assumes the interchangeability of the partial derivatives:

$$\frac{d}{dt}\frac{\partial x_n}{\partial q_k} = \sum_{l=1}^{f} \frac{\partial^2 x_n}{\partial q_l\,\partial q_k}\,\dot{q}_l + \frac{\partial^2 x_n}{\partial t\,\partial q_k}$$
$$= \frac{\partial}{\partial q_k}\left(\sum_{l=1}^{f} \frac{\partial x_n}{\partial q_l}\,\dot{q}_l + \frac{\partial x_n}{\partial t}\right) = \frac{\partial}{\partial q_k}\frac{dx_n}{dt} \qquad (9.23)$$

By

$$Q_k = \sum_{n=1}^{3N} F_n\,\frac{\partial x_n}{\partial q_k} \qquad (9.24)$$

we define the *generalized forces* Q_k. We now substitute (9.22) with (9.20) and (9.24) into the equations of motion (9.12):

$$\frac{d}{dt}\left(\frac{\partial T}{\partial \dot{q}_k}\right) - \frac{\partial T}{\partial q_k} = Q_k \quad (k = 1, 2, \ldots, f) \qquad (9.25)$$

For now, we restrict ourselves to forces F_n, which can be represented by a potential:

$$F_n = -\frac{\partial U(x)}{\partial x_n} \qquad (9.26)$$

In $U(x)$ we insert the transformation $x_n = x_n(q, t)$ and obtain

$$U(q, t) = U(q_1, \ldots, q_f, t) = U(x_1(q, t), \ldots, x_n(q, t)) \qquad (9.27)$$

The use of the same letter U for the different functions $U(x)$ and $U(q, t)$ is done in the same sense as for the kinetic energy T. The generalized forces can now be expressed as the derivative of $U(q, t)$:

$$Q_k = \sum_{n=1}^{3N} F_n\,\frac{\partial x_n}{\partial q_k} = -\sum_{n=1}^{3N} \frac{\partial U(x)}{\partial x_n}\,\frac{\partial x_n}{\partial q_k} = -\frac{\partial U(q, t)}{\partial q_k} \qquad (9.28)$$

This relation makes the designation "generalized force" plausible. Considering $\partial U/\partial \dot{q}_k = 0$ we can transform (9.25) into

$$\frac{d}{dt}\,\frac{\partial (T - U)}{\partial \dot{q}_k} = \frac{\partial (T - U)}{\partial q_k} \qquad (9.29)$$

For velocity-dependent potentials, (9.29) is equivalent to (9.25), if

$$Q_k = -\frac{\partial U(q,\dot{q},t)}{\partial q_k} + \frac{d}{dt}\frac{\partial U(q,\dot{q},t)}{\partial \dot{q}_k} \tag{9.30}$$

For $U = U(q,t)$ this reduces to the already familiar form.

We now introduce the *Lagrange function* $\mathcal{L}$ of the non-relativistic mechanics,

$$\mathcal{L}(q,\dot{q},t) = T(q,\dot{q},t) - U(q,t) \quad \text{Lagrange function} \tag{9.31}$$

The Lagrange function is defined as the difference between kinetic and potential energy where T and U are written as functions of the generalized coordinates. Since one can choose different generalized coordinates for a given system, the function $\mathcal{L}(q,\dot{q},t)$ is not uniquely fixed. As we will see later, additional terms to $\mathcal{L}$ are possible, which do not change the equations of motion. In this respect, $\mathcal{L}$ is a theoretical quantity. In contrast, the kinetic energy and the potential energy are physical, i.e. measurable quantities. With the Lagrangian function $\mathcal{L}$ the equations (9.29) receive their final shape:

Lagrange's equations second kind:

$$\frac{d}{dt}\frac{\partial \mathcal{L}(q,\dot{q},t)}{\partial \dot{q}_k} = \frac{\partial \mathcal{L}(q,\dot{q},t)}{\partial q_k} \quad (k = 1,\ldots,f) \tag{9.32}$$

These equations represent a system of f second-order differential equations for the (generalized) trajectories $q_k(t)$. The second kind of Lagrangian equations are the preferred equations for the solution of problems in mechanics. This is for a number of reasons:

1. Compared to the Lagrangian equations of first kind we have to deal with only $f = 3N - R$ instead of $3N + R$ equations. In this approach, we do not calculate the constraint forces.
2. For complex systems, the establishment of the Lagrangian function is *far easier* than the equations of motion themselves. This is mainly due to the fact that the Lagrange function is a single *scalar* quantity. In general, it is much simpler to set up this quantity than the equations of motion.
3. In many cases, the Lagrange function is a particularly simple function of the variables under consideration. This is an important

point of view, if one wants to develop new physical theories (in particular field theories).

System and system state

The part of the world to be described we call *system*. This can be, for example, a double pendulum, the solar system, or a hydrogen atom. In the context of (point) mechanics, we restrict ourselves to systems of mass points.

Such a system always relates to an extremely small section of reality. If we examine the orbit of the earth we disregard its own rotation and all other processes on the earth. The system "sun–earth" is then replaced by the model system "mass point in a gravitational potential". This example shows that an "extremely small section" does not necessarily mean spatial extension; rather only a few degrees of freedom are treated explicitly. The considered system does not have to be closed; the influence of the environment can be taken into account by external forces.

Which degrees of freedom one wants to treat explicitly is a question of the purpose, of the computational possibilities, and of the goals pursued. In this context, a mechanical (model) system is defined by the choice of the generalized coordinates and by the specification of the Lagrange function.

The second kind of Lagrange equations are second order differential equations for the coordinates $q_1(t), \ldots, q_f(t)$. The solution is determined by initial conditions, i.e. $2f$ values for $q_i(0)$ and $\dot{q}_i(0)$. These values define the *system state* at time $t = 0$. The Lagrangian equations then determine the state of the system at later times, i.e. the $2f$ values $q_1, \ldots, q_f$ and $\dot{q}_1, \ldots, \dot{q}_f$ at time t.

The remarks about physical systems apply in general, those about the system state for all mechanical systems (with modifications for continuous media, Part VIII).

Constants of Motion

The equations of motion may lead to solutions for which certain quantities (like, for example, the energy) are conserved. These quantities are thus *constants of motion*.

The Lagrange function $\mathcal{L}(q_1, \ldots, q_f, \dot{q}_1, \ldots, \dot{q}_f, t)$ can depend on the generalized coordinates and velocities, and on the time. In special cases, $\mathcal{L}$ can be independent of one or more coordinates, or of the time; we study the consequences of this. The case where $\mathcal{L}$ is independent of a particular $\dot{q}_k$ is of no interest; because then there would be no dynamics for this degree of freedom.

Energy conservation

Using the equations of motion, we calculate the time derivatives of $\sum(\partial\mathcal{L}/\partial\dot{q}_k)\,\dot{q}_k$ and of $\mathcal{L}$:

$$\begin{aligned}\frac{d}{dt}\sum_{k=1}^{f}\frac{\partial\mathcal{L}}{\partial\dot{q}_k}\,\dot{q}_k &= \sum_{k=1}^{f}\dot{q}_k\,\frac{d}{dt}\frac{\partial\mathcal{L}}{\partial\dot{q}_k} + \sum_{k=1}^{f}\frac{\partial\mathcal{L}}{\partial\dot{q}_k}\,\ddot{q}_k \\ &= \sum_{k=1}^{f}\frac{\partial\mathcal{L}}{\partial q_k}\,\dot{q}_k + \sum_{k=1}^{f}\frac{\partial\mathcal{L}}{\partial\dot{q}_k}\,\ddot{q}_k \qquad (9.33)\end{aligned}$$

$$\frac{d\mathcal{L}}{dt} = \sum_{k=1}^{f}\frac{\partial\mathcal{L}}{\partial q_k}\,\dot{q}_k + \sum_{k=1}^{f}\frac{\partial\mathcal{L}}{\partial\dot{q}_k}\,\ddot{q}_k + \frac{\partial\mathcal{L}}{\partial t} \qquad (9.34)$$

From the last two equations we get

$$\frac{d}{dt}\left(\sum_{k=1}^{f}\frac{\partial\mathcal{L}}{\partial\dot{q}_k}\,\dot{q}_k - \mathcal{L}\right) = -\frac{\partial\mathcal{L}}{\partial t} \qquad (9.35)$$

From this follows the conservation law:

$$\frac{\partial\mathcal{L}}{\partial t} = 0 \quad\longrightarrow\quad \sum_{k=1}^{f}\frac{\partial\mathcal{L}}{\partial\dot{q}_k}\,\dot{q}_k - \mathcal{L} = \text{const.} \qquad (9.36)$$

If constraints do not depend explicitly on time, then this is also true for (9.4), $x_n = x_n(q)$, and the kinetic energy is of the form (9.19), $T = \sum m_{ik}(q)\,\dot{q}_i\,\dot{q}_k$. Moreover, if $U = U(q,t)$ does not depend on the velocities we get

$$\sum_{k=1}^{f}\frac{\partial\mathcal{L}}{\partial\dot{q}_k}\,\dot{q}_k = \sum_{k=1}^{f}\frac{\partial T(q,\dot{q})}{\partial\dot{q}_k}\,\dot{q}_k = 2\,T(q,\dot{q}) \qquad \begin{pmatrix} x_n = x_n(q) \\ U = U(q,t)\end{pmatrix} \qquad (9.37)$$

Then the conservation law (9.36) becomes the *energy conservation law*:

$$\left.\begin{aligned} \partial\mathcal{L}/\partial t = 0 \\ x_n = x_n(q) \end{aligned}\right\} \quad \longrightarrow \quad E = T + U = \text{const.} \tag{9.38}$$

Cyclic coordinate

If a generalized coordinate q_k does not occur explicitly in the Lagrange function,

$$\frac{\partial\mathcal{L}}{\partial q_k} = 0 \tag{9.39}$$

this coordinate is called *cyclic*. From the Lagrange equations, it follows immediately that the corresponding *generalized momentum* p_k,

$$p_k = \frac{\partial\mathcal{L}}{\partial\dot{q}_k} \quad \text{(generalized momentum)} \tag{9.40}$$

is conserved:

$$\frac{\partial\mathcal{L}}{\partial q_k} = 0 \quad \longrightarrow \quad p_k = \text{const.} \tag{9.41}$$

A trivial example is the free particle: The Lagrangian function $\mathcal{L} = T = m\dot{\boldsymbol{r}}^2/2$ does not depend on $\boldsymbol{r}$, therefore the momentum $\boldsymbol{p} = m\dot{\boldsymbol{r}}$ is a constant of motion.

Conservation laws like (9.36), (9.38), or (9.41) are of the form $Q(\dot{x}, x, t) = \text{const.}$ Thus, they represent differential equations of the first order, i.e. first integrals of the equations of motion. They can facilitate the solution of a problem (solving first-order differential equation is easier than solving second-order differential equations).

Electromagnetic Forces

In the derivation of the second of kind, Lagrangian equations, we have assumed a potential of the form $U(q, t)$. Thus, electromagnetic and frictional forces are excluded for the time being. We investigate how these forces can be treated in the framework of the second kind of Lagrange equations.

Following (9.30), we can also admit a potential $U(q,\dot{q},t)$, if the forces Q_k are given by

$$Q_k = -\frac{\partial U}{\partial q_k} + \frac{d}{dt}\frac{\partial U}{\partial \dot{q}_k} \tag{9.42}$$

This generalization of $Q_k = -\partial U/\partial q_k$ is sufficient to include electromagnetic forces.

The electromagnetic fields $\boldsymbol{E}$ and $\boldsymbol{B}$ can be expressed by the scalar and vector potentials, Φ and $\boldsymbol{A}$, respectively:

$$\boldsymbol{E}(\boldsymbol{r},t) = -\text{grad}\,\Phi(\boldsymbol{r},t) - \frac{1}{c}\frac{\partial \boldsymbol{A}(\boldsymbol{r},t)}{\partial t} \tag{9.43}$$

$$\boldsymbol{B}(\boldsymbol{r},t) = \text{curl}\,\boldsymbol{A}(\boldsymbol{r},t) \tag{9.44}$$

For the potential

$$U(\boldsymbol{r},\dot{\boldsymbol{r}},t) = q\,\Phi(\boldsymbol{r},t) - \frac{q}{c}\,\boldsymbol{A}(\boldsymbol{r},t)\cdot\dot{\boldsymbol{r}} \tag{9.45}$$

the forces (9.42) result in (Exercise 9.3)

$$\boldsymbol{F} = \boldsymbol{F}(\boldsymbol{r},\dot{\boldsymbol{r}},t) = \sum_{k=1}^{3} Q_k\,\boldsymbol{e}_k = q\left(\boldsymbol{E}(\boldsymbol{r},t) + \frac{1}{c}\,\dot{\boldsymbol{r}}\times\boldsymbol{B}(\boldsymbol{r},t)\right) \tag{9.46}$$

Her we use the Cartesian coordinates so that the position vector reads $\boldsymbol{r} = \sum x_i\,\boldsymbol{e}_i$. The result is the well-known Lorentz force, which is studied and justified in more detail in electrodynamics. The (non-relativistic) Lagrangian function for a charged particle in the electromagnetic field is thus:

$$\mathcal{L}(\dot{\boldsymbol{r}},\boldsymbol{r},t) = \frac{m}{2}\,\dot{\boldsymbol{r}}^2 - q\,\Phi(\boldsymbol{r},t) + \frac{q}{c}\,\dot{\boldsymbol{r}}\cdot\boldsymbol{A}(\boldsymbol{r},t) \tag{9.47}$$

Friction Forces

For a realistic description of mechanical problems, the introduction of frictional forces can be useful. In Cartesian coordinates, such forces can often be described by the approach

$$F_{\text{diss},n} = -\gamma_n\,\dot{x}_n \tag{9.48}$$

No potential can be assigned to these forces. We must therefore first go back to (9.12). Used there, (9.48) leads to the generalized, dissipative forces

$$Q_{\text{diss},k} = \sum_{n=1}^{3N} F_{\text{diss},n} \frac{\partial x_n}{\partial q_k} \tag{9.49}$$

These forces are to be added to the right-hand side of (9.25). A significant advantage of the Lagrange function is, that the system is represented by a single scalar function $\mathcal{L}(q, \dot{q}, t)$. Allowing for an additional term in the Lagrangian equations, we can use this advantage also for the frictional forces. We define the *Rayleigh dissipation function* D by

$$D(\dot{x}) = \sum_{n=1}^{3N} \frac{\gamma_n}{2} \dot{x}_n^2 \quad \text{and} \quad D(q, \dot{q}, t) = \sum_{n=1}^{3N} \frac{\gamma_n}{2} [\dot{x}_n(q, \dot{q}, t)]^2 \tag{9.50}$$

The function $D(q, \dot{q}, t)$ is obtained by substituting (9.15) into $D(\dot{x})$. Using (9.48) and (9.50), we can obtain the generalized force $Q_{\text{diss},k}$ expressed by D:

$$Q_{\text{diss},k} = -\sum_{n=1}^{3N} \frac{\partial D}{\partial \dot{x}_n} \frac{\partial x_n}{\partial q_k} \overset{(9.16)}{=} -\sum_{n=1}^{3N} \frac{\partial D}{\partial \dot{x}_n} \frac{\partial \dot{x}_n}{\partial \dot{q}_k} = -\frac{\partial D(q, \dot{q}, t)}{\partial \dot{q}_k} \tag{9.51}$$

By the modified Lagrangian equations

$$\frac{d}{dt} \frac{\partial \mathcal{L}}{\partial \dot{q}_i} - \frac{\partial \mathcal{L}}{\partial q_i} + \frac{\partial D}{\partial \dot{q}_i} = 0 \tag{9.52}$$

the frictional forces can now be taken into account.

According to (3.19), $-\boldsymbol{F}_{\text{diss}} \cdot \dot{\boldsymbol{r}}$ is equal to the power delivered by a particle. For the frictional force $\boldsymbol{F}_{\text{diss}} = -\gamma \dot{\boldsymbol{r}}$, this becomes $\gamma \dot{\boldsymbol{r}}^2$. In this case, Rayleigh's dissipation function is equal to half the power dissipated into other energy forms (like heat).

Exercises

9.1 *Motion in a spherically symmetric potential*

The motion of a particle in a spherically symmetric potential U shall be described with spherical coordinates, i.e. $q_1 = r$, $q_2 = \theta$, $q_3 = \phi$ and $U = U(r, t)$. Set up the Lagrange function, and determine any cyclic coordinate and the corresponding associated constant of motion.

9.2 *Form of kinetic energy*

Let the kinetic energy be of the form $T = T(q, \dot{q}) = \sum_{i,k} m_{ik}(q)\, \dot{q}_i\, \dot{q}_k$ where $m_{ik} = m_{ki}$. Show $\sum_n \dot{q}_n\, (\partial T/\partial \dot{q}_n) = 2T$.

9.3 *Particles in an electromagnetic field*

Set up the Lagrange equations for the Lagrange function

$$\mathcal{L}(\boldsymbol{r}, \dot{\boldsymbol{r}}, t) = \frac{m}{2}\, \dot{\boldsymbol{r}}^2 - q\, \Phi(\boldsymbol{r}, t) + \frac{q}{c}\, \dot{\boldsymbol{r}} \cdot \boldsymbol{A}(\boldsymbol{r}, t)$$

Put these equations into the form $m\, \ddot{\boldsymbol{r}} = \boldsymbol{F}$ and show

$$\boldsymbol{F} = q \left(\boldsymbol{E} + \frac{\dot{\boldsymbol{r}}}{c} \times \boldsymbol{B} \right) \tag{9.53}$$

Use $\boldsymbol{E} = -\mathrm{grad}\, \Phi - (\partial \boldsymbol{A}/\partial t)/c$ and $\boldsymbol{B} = \mathrm{curl}\, \boldsymbol{A}$.

By $\boldsymbol{E} = E_0\, \boldsymbol{e}_z$ and $\boldsymbol{B} = B_0\, \boldsymbol{e}_z$ a homogeneous constant electromagnetic field is given. Calculate and discuss the trajectory of a particle (mass m, charge q) for the initial conditions $\boldsymbol{r}(0) = 0$ and $\dot{\boldsymbol{r}}(0) = v_0\, \boldsymbol{e}_x$.

Chapter 10

Applications II

We apply the Lagrange equations of Second kind to all examples (inclined plane, mass point on the rotating rod, Atwood's falling machine) for which we have solved the Lagrangian equations of the first kind. For the double pendulum and the curvilinear coordinates, it is demonstrated that the solution via the Lagrangian function is much simpler than directly setting up the equations of motion. Finally, we will examine the Lagrangian equations of the second kind for a new example (mass point on a circular cone).

The Lagrange equations of second kind are:

$$\frac{d}{dt}\frac{\partial \mathcal{L}(q,\dot{q},t)}{\partial \dot{q_k}} = \frac{\partial \mathcal{L}(q,\dot{q},t)}{\partial q_k}, \quad (k = 1, \ldots, f) \tag{10.1}$$

Its application to concrete problems can be divided into the following steps:

1. Choice of the generalized coordinates $q = (q_1, \ldots, q_f)$ and specification of the transformation $x_n = x_n(q,t)$ to Cartesian coordinates.
2. Determination of the Lagrange function $\mathcal{L}(q,\dot{q},t)$.
3. Establishment of the equations of motion.
4. Determination of the constants of motion (conservation laws).
5. Solution of the equations of motion, possibly using conservation law.
6. Determination of integration constants.
7. Discussion of the solution.

We first consider the three examples already studied in Chapter 8. For these known cases, this is done in a relatively concise form. Then we highlight the advantages of the present formulation for the example of the double pendulum and for curvilinear coordinates. Finally, we treat systematically and in detail a new example.

Inclined Plane

For the inclined plane, in Figure 8.1, we choose the path length s as the generalized coordinate. Then the transformation to Cartesian coordinates reads

$$x = x(s) = s\,\cos\alpha, \quad y = y(s) = 0, \quad z = z(s) = s\,\sin\alpha \tag{10.2}$$

From $T = m\,(\dot{x}^2 + \dot{y}^2 + \dot{z}^2)/2$ and $U = m\,g\,z$ follows:

$$\mathcal{L}(s, \dot{s}) = T - U = \frac{m}{2}\,\dot{s}^2 - m\,g\,s\,\sin\alpha \tag{10.3}$$

The Lagrange equation

$$m\,\ddot{s}(t) = -m\,g\,\sin\alpha \tag{10.4}$$

is easy to integrate:

$$s(t) = -\frac{g}{2}\,t^2\,\sin\alpha + v_0\,t + s_0 \tag{10.5}$$

Since neither $\mathcal{L}$ nor the transformations (10.2) depend explicitly on the time, the energy is conserved:

$$E = T + U = \frac{m}{2}\,\dot{s}(t)^2 + m\,g\,s(t)\,\sin\alpha = \text{const.} \tag{10.6}$$

Since the Lagrangian equation (10.4) is trivial to integrate, the conservation law is of no practical advantage in this case.

Mass Point on a Rotating Rod

For the mass point on the rotating rod, in Figure 8.2, we select the distance ρ from the center as the generalized coordinate. Then

$$x = x(\rho, t) = \rho\,\cos(\omega t), \quad y = y(\rho, t) = \rho\sin(\omega t), \quad z = 0 \tag{10.7}$$

From $\mathcal{L} = T = m(\dot{x}^2 + \dot{y}^2 + \dot{z}^2)/2$ and $U = 0$ we get

$$\mathcal{L}(\rho, \dot{\rho}) = T - U = \frac{m}{2}\left(\dot{\rho}^2 + \omega^2 \rho^2\right) \tag{10.8}$$

This leads to the equation of motion

$$\ddot{\rho}(t) = \omega^2\, \rho(t) \tag{10.9}$$

The formulation of the Lagrangian equations of second kind is much simpler than that of the Lagrangian equations of first kind (Chapter 8). The solution of (10.9) was given in (8.37).

Since $\partial\mathcal{L}/\partial t = 0$, the conservation law (9.36) holds

$$\frac{\partial \mathcal{L}}{\partial \dot{\rho}}\, \dot{\rho} - \mathcal{L} = \frac{m}{2}\left(\dot{\rho}^2 - \omega^2 \rho^2\right) = \text{const.} \tag{10.10}$$

Since $\partial x_n/\partial t \neq 0$, this is not equal to the energy. The energy

$$E = T + U = T = \frac{m}{2}\left(\dot{\rho}^2 + \omega^2 \rho^2\right) \neq \text{const.} \tag{10.11}$$

is not conserved despite $\partial\mathcal{L}/\partial t = 0$. The conservation law (10.10) can, however, be expressed (analogously to the energy conservation) in the form $T_\rho + U_{\text{eff}} = \text{const.}$ with the effective potential

$$U_{\text{eff}}(\rho) = -\frac{m}{2}\, \omega^2 \rho^2 \quad \text{and} \quad T_\rho = \frac{m}{2}\, \dot{\rho}^2 \tag{10.12}$$

The contribution U_{eff} is a part of the kinetic energy (in the inertial frame). For a co-rotating observer, who sees only the ρ-motion, it appears like the potential of an external force field.

Atwood's Falling Machine

Atwood's falling machine, Figure 8.3, is a system with just one degree of freedom. As a generalized coordinate, we choose $q = x_1$. Then the transformation to the Cartesian coordinates is

$$x_1 = q, \quad x_2 = l - q, \quad y_1 = -R, \quad y_2 = R, \quad z_1 = z_2 = 0 \tag{10.13}$$

The kinetic energy is

$$T = \frac{1}{2}(m_1\, \dot{x_1}^2 + m_2\, \dot{x_2}^2) = \frac{1}{2}\,(m_1 + m_2)\, \dot{q}^2 \tag{10.14}$$

The potential energy is

$$U = -m_1 g\, x_1 - m_2 g\, x_2 = g\,(m_2 - m_1)\, q + \text{const.} \tag{10.15}$$

The constant is of no influence in the equations of motion and can be omitted. Thus,

$$\mathcal{L}(q, \dot{q}) = \frac{m_1 + m_2}{2}\, \dot{q}^2 - g\,(m_2 - m_1)\, q \tag{10.16}$$

From this follow the Lagrangian equations

$$(m_1 + m_2)\, \ddot{q}(t) = (m_1 - m_2)\, g \tag{10.17}$$

in agreement with (8.43).

Double Pendulum

In the three examples discussed so far, one could easily establish the equations of motion for the generalized coordinates directly. In the next example, the double pendulum in Figure 9.1, it can be seen that the "detour" via the Lagrange function is an essential *simplification.*

In (9.10), we have introduced the angles φ_1 and φ_2 as suitable generalized coordinates of the double pendulum:

$$\begin{aligned} x_1 &= l_1 \sin\varphi_1 \qquad & x_2 &= l_1 \sin\varphi_1 + l_2 \sin\varphi_2 \\ y_1 &= -l_1 \cos\varphi_1 \qquad & y_2 &= -l_1 \cos\varphi_1 - l_2 \cos\varphi_2 \\ z_1 &= 0 & z_2 &= 0 \end{aligned} \tag{10.18}$$

From this follows the kinetic energy

$$\begin{aligned} T &= \frac{m_1}{2}\,(\dot{x}_1^2 + \dot{y}_1^2 + \dot{z}_1^2) + \frac{m_2}{2}\,(\dot{x}_2^2 + \dot{y}_2^2 + \dot{z}_2^2) \\ &= \frac{m_1}{2}\, l_1^2\, \dot{\varphi}_1^2 + \frac{m_2}{2}\,[l_1^2\, \dot{\varphi}_1^2 + l_2^2\, \dot{\varphi}_2^2 + 2\, l_1\, l_2 \cos(\varphi_1 - \varphi_2)\, \dot{\varphi}_1\, \dot{\varphi}_2] \end{aligned} \tag{10.19}$$

Together with the potential energy $U = m_1 g\, y_1 + m_2 g\, y_2$ we get

$$\begin{aligned} \mathcal{L} = {} & \frac{m_1 + m_2}{2}\, l_1^2\, \dot{\varphi}_1^2 + \frac{m_2}{2}\, l_2^2\, \dot{\varphi}_2^2 + m_2\, l_1\, l_2\, \dot{\varphi}_1\, \dot{\varphi}_2 \cos(\varphi_1 - \varphi_2) \\ & + (m_1 + m_2)\, g\, l_1 \cos\varphi_1 + m_2\, g\, l_2 \cos\varphi_2 \end{aligned} \tag{10.20}$$

Herewith, the equations of motion (10.1) can be written down. A direct establishment of the equations of motion for $\varphi_1(t)$ and

$\varphi_2(t)$ is, however, difficult (consider, for example, the term with $\dot{\varphi}_1\,\dot{\varphi}_2$). Our procedure consists of rather simple steps: The choice of coordinates (10.18) is obvious and simple, the steps to (10.20) and the equations of motion are straight-forward. If one proceeds now to the triple pendulum, then the way shown here is straight-forward, too. In contrast to this, attempting to construct the equations of motion directly from Newton's laws would be quite cumbersome.

For small oscillations of the double pendulum, the equations of motion are to be solved in Exercise 10.1.

Curvilinear Coordinates

The equation of motion of a point mass shall be written down in *curvilinear coordinates*. This task can be solved most simply with the help of the Lagrange function.

As generalized coordinates, we choose the desired curvilinear coordinates (such as cylindrical or spherical coordinates). The definition of the generalized coordinates is given by the transformation between the Cartesian and the curvilinear coordinates. The number of Cartesian coordinates is equal to that of the generalized (curvilinear) coordinates (no constraint). The transformation to arbitrary curvilinear coordinates is of the form:

$$x_n = x_n(q_1, \ldots, q_N), \quad n = 1, \ldots, N \tag{10.21}$$

Then

$$dx_n = \sum_{k=1}^{N} \frac{\partial x_n}{\partial q_k}\, dq_k \tag{10.22}$$

and

$$ds^2 = \sum_{n=1}^{N} (dx_n)^2 = \sum_{n=1}^{N} \sum_{i,k=1}^{N} \frac{\partial x_n}{\partial q_i} \frac{\partial x_n}{\partial q_k}\, dq_i\, dq_k = \sum_{i,k=1}^{N} g_{ik}(q)\, dq_i\, dq_k \tag{10.23}$$

Here, we have introduced the functions $g_{ik}(q) = \sum_n (\partial x_n/\partial q_i)(\partial x_n/\partial q_k)$. The *line element* ds is the distance between adjacent points of space, it thereby defines the *metric* of the space. The quantities g_{ik} are therefore also called metric tensors.

We now consider *one* particle, i.e. $N = 3$ and $(x_1, x_2, x_3) = (x, y, z)$. We further restrict ourselves to *orthogonal* coordinates, which are defined by the fact that the metric tensor is of the form $g_{ik} = h_i(q)^2\,\delta_{ik}$. These coordinates include, among others, the Cartesian coordinates and spherical, cylindrical, elliptic, or hyperbolic coordinates. For the first three examples, the square of the line element is

$$ds^2 = \sum_{i,k=1}^{3} g_{ik}(q)\,dq_i\,dq_k = \begin{cases} dx^2 + dy^2 + dz^2 & \text{(Cartesian)} \\ d\rho^2 + \rho^2\,d\varphi^2 + dz^2 & \text{(cylindrical)} \\ dr^2 + r^2\,d\theta^2 + r^2\,\sin^2\!\theta\,d\phi^2 & \text{(spherical)} \end{cases} \tag{10.24}$$

The Lagrange function $\mathcal{L}$ of a free particle reads

$$\mathcal{L}(q, \dot{q}) = T(q, \dot{q}) = \frac{m}{2}\sum_{n=1}^{3}\left(\frac{dx_n}{dt}\right)^2 = \frac{m}{2}\sum_{i,k=1}^{3} g_{ik}(q)\,\dot{q}_i\,\dot{q}_k \tag{10.25}$$

For the cases listed in (10.24), this becomes.

$$\mathcal{L}(q, \dot{q}) = \begin{cases} \frac{m}{2}(\dot{x}^2 + \dot{y}^2 + \dot{z}^2) & \text{(Cartesian)} \\ \frac{m}{2}(\dot{\rho}^2 + \rho^2\,\dot{\varphi}^2 + \dot{z}^2) & \text{(cylindrical)} \\ \frac{m}{2}(\dot{r}^2 + r^2\,\dot{\theta}^2 + r^2\sin^2\!\theta\;\dot{\phi}^2) & \text{(spherical)} \end{cases} \tag{10.26}$$

The equations of motion follow from

$$\frac{d}{dt}\frac{\partial\mathcal{L}}{\partial\dot{q}_k} = \frac{\partial\mathcal{L}}{\partial q_k} \tag{10.27}$$

The procedure given here is much simpler than, say, the derivation (1.8)–(1.14) for polar coordinates. When writing down the equation of motion directly, the position dependence of the basis vectors must be taken into account (as discussed in Chapter 1). In contrast, the Lagrange function is the scalar kinetic energy, which is easy to determine for arbitrary coordinates.

Mass Point on a Circular Cone

We consider a mass point moving frictionless in the gravitational field on a circular cone, Figure 10.1. This can be thought of as

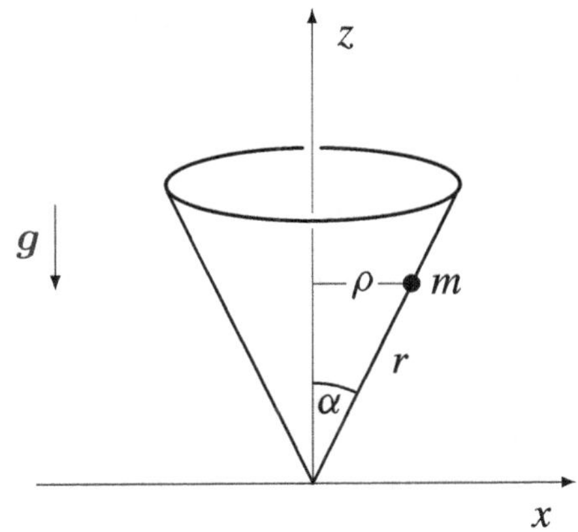

Figure 10.1 In the gravitational field a particle glides frictionless on a circular cone.

being realized by a small ball rolling inside a cone-shaped funnel; the rotational energy of the sphere is neglected. For this new example, we give explicitly all steps up to the solution as listed at the beginning of this chapter.

Choice of generalized coordinates: The system is subject to the constraint

$$g(x, y, z) = x^2 + y^2 - z^2 \tan^2\alpha = 0 \tag{10.28}$$

Given one constraint and three Cartesian coordinates, two generalized coordinates have to be chosen. In question are, for example,

$$(x, y), \quad (x, z), \quad (r, \phi), \quad (\rho, \phi) \quad \text{or} \quad (z, \phi) \tag{10.29}$$

Particularly suitable are generalized coordinates, which reflect the symmetry of the system. In our case, this is the coordinate ϕ, since the problem is symmetric with respect to rotations around the z-axis. As the second coordinate, we choose r; equally suitable would be ρ or z. Thus, the transformation to Cartesian coordinates is

$$\begin{aligned} x &= x(r, \phi) = r \sin\alpha \cos\phi \\ y &= y(r, \phi) = r \sin\alpha \sin\phi \\ z &= z(r, \phi) = r \cos\alpha \end{aligned} \tag{10.30}$$

Note that α is a parameter, not a coordinate.

Lagrange function: From the transformation equations follows:

$$\begin{aligned}\dot{x} &= \dot{x}(r,\phi,\dot{r},\dot{\phi}) = \dot{r}\,\sin\alpha\,\cos\phi - r\,\sin\alpha\,\sin\phi\,\dot{\phi}\\ \dot{y} &= \dot{y}(r,\phi,\dot{r},\dot{\phi}) = \dot{r}\,\sin\alpha\,\sin\phi + r\,\sin\alpha\,\cos\phi\,\dot{\phi}\\ \dot{z} &= \quad \dot{z}(\dot{r}) \quad = \dot{r}\,\cos\alpha\end{aligned} \tag{10.31}$$

The kinetic and the potential energy are written as functions of the generalized coordinates:

$$T = \frac{m}{2}\,(\dot{x}^2 + \dot{y}^2 + \dot{z}^2) = \frac{m}{2}\,(\dot{r}^2 + r^2\,\dot{\phi}^2\sin^2\alpha) \tag{10.32}$$

$$U = m\,g\,z = m\,g\,r\cos\alpha \tag{10.33}$$

Thus, the Lagrange function is

$$\mathcal{L}(r,\dot{r},\dot{\phi}) = T - U = \frac{m}{2}\,(\dot{r}^2 + r^2\,\dot{\phi}^2\sin^2\alpha) - m\,g\,r\cos\alpha \tag{10.34}$$

Equations of motion:

$$\frac{d}{dt}\frac{\partial\mathcal{L}}{\partial\dot{r}} - \frac{\partial\mathcal{L}}{\partial r} = m\,\ddot{r} - m\,r\,\dot{\phi}^2\sin^2\alpha + m\,g\cos\alpha = 0 \tag{10.35}$$

$$\frac{d}{dt}\frac{\partial\mathcal{L}}{\partial\dot{\phi}} - \frac{\partial\mathcal{L}}{\partial\phi} = \frac{d}{dt}(\,m\,r^2\,\dot{\phi}\,\sin^2\alpha) = 0 \tag{10.36}$$

Constants of motion: The Lagrange function $\mathcal{L}(r,\dot{r},\dot{\phi})$ does not depend explicitly on ϕ or on t. The conserved quantity belonging to the cyclic coordinate results directly from the equation of motion (10.36), provided that the derivative with respect to time has not yet been carried out:

$$\ell = m\,r^2\,\dot{\phi}\,\sin^2\alpha = \text{const.} \tag{10.37}$$

This is the z-component of the angular momentum. Its conservation follows from the invariance of the present problem with respect to rotations around the z-axis; for this, the potential *and* the constraint must be rotationally invariant.

From $\partial\mathcal{L}/\partial t = 0$ and $\partial x_n/\partial t = 0$ follows the conservation of energy

$$E = T + U = \frac{m}{2}\,(\dot{r}^2 + r^2\,\dot{\phi}^2\,\sin^2\alpha) + m\,g\,r\cos\alpha = \text{const.} \tag{10.38}$$

Solution of the equation of motion: Each constant of motion represents a first integral of the equations of motion; thus it can replace one of the equations of motion. Thus, (10.37) trivially replaces (10.36). Using (10.37), we eliminate $\dot{\phi}$ from (10.38) and obtain

$$E = \frac{m\,\dot{r}^2}{2} + U_{\text{eff}}(r), \quad U_{\text{eff}}(r) = \frac{\ell^2}{2m r^2 \sin^2\alpha} + m g r \cos\alpha \tag{10.39}$$

The two equations (10.37) and (10.39) are differential equations of first order. They can be used to replace the equations of motion (10.35) and (10.36) (both differential equations of second order).

Equation (10.39) is equivalent to the problem discussed in (2.21). As there, we solve for $dr/dt = f(r)$ and integrate according to $\int dt = \int dr/f(r)$,

$$\int_{t_0}^{t} dt' = t - t_0 = \int_{r_0}^{r} dr' \sqrt{\frac{m}{2[E - U_{\text{eff}}(r')]}} \tag{10.40}$$

The solution of the integral is a function $t = t(r)$, which may be resolved to $r = r(t)$. As an elliptic integral, however, (10.40) cannot be solved elementarily, but it could be solved numerically. From the solution $r(t)$ results then also $\phi(t)$ by the integration of (10.37):

$$\phi(t) - \phi_0 = \frac{\ell}{m \sin^2\alpha} \int_{t_0}^{t} \frac{dt'}{r(t')^2} \tag{10.41}$$

Integration constants: For the complete determination of the solution the four initial conditions

$$r(t_0) = r_0, \quad \dot{r}(t_0) = \dot{r}_0, \quad \phi(t_0) = \phi_0, \quad \dot{\phi}(t_0) = \dot{\phi}_0 \tag{10.42}$$

must be given at some time t_0. The four integration constants of the solution of (10.35) and (10.36) can be determined from the four numbers r_0, $\dot{r}_0$, ϕ_0, and $\dot{\phi}_0$. Alternatively, the solution is fixed by following integration constants:

$$\ell,\ E,\ \phi_0,\ r_0 \quad \text{(integration constants)} \tag{10.43}$$

With each conservation law, an integration constant is obtained.

Discussion of the solution: Even without an explicit solution we can discuss the form of possible trajectories with the help of the energy

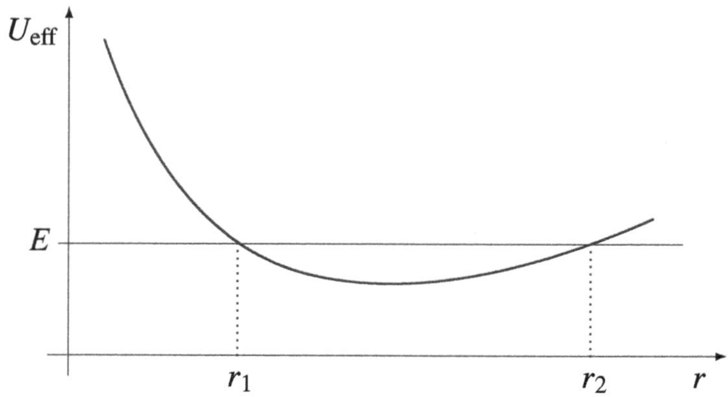

Figure 10.2 Graphical discussion of the particle's motion on the circular cone (Figure 10.1). The energy of the particle is conserved, $E = m\dot{r}^2/2 + U_{eff}(r) =$ const. The equation $E = U_{eff}(r)$ has two solutions, r_1 and r_2; the solution is restricted to $r_1 \leq r \leq r_2$. The kinetic energy $m\dot{r}^2/2 \geq 0$ is equal to the vertical distance between the horizontal $E =$ const. and $U_{eff}(r)$. The particle oscillates between the turning points r_1 and r_2.

expression (10.39). For $\ell \neq 0$, $U_{eff}(r)$ is of the form $a/r^2 + br$. As sketched in Figure 10.2, $U_{eff}(r)$ has a minimum and goes to infinity for $r \to 0$ and $r \to \infty$. By E and ℓ the two turning points r_1 and r_2 are determined, Figure 10.2. The solution $r(t)$ oscillates between r_1 and r_2. At the same time, because of $\dot{\phi} \neq 0$ a motion in ϕ direction occurs. The mass point glides up and down in the cone on a wavy line, which is limited by the circles at $z_1 = r_1 \cos\alpha$ and $z_2 = r_2 \cos\alpha$. This wavy line is in general not closed in itself.

Exercises

10.1 *Small oscillations of the double pendulum*

Formulate the constraint conditions for the plane double pendulum, and introduce the angles φ_1 and φ_2 as the generalized coordinates. Set up the Lagrangian equations. Restrict to small oscillations and solve the equations using the ansatz

$$\begin{pmatrix} \varphi_1(t) \\ \varphi_2(t) \end{pmatrix} = \begin{pmatrix} a_1 \\ a_2 \end{pmatrix} \exp(\mathrm{i}\omega t)$$

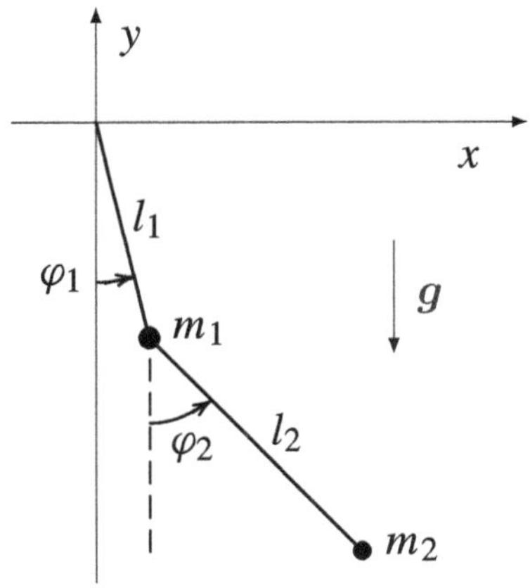

Discuss the solution for the following three cases: (i) $m_1 \ll m_2$, (ii) $m_1 \gg m_2$, and (iii) $m_1 = m_2 = m$, $l_1 = l_2 = l$.

10.2 *Dumbbell with friction force*

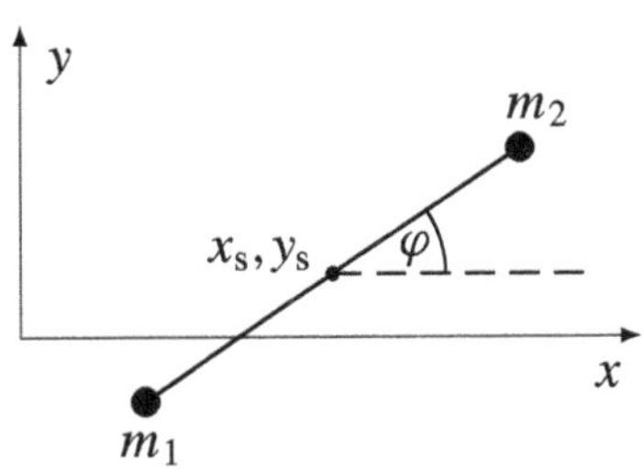

Two point masses are connected by a massless rod of length L and can move in the x–y-plane. Both masses are subject to a frictional force, which is proportional to their velocity.

Set up the Lagrange function. Use as generalized coordinates q_k the center-of-mass coordinates x_s, y_s of the dumbbell and the angle φ between the dumbbell and the x-axis. Calculate the generalized frictional forces Q_k. Give the general solution of the equations of motion.

Chapter 11

Space–Time Symmetries

Space and time of an inertial frame show certain symmetries. We investigate the consequences of these space-time symmetries. For closed systems, these symmetries lead to conservation laws.

Introduction

By *symmetry*, we understand the *invariance* under a certain *operation*. The words "noon" or "madam" do not change when read from the back; the special symmetry of this word consists in the invariance under the operation that reverses the order of the letters. A chessboard is invariant under the operation "exchanging light and dark squares, and rotating by $\pi/2$". An (infinite) crystal is invariant under spatial displacement by one lattice constant. A sphere is invariant under arbitrary rotations around the center. One distinguishes between discrete symmetries (reflection, displacement by one lattice constant, rotation by $\pi/2$) and continuous symmetries, which depend on a continuous parameter (e.g. a rotation angle).

In physics, the objects whose symmetry is studied, are physical systems. The systems and their dynamics are described mathematically in models. The symmetry of the systems should therefore be reflected in the mathematical objects (such as the equation of motion or the Lagrange function) that describe the system. In Chapter 5, we have investigated under which operations (which were called transformations there) Newton's first law is form invariant. This led to the general Galilean transformation (5.9). In the previous chapters, we have studied the Lagrange function $\mathcal{L}$, which characterizes the system under consideration; from $\mathcal{L}$ follow in particular the equations of motion. In this chapter, we investigate how the Lagrange function behaves under active Galilean transformations. As a result, we obtain the statement that from any invariance of the Lagrangian function (i.e. from any symmetry of the system) a conservation law follows.

By inertial system (IS) we mean an inertial reference frame with coordinates x_1, x_2, x_3, and t. We consider the operation

$$x_i \to x_i^*, \quad t \to t^* \tag{11.1}$$

where x_i^*, t^* and x_i, t are connected by a Galilean transformation:

$$x_i^* = \sum_{j=1}^{3} \alpha_{ij}\, x_j - v_i\, t - a_i, \quad t^* = t - t_0 \tag{11.2}$$

As already discussed in Chapter 5, a Galilean transformation can be considered from the following points of view:

(i) Passive transformation: The same physical system is viewed from two different IS. Then (11.2) links the coordinates, which the same event has in the two inertial systems.
(ii) Active transformation: Within one IS, two physical systems are considered which are connected by the transformation. Then (11.2) connects the coordinates of an event in the one system with those of the corresponding event in the transformed system.

In what follows, we study active transformations, such as the displacement or rotation of a physical system. Thereby we use the terms transformation and operation synonymously; in the narrower sense the term transformation refers to the equation (11.1) with (11.2), and the term operation to the displacement or rotation described by it.

By (11.1) with (11.2) the following operations can be described:

1. Time shift by a constant amount t_0.
2. Spatial displacement by a constant vector a_i.
3. Spatial rotation by three constant angles (contained in α_{ij}).
4. Spatial displacement by the time-dependent vector $v_i\, t$.

External influences generally cause the system is not invariant under the indicated operations. Thus, for example, the billiard game is decisively changed, if the plate is tilted a little bit; the gravity field of the earth distinguishes the vertical direction. But the billiard game is invariant under rotations around the vertical axis or under spatial displacements in horizontal direction; and it is also possible to play billiards in a train moving with constant velocity (no acceleration). In contrast to this, a three-dimensional billiard inside a cube (without a gravitational field) would be a system that is invariant under all four operations. This is also true for our solar system if we neglect the gravitational field of our galaxy. A system, which is not subject to any external influences (or for which they can be neglected), is called *closed*.

All facts of experience are compatible with the assumption that closed systems function in the same way under the following conditions:

1. Today, tomorrow, or in 10^8 years.
2. Here or at a distance of 10^8 lightyears.
3. As it is, or somehow rotated.
4. As it is, or moving with constant velocity relative to it.

"To function in the same way" means "to run according to the same laws". In this sense, closed systems are invariant under the operations under consideration. Since these symmetries apply to *all* closed systems, they represent properties of space and time. The symmetries are called:

1. Homogeneity of time.
2. Homogeneity of space.
3. Isotropy of space.
4. Relativity of space–time.

Our discussion refers to inertial frames or systems (IS). The just mentioned homogeneity, isotropy, and relativity properties apply only to the space–time of inertial systems (IS). Thus, for example, only for an IS, all directions are equivalent. For example, if a relative to IS rotating system has a distinguished direction (the axis of rotation relative to IS); then the isotropy of space is invalid. In contrast to the first three symmetries, the fourth one involves space *and* time; the corresponding transformation connects space and time coordinates.

The statement that the space and time of an IS has enumerated symmetries is an *assumption*, which is in agreement with all facts of experience. In particular, there are far-reaching consequences (like the corresponding conservation laws), which are well-verified experimentally. There are two reservations to be made: Although the relativity of space–time (principle of relativity) applies in general, the transformation (11.2) is only correct, if the considered relative velocity is small compared to the velocity of light. And on cosmic scales (for instance, over distances of 10^{10} light years) there is no global IS; local IS (for instance, in the range of the Milky Way) is possible.

The listed space–time symmetries were discussed in Chapter 5 for Newton's first law. We now apply this discussion to more general mechanical systems, which we describe by a Lagrangian function. The most general system we have treated so far is a system of mass points with the Lagrange function

$$\mathcal{L}(\boldsymbol{r}_1, \ldots, \boldsymbol{r}_N, \dot{\boldsymbol{r}}_1, \ldots, \dot{\boldsymbol{r}}_N, t) = \frac{1}{2} \sum_{\nu=1}^{N} m_\nu \, \dot{\boldsymbol{r}}_\nu^{\,2} - U(\boldsymbol{r}_1, \ldots, \boldsymbol{r}_N, t) \tag{11.3}$$

The position vector of the νth mass point can be represented by Cartesian coordinates, $\boldsymbol{r}_\nu := (x_1^\nu, x_2^\nu, x_3^\nu)$. For these Cartesian coordinates the transformation (11.1) is considered. Let there be no constraints; in principle, constraints can be replaced by suitable potentials. The potential U generally includes the contributions of internal and external forces; an example of U is (4.36).

The system described by (11.3) could be, for example, our solar system with all planets and moons, or also a polyatomic molecule, which executes rotations and oscillations.

As a special case of (11.3), we consider a *closed* system. This means, for example, that for the solar system, we neglect the gravitational field of the galaxy, or that we do not consider external electromagnetic fields for the molecule. For the internal forces, we assume the potential form discussed in Chapter 4. The Lagrange function of the closed system is then

$$\mathcal{L}_0(\boldsymbol{r}_1,\ldots,\boldsymbol{r}_N,\dot{\boldsymbol{r}}_1,\ldots,\dot{\boldsymbol{r}}_N) = \frac{1}{2}\sum_{\nu=1}^{N} m_\nu\,\dot{\boldsymbol{r}}_\nu^{\,2} - \sum_{\nu=2}^{N}\sum_{\mu=1}^{\nu-1} U_{\nu\mu}(|\boldsymbol{r}_\nu - \boldsymbol{r}_\mu|) \tag{11.4}$$

The closed system described by $\mathcal{L}_0$ has all the space–time symmetries discussed here. We investigate how this can be seen from the Lagrangian function $\mathcal{L}_0$, and identify the associated conservation laws.

Homogeneity of Time

We write the respective transformations in dependence on a parameter ϵ, where $\epsilon = 0$ is the identity transformation. In this way, we obtain a standard form of the invariance condition.

We start with temporal shifts, i.e. with the transformation

$$\boldsymbol{r}_\nu \to \boldsymbol{r}_\nu^* = \boldsymbol{r}_\nu, \quad t \;\to\; t^* = t + \epsilon \tag{11.5}$$

This is a special case of (11.1) and (11.2). From $\boldsymbol{r}_\nu^* = \boldsymbol{r}_\nu$ and $dt^* = dt$, it follows $d\boldsymbol{r}_\nu^*/dt^* = d\boldsymbol{r}_\nu/dt$ or $\dot{\boldsymbol{r}}^* = \dot{\boldsymbol{r}}$. For the general many-particle Lagrangian function (14.4), the following holds:

$$\mathcal{L}^* = \mathcal{L}(\ldots,\boldsymbol{r}_\nu^*,\ldots,\dot{\boldsymbol{r}}_\nu^*,\ldots,t^*) = \mathcal{L}(\ldots,\boldsymbol{r}_\nu,\ldots,\dot{\boldsymbol{r}}_\nu,\ldots,t+\epsilon) \tag{11.6}$$

Using the equations of motion, $d\mathcal{L}^*/d\epsilon$ can be transformed into a total time derivative:

$$\left(\frac{d\mathcal{L}^*}{d\epsilon}\right)_{\epsilon=0} = \frac{\partial\mathcal{L}}{\partial t} \overset{(9.35)}{=} \frac{d}{dt}\left(\mathcal{L} - \sum_{\nu=1}^{N}\frac{\partial\mathcal{L}}{\partial\dot{\boldsymbol{r}}_\nu}\cdot\dot{\boldsymbol{r}}_\nu\right) \tag{11.7}$$

The partial derivatives with respect to vectors are defined as in (4.20).

The Lagrangian function $\mathcal{L}_0$ of a closed system (11.4) does not explicitly depend on time; therefore $d\mathcal{L}_0^*/d\epsilon = 0$. Hereby, we obtain

$$\left(\frac{d\mathcal{L}_0^*}{d\epsilon}\right)_{\epsilon=0} = 0 \quad \overset{(11.7)}{\Longrightarrow} \quad \mathcal{L}_0 - \sum_{\nu=1}^{N} \frac{\partial \mathcal{L}_0}{\partial \dot{\boldsymbol{r}}_\nu} \cdot \dot{\boldsymbol{r}}_\nu = -(T+U) = \text{const.} \tag{11.8}$$

Herein, we used $\mathcal{L}_0 = T - U$ and $\sum(\partial\mathcal{L}_0/\partial\dot{\boldsymbol{r}}_\nu)\cdot\dot{\boldsymbol{r}}_\nu = 2T$. On the left side of (11.8) is the symmetry or invariance condition, and on the right side is the associated conservation law. This scheme is repeated for all the symmetries considered in the following.

We have established the relationship between invariance with respect to (11.5) and the conservation of energy for a relatively general mechanical system. This connection is in fact valid for all fields of physics: the energy of a closed system is identified as that conservation quantity that results from the homogeneity of time:

$$\text{Homogeneity of time} \quad \longrightarrow \quad \text{Energy conservation} \tag{11.9}$$

Homogeneity of Space

We consider a spatial displacement with ϵ in $\boldsymbol{n}$ direction:

$$\boldsymbol{r}_\nu \to \boldsymbol{r}_\nu^* = \boldsymbol{r}_\nu + \epsilon\,\boldsymbol{n}, \quad t \to t^* = t \tag{11.10}$$

This is a special case of (11.1) and (11.2). The unit vector $\boldsymbol{n}$ has an arbitrary constant direction; it could be, for example, $\boldsymbol{e}_x$, $\boldsymbol{e}_y$, or $\boldsymbol{e}_z$. For a general Lagrangian function, the transformation (11.10) yields

$$\mathcal{L}^* = \mathcal{L}(\ldots, \boldsymbol{r}_\nu^*, \ldots, \dot{\boldsymbol{r}}_\nu^*, \ldots, t^*) = \mathcal{L}(\ldots, \boldsymbol{r}_\nu + \epsilon\,\boldsymbol{n}, \ldots, \dot{\boldsymbol{r}}_\nu, \ldots, t) \tag{11.11}$$

Since $d\boldsymbol{r}_\nu^* = d\boldsymbol{r}_\nu$, the velocities do not change. Using the equations of motion, $d\mathcal{L}^*/d\epsilon$ can be transformed into a total time derivative:

$$\begin{aligned}\left(\frac{d\mathcal{L}^*}{d\epsilon}\right)_{\epsilon=0} &= \sum_{\nu=1}^{N} \frac{\partial \mathcal{L}}{\partial \boldsymbol{r}_\nu} \cdot \boldsymbol{n} = \frac{d}{dt}\sum_{\nu=1}^{N} \frac{\partial \mathcal{L}}{\partial \dot{\boldsymbol{r}}_\nu} \cdot \boldsymbol{n} \\ &= \frac{d}{dt}\sum_{\nu=1}^{N} \boldsymbol{p}_\nu \cdot \boldsymbol{n} = \frac{d(\boldsymbol{n}\cdot\boldsymbol{P})}{dt}\end{aligned} \tag{11.12}$$

Here, $\boldsymbol{P} = \sum \boldsymbol{p}_\nu = \sum \partial\mathcal{L}/\partial\dot{\boldsymbol{r}}_\nu$ is the center-of-mass momentum (i.e. the total momentum of the system).

The coordinate differences $\boldsymbol{r}_\nu^* - \boldsymbol{r}_\mu^* = \boldsymbol{r}_\nu - \boldsymbol{r}_\mu$ do not depend on ϵ. Therefore, the Lagrangian function $\mathcal{L}_0$ of a closed system (11.4) is invariant under the transformation (11.10):

$$\left(\frac{d\mathcal{L}_0^*}{d\epsilon}\right)_{\epsilon=0} = 0 \quad \overset{(11.12)}{\Longrightarrow} \quad \boldsymbol{P} = \text{const.} \tag{11.13}$$

Since $\boldsymbol{n}$ is arbitrary, the constancy of $\boldsymbol{n} \cdot \boldsymbol{P}$ implies that of $\boldsymbol{P}$.

Again, in general: The total momentum of a closed system is identified as the conservation quantity which results from the homogeneity of the space:

$$\text{Homogeneity of space} \quad \longrightarrow \quad \text{Momentum conservation} \tag{11.14}$$

Isotropy of Space

We consider an infinitesimal rotation about a constant angle ϵ; let the axis of rotation be denoted by the constant unit vector $\boldsymbol{n}$. Then the transformation (Figure 11.1) is

$$\boldsymbol{r}_\nu \to \boldsymbol{r}_\nu^* = \boldsymbol{r}_\nu + (\boldsymbol{n} \times \boldsymbol{r}_\nu)\,\epsilon \qquad t \to t^* = t \tag{11.15}$$

This is a special case of (11.1) and (11.2). Since ϵ and $\boldsymbol{n}$ are time-independent, it follows for the velocities

$$\dot{\boldsymbol{r}}_\nu \to \dot{\boldsymbol{r}}_\nu^* = \dot{\boldsymbol{r}}_\nu + (\boldsymbol{n} \times \dot{\boldsymbol{r}}_\nu)\epsilon \tag{11.16}$$

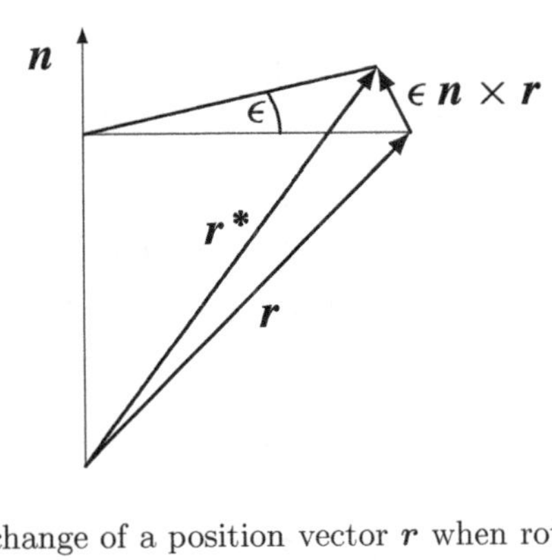

Figure 11.1 The change of a position vector $\boldsymbol{r}$ when rotated by the angle ϵ is $(\boldsymbol{n} \times \boldsymbol{r})\,\epsilon$, where the unit vector $\boldsymbol{n}$ points in the direction of the axis of rotation.

For a general Lagrange function, this transformation yields

$$\begin{aligned}\mathcal{L}^* &= \mathcal{L}(\ldots, \boldsymbol{r}_\nu^*, \ldots, \dot{\boldsymbol{r}}_\nu^*, \ldots, t^*) \\ &= \mathcal{L}(\ldots, \boldsymbol{r}_\nu + (\boldsymbol{n} \times \boldsymbol{r}_\nu)\,\epsilon, \ldots, \dot{\boldsymbol{r}}_\nu + (\boldsymbol{n} \times \dot{\boldsymbol{r}}_\nu)\,\epsilon, \ldots, t) \qquad (11.17)\end{aligned}$$

Using the equations of motion, $d\mathcal{L}^*/d\epsilon$ can be transformed into a total time derivative:

$$\begin{aligned}\left(\frac{d\mathcal{L}^*}{d\epsilon}\right)_{\epsilon=0} &= \sum_{\nu=1}^{N} \frac{\partial \mathcal{L}}{\partial \boldsymbol{r}_\nu} \cdot (\boldsymbol{n} \times \boldsymbol{r}_\nu) + \sum_{\nu=1}^{N} \frac{\partial \mathcal{L}}{\partial \dot{\boldsymbol{r}}_\nu} \cdot (\boldsymbol{n} \times \dot{\boldsymbol{r}}_\nu) \\ &= \sum_{\nu=1}^{N} \left(\frac{d\boldsymbol{p}_\nu}{dt} \cdot (\boldsymbol{n} \times \boldsymbol{r}_\nu) + \boldsymbol{p}_\nu \cdot \frac{d\,(\boldsymbol{n} \times \boldsymbol{r}_\nu)}{dt} \right) \\ &= \frac{d}{dt} \sum_{\nu=1}^{N} \boldsymbol{n} \cdot (\boldsymbol{r}_\nu \times \boldsymbol{p}_\nu) = \frac{d(\boldsymbol{n} \cdot \boldsymbol{L})}{dt} \qquad (11.18)\end{aligned}$$

Here, $\boldsymbol{L} = \sum_\nu \boldsymbol{r}_\nu \times \boldsymbol{p}_\nu$ is the total angular momentum of the system. The Lagrange function $\mathcal{L}_0$ from (11.4) depends only on $\dot{\boldsymbol{r}}_\nu^{\,2}$ and on the distances $|\boldsymbol{r}_\nu - \boldsymbol{r}_\mu|$. These quantities do not change under the considered transformation:

$$(\dot{\boldsymbol{r}}_\nu^*)^2 = \dot{\boldsymbol{r}}_\nu^{\,2} + \mathcal{O}(\epsilon^2), \quad (\boldsymbol{r}_\nu^* - \boldsymbol{r}_\mu^*)^2 = (\boldsymbol{r}_\nu - \boldsymbol{r}_\mu)^2 + \mathcal{O}(\epsilon^2) \qquad (11.19)$$

The terms $\mathcal{O}(\epsilon^2)$ occur because the transformation (11.15) describes the rotation correctly only up to the first order in ϵ. The invariance condition $(d\mathcal{L}_0^*/d\epsilon)_{\epsilon=0} = 0$ is again satisfied in any case. Thus, we obtain

$$\left(\frac{d\mathcal{L}_0^*}{d\epsilon}\right)_{\epsilon=0} = 0 \quad \overset{(11.18)}{\Longrightarrow} \quad \boldsymbol{L} = \text{const.} \qquad (11.20)$$

Since $\boldsymbol{n}$ is arbitrary, the constancy of $\boldsymbol{n} \cdot \boldsymbol{L}$ implies that of $\boldsymbol{L}$.

Again, in general: The total angular momentum of a closed system is identified as that conservation quantity which results from the isotropy of the space:

$$\text{Isotropy of space} \quad \longrightarrow \quad \text{Conservation of angular momentum} \qquad (11.21)$$

Relativity of Space–Time

We consider the transformation

$$\boldsymbol{r}_\nu \to \boldsymbol{r}_\nu^* = \boldsymbol{r}_\nu + \epsilon\, \boldsymbol{n}\, t \quad t \to\ t^* = t \tag{11.22}$$

between two systems moving relative to each other with the infinitesimal and constant velocity $\epsilon \boldsymbol{n}$. The transformation (11.22) is a special case of (11.1) and (11.2). From (11.22) follows:

$$\dot{\boldsymbol{r}}_\nu \to \dot{\boldsymbol{r}}_\nu^* = \dot{\boldsymbol{r}}_\nu + \epsilon\, \boldsymbol{n} \tag{11.23}$$

For a general Lagrange function, this transformation yields

$$\begin{aligned} \mathcal{L}^* &= \mathcal{L}(\ldots, \boldsymbol{r}_\nu^*, \ldots, \dot{\boldsymbol{r}}_\nu^*, \ldots, t^*) \\ &= \mathcal{L}(\ldots, \boldsymbol{r}_\nu + \epsilon\, \boldsymbol{n}\, t, \ldots, \dot{\boldsymbol{r}}_\nu + \epsilon\, \boldsymbol{n}, \ldots, t) \end{aligned} \tag{11.24}$$

Using the equations of motion, $d\mathcal{L}^*/d\epsilon$ can be transformed into a total time derivative:

$$\begin{aligned} \left(\frac{d\mathcal{L}^*}{d\epsilon}\right)_{\epsilon=0} &= \sum_{\nu=1}^{N} \frac{\partial \mathcal{L}}{\partial \boldsymbol{r}_\nu} \cdot \boldsymbol{n}\, t + \sum_{\nu=1}^{N} \frac{\partial \mathcal{L}}{\partial \dot{\boldsymbol{r}}_\nu} \cdot \boldsymbol{n} \\ &= \boldsymbol{n} \cdot \sum_{\nu=1}^{N} (\dot{\boldsymbol{p}}_\nu\, t + \boldsymbol{p}_\nu) = \frac{d(\boldsymbol{n} \cdot \boldsymbol{P} t)}{dt} \end{aligned} \tag{11.25}$$

Here, $\boldsymbol{P} = \sum_\nu \boldsymbol{p}_\nu$ is the center-of-mass momentum.

The Lagrange function $\mathcal{L}_0$ from (11.4) depends only on $\dot{\boldsymbol{r}}_\nu^{\,2}$ and on the distances $|\boldsymbol{r}_\nu - \boldsymbol{r}_\mu|$. The distances are invariant under (11.22); therefore $U^* = U$ holds for the potential energy. In the kinetic energy, on the other hand, consider (11.23):

$$\begin{aligned} \mathcal{L}_0^* &= \frac{1}{2} \sum_{\nu=1}^{N} m_\nu\, (\dot{\boldsymbol{r}}_\nu^*)^2 - U^* = T + \sum_{\nu=1}^{N} m_\nu\, \dot{\boldsymbol{r}}_\nu \cdot \boldsymbol{n}\, \epsilon + \mathcal{O}(\epsilon^2) - U \\ &= \mathcal{L}_0 + \epsilon\, \frac{d}{dt} \sum_{\nu=1}^{N} m_\nu\, \boldsymbol{n} \cdot \boldsymbol{r}_\nu + \mathcal{O}(\epsilon^2) = \mathcal{L}_0 + \epsilon M\, \frac{d(\boldsymbol{n} \cdot \boldsymbol{R})}{dt} + \mathcal{O}(\epsilon^2) \end{aligned} \tag{11.26}$$

In the last step, we introduced the center-of-mass coordinate $\boldsymbol{R} = \sum_\nu m_\nu \boldsymbol{r}_\nu / M$ and the total mass $M = \sum_\nu m_\nu$.

According to (11.26), the Lagrange function $\mathcal{L}_0$ is not invariant under the transformation (11.22). The additional term appearing under the transformation is a total time derivative. One now easily convinces oneself (Exercise 11.1) that such an additional term does not change the equations of motion. The equations of motion and therefore the system under consideration are invariant under the transformation (11.22).

From (11.26) follows:

$$\left(\frac{d\mathcal{L}_0^*}{d\epsilon}\right)_{\epsilon=0} = M\,\frac{d(\boldsymbol{n}\cdot\boldsymbol{R})}{dt} \tag{11.27}$$

Unlike the cases discussed above, $d\mathcal{L}_0^*/d\epsilon$ does not vanish. However, if we consider (11.25) for $\mathcal{L}_0^*$ and combine it with (11.27), one obtains $d(\boldsymbol{n}\cdot\boldsymbol{P}t)/dt = M\,d(\boldsymbol{n}\cdot\boldsymbol{R})/dt$. With $\boldsymbol{P} = M\dot{\boldsymbol{R}}$ and taking into account the arbitrariness of $\boldsymbol{n}$, this gives $d(\dot{\boldsymbol{R}}t - \boldsymbol{R})/dt = 0$ or

$$\dot{\boldsymbol{R}}t - \boldsymbol{R} = \text{const.} \tag{11.28}$$

This, like the other conservation laws, is a first order differential equation. Its integration leads to $\boldsymbol{R} = \boldsymbol{A}t + \boldsymbol{B}$ with the integration constants $\boldsymbol{A}$ and $\boldsymbol{B}$. This statement follows also from (11.13), so in this case, we do not obtain a new conserved quantity.

A closed system is invariant under Galilean transformations. The Galilean transformations form a 10-parameter group. For each of the 10 invariances or symmetries, we have assigned a conservation quantity, namely the energy E, the momentum $\boldsymbol{P}$, the angular momentum $\boldsymbol{L}$ and the magnitude $\dot{\boldsymbol{R}}t - \boldsymbol{R}$.

Exercises

11.1 *Total time derivative in the Lagrange function*

Two Lagrange functions differ by the total time derivative of an arbitrary function $f(q,t)$ of the coordinates and of time:

$$\mathcal{L}^*(q,\dot{q},t) = \mathcal{L}(q,\dot{q},t) + \frac{d}{dt}\,f(q,t) \tag{11.29}$$

Show that the Lagrange functions $\mathcal{L}^*$ and. $\mathcal{L}$ yield the same equations of motion.

PART III
Variational Principles

Chapter 12

Variation without Constraint

Part III introduces the variational calculus (Chapters 12 and 13). In Chapter 14, the basic laws of mechanics are formulated as a variational principle (Hamilton's principle). Chapter 15 deals with the connection between symmetries and conservation laws in a general form (Noether's theorem).

Problem

A *function* $y = y(x)$ is a prescription that assigns to each x-value a number (the y-value). In the calculus of variations, one considers *functionals*, which assigns to each function a number (the value of the functional). For example,

$$J = J[y] = \int_1^2 ds = \int_{x_1}^{x_2} dx\,\sqrt{1 + y'(x)^2} \tag{12.1}$$

is the functional that calculates the distance along the curve $y = y(x)$ between the two points $(x_1, y(x_1))$ and $(x_2, y(x_2))$. This functional assigns to each curve $y(x)$ connecting the two points, a number (the path length). To distinguish it from functions we enclose the argument y of the functional in square brackets.

When discussing a function $y = y(x)$, a common question is, for which argument values the function is maximal or minimal. The necessary condition for a local extremum of a differentiable function $y = y(x)$ is

$$\frac{dy(x)}{dx} = 0 \tag{12.2}$$

At the points x that satisfy this condition, the function is *stationary.* A more detailed examination then shows whether there is a minimum or maximum or a turning point with a horizontal tangent at this point.

We consider the analogous question: For which function is a certain functional extremal? As an answer, we will set up a condition for the stationarity of the functional. This is then, as in (12.2), a necessary condition for the functional to become *maximal* or *minimal*. For simplicity, we adopt the formulation "$J[y]$ = minimal" in the following.

We restrict the wanted functions $y(x)$ by boundary values:

$$y(x_1) = y_1 \quad \text{and} \quad y(x_2) = y_2 \tag{12.3}$$

We can now ask: For which curve is the path distance between two given points minimal? This means mathematically: For which function $y(x)$ the functional (12.1) becomes minimal?

A starting point of the calculus of variations was the problem of the *brachistochrone*, Figure 12.1, which was formulated in 1696 by Jakob Bernoulli: A mass point slides frictionless and under the influence of gravity along a curve $y(x)$ from P_1 to P_2. Let its initial velocity be zero. Now, $y(x)$ is to be determined in such a way that the time T, which the particle needs for the path from P_1 to P_2 becomes minimal. The result can be viewed as an optimal packet slide. The time T to be minimized is a functional of $y(x)$.

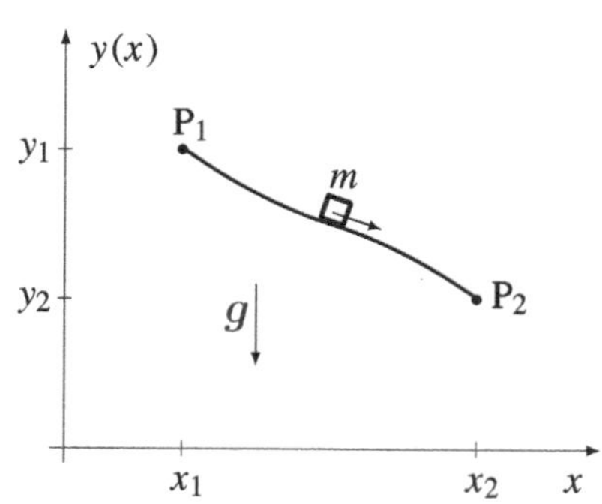

Figure 12.1 On which curve a frictionless gliding body in the gravitational field comes fastest from P_1 to P_2? The solution curve is called brachistochrone.

Due to energy conservation, the kinetic energy of the mass is equal to $mv^2/2 = mg\,(y_1 - y)$. Thus, we obtain

$$J[y] = T = \int_1^2 \frac{ds}{v} = \int_{x_1}^{x_2} dx\,\sqrt{\frac{1+y'(x)^2}{2g\,(y_1 - y(x))}} \tag{12.4}$$

We introduce another physical problem: By rotation around the x-axis, the points P_1 and P_2 in Figure 12.1 describe circular rings (which could be realized as wire rings). By this rotation, the curve $y(x)$ becomes an area of revolution A. This area might be realized by a skin of soapy water, which is stretched between two wire rings. The surface tension implies that this surface is minimal (in equilibrium, no gravity field). Therefore, the following functional must be minimal:

$$J[y] = A = 2\pi \int_{x_1}^{x_2} ds\; y = 2\pi \int_{x_1}^{x_2} dx\; y\,\sqrt{1+y'^2} \tag{12.5}$$

From $J[y]$ = minimal, $y(x)$ and thus the position of the soap skin can be determined (Exercise 12.2).

Euler–Lagrange Equation

The general problem is: which function $y(x)$ makes the functional

$$J = J[y] = \int_{x_1}^{x_2} dx\; F(y, y', x) \tag{12.6}$$

minimal? Here, the function $F(y, y', x)$ and the boundary values

$$y(x_1) = y_1, \quad y(x_2) = y_2 \tag{12.7}$$

are taken as given.

Now, let $y(x)$ be the function that minimizes the functional J. Then $J[y + \delta y]$ with an arbitrarily small deviation δy is larger than $J[y]$. An infinitesimal deviation $\delta y(x)$ from $y(x)$ is written as

$$\delta y(x) = \epsilon\,\eta(x) \tag{12.8}$$

Here, the factor ϵ is infinitesimal. The function $\eta(x)$ is restricted by

$$\eta(x_1) = \eta(x_2) = 0 \tag{12.9}$$

because $y + \delta y$ must pass through the boundary points. We assume that $\eta(x)$ is defined in the whole interval and is differentiable; otherwise $\eta(x)$ is arbitrary.

The functional $J[y + \epsilon\eta]$ is a function of ϵ, which we denote by $J(\epsilon)$. For the searched function $y(x)$, $J(\epsilon)$ must be minimal at $\epsilon = 0$:

$$J[y + \epsilon\eta] \;\text{ is minimal at } \epsilon = 0 \quad \text{for any } \eta(x) \tag{12.10}$$

If this were not true for some $\eta_0(x)$, then $J[y + \epsilon\eta_0]$ could be made smaller than $J[y]$ for $\epsilon \neq 0$. A possible deviation $\epsilon\,\eta(x)$ and the resulting $J[y + \epsilon\eta]$ are sketched in Figure 12.2.

The condition (12.10) implies

$$\left(\frac{d\,J[y + \epsilon\eta]}{d\epsilon}\right)_{\epsilon=0} = 0 \quad \text{for any } \eta(x) \tag{12.11}$$

This is the condition for stationarity, and thus — just like (12.2) — a necessary condition for a local extremum. We evaluate (12.11), by expanding $J[y + \epsilon\eta]$ into powers of ϵ,

$$\begin{aligned} J[y + \epsilon\eta] &= \int_{x_1}^{x_2} dx\; F(y + \epsilon\eta, y' + \epsilon\eta', x) \\ &= \int_{x_1}^{x_2} dx\; \big\{F(y, y', x) + F_y(y, y', x)\,\epsilon\,\eta(x) \\ &\qquad + F_{y'}(y, y', x)\,\epsilon\,\eta'(x)\big\} + \mathcal{O}(\epsilon^2) \end{aligned} \tag{12.12}$$

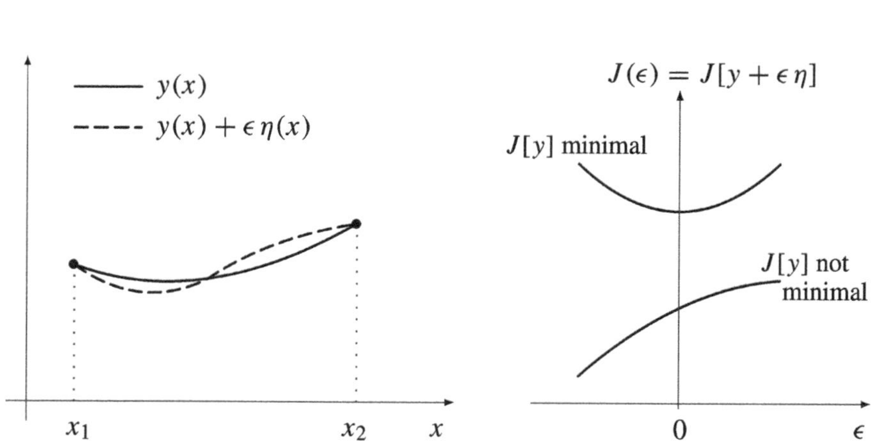

Figure 12.2 For a function $y(x)$ with fixed boundary points (•) a deviating function $y(x) + \epsilon\,\eta(x)$ is considered (left). If $y(x)$ is the sought solution then $J(\epsilon) = J[y + \epsilon\eta]$ has a minimum at $\epsilon = 0$ (right).

We assume that $F(y, y', x)$ is differentiable, and use the notation

$$F_y = \frac{\partial F}{\partial y} \quad \text{and} \quad F_{y'} = \frac{\partial F}{\partial y'} \tag{12.13}$$

From (12.11) follows:

$$\begin{aligned} 0 = \left(\frac{d\,J[y+\epsilon\eta]}{d\epsilon}\right)_{\epsilon=0} &= \int_{x_1}^{x_2} dx\,\left\{F_y\,\eta(x) + F_{y'}\,\eta'(x)\right\} \\ &\overset{\text{p.i.}}{=} \underbrace{F_{y'}\,\eta(x)\,\Big|_{x_1}^{x_2}}_{=0} + \int_{x_1}^{x_2} dx\,\left(F_y - \frac{dF_{y'}}{dx}\right)\eta(x) \quad \text{for any } \eta(x) \end{aligned} \tag{12.14}$$

In the step from the first to the second line we performed a partial integration (p.i.). Since $\eta(x)$ is arbitrary, the bracket expression in the last integrand must be zero. If the expression were positive at any point (or negative), it would be so also in a certain environment. Then an $\eta(x)$, which is positive in this environment and otherwise equal to zero, would lead to a non-zero integral. Therefore, $F_y - dF_{y'}/dx = 0$ or, more explicitly,

$$\frac{d}{dx}\frac{\partial F(y,y',x)}{\partial y'} = \frac{\partial F(y,y',x)}{\partial y} \quad \text{Euler–Lagrange equation} \tag{12.15}$$

This *Euler–Lagrange equation* is a differential equation of the second order for the function $y(x)$; it is linear in y''. The Euler–Lagrange equation is equivalent to the stationarity of the functional $J[y]$. It is therefore a necessary condition for an extremum; a more detailed investigation can show whether in the specific case actually a minimum or a maximum is present.

The $\delta y(x) = \epsilon\eta(x)$ introduced in (12.8) is called also a "variation of $y(x)$". For the wanted $y(x)$ the functional J is stationary, which means J does not change at infinitesimal changes δy of $y(x)$; the variation δJ of J vanishes. In a common short notation, we sketch once more the derivation of the Euler–Lagrange equations:

$$\begin{aligned} \delta J = J[y+\delta y] - J[y] &= \int_{x_1}^{x_2} dx\,\left(F_y\,\delta y + F_{y'}\,\delta y'\right) \\ &= \int_{x_1}^{x_2} dx\,\left(F_y - \frac{dF_{y'}}{dx}\right)\delta y = 0 \end{aligned} \tag{12.16}$$

From the arbitrariness of δy follows again the Euler–Lagrange equations. From this we see that the Euler–Lagrange equations and $\delta J = 0$ are equivalent:

$$\frac{dF_{y'}}{dx} = F_y \longleftrightarrow \delta J = 0 \tag{12.17}$$

Shortest path

As an example, we determine the shortest path between two given points, i.e. the function $y(x)$ for which (12.1) becomes minimal. For

$$F = \sqrt{1 + y'(x)^2} \tag{12.18}$$

we obtain the Euler–Lagrange equation

$$\frac{d}{dx} \frac{y'(x)}{\sqrt{1 + y'(x)^2}} = 0 \tag{12.19}$$

The integration of this equation yields

$$y' = \text{const.} \quad y = ax + b \tag{12.20}$$

The constants are defined by the given boundary points. Thus, we get a straight line as the shortest path between two points (no surprise).

Solving the Euler–Lagrange Equation

With the correspondence

$$y(x) \longleftrightarrow q(t) \quad \text{and} \quad F(y, y', x) \longleftrightarrow \mathcal{L}(q, \dot{q}, t) \tag{12.21}$$

one sees that the Euler–Lagrange equation of variational calculus and the Lagrange equation of mechanics have the same structure. The solution strategies are therefore the same. In particular, we can carry over the known conservation laws:

Momentum conservation: The integrand F does not explicitly depend on y, i.e. $F = F(y', x)$. From $F_y = 0$ and the Euler–Lagrange equation we obtain

$$F_{y'}(y', x) = c = \text{const.} \tag{12.22}$$

We may resolve this first-order differential equation to $y' = g(x, c)$. Then a further integration yields the solution

$$y(x) = \int dx\; g(x, c) + \text{const.} \tag{12.23}$$

Conservation of energy: The integrand F does not explicitly depend on x, i.e. $F = F(y, y')$. From $F_x = 0$ and the Euler–Lagrange equation then follows (analogous to (9.35))

$$F_{y'}(y, y')\, y' - F(y, y') = c = \text{const.} \tag{12.24}$$

We may resolve this first-order differential equation to $y' = h(y, c)$. Then a further integration yields

$$x = x(y) = \int \frac{dy}{h(y, c)} + \text{const.} \tag{12.25}$$

Generalizations

We derived the Euler–Lagrange equation for a function $y(x)$ which depends on one variable. The form of the functional was assumed as $F = F(y, y', x)$. A variation of the boundary values was not allowed. We present now some generalizations.

Several functions

We consider a functional

$$J = J[y_1, \ldots, y_N] = \int_{x_1}^{x_2} dx\; F(y_1, \ldots, y_N, y_1', \ldots, y_N', x) \tag{12.26}$$

that depends on N functions $y_1(x), y_2(x), \ldots, y_N(x)$ with fixed boundary values

$$y_i(x_1) = y_{i1}, \quad y_i(x_2) = y_{i2} \quad (i = 1, \ldots, N) \tag{12.27}$$

We look for the functions $y_i(x)$ for which J becomes stationary. For these functions y_i the function

$$J(\epsilon_1, \epsilon_2, \ldots, \epsilon_N) = J[y_1 + \epsilon_1\, \eta_1, \ldots, y_N + \epsilon_N\, \eta_N] \tag{12.28}$$

must be stationary at $\epsilon_1 = \epsilon_2 = \cdots = \epsilon_N = 0$, i.e.

$$\left(\frac{\partial J(\epsilon_1, \ldots, \epsilon_N)}{\partial \epsilon_i} \right)_{\epsilon_i = 0} = 0 \quad (i = 1, \ldots, N) \tag{12.29}$$

This yields the N Euler–Lagrange equations

$$\frac{d}{dx} \frac{\partial F(y, y', x)}{\partial y_i'} - \frac{\partial F(y, y', x)}{\partial y_i} = 0 \quad (i = 1, \ldots, N) \tag{12.30}$$

Here, y and y' in the argument stand for $y_1, \ldots, y_N$ and $y_1', \ldots, y_N'$. These are N second-order differential equations for the N wanted functions $y_1(x), \ldots, y_N(x)$. In mechanics, this generalization is needed for a system of mass points.

Several arguments

We consider many arguments $x_1, \ldots, x_N$ instead of x only. We look for the function $y(x_1, \ldots, x_N)$ for which the functional

$$J = J[y] = \int_B dx_1 \cdots \int_B dx_N \; F\left(y, \frac{\partial y}{\partial x_1}, \ldots, \frac{\partial y}{\partial x_N}, x_1, \ldots, x_N \right) \tag{12.31}$$

is extremal. The function $y(x_1, \ldots, x_N)$ shall be fixed at the boundary of the integration range B. For the variation $\delta y = \epsilon\, \eta(x_1, \ldots, x_N)$ the functional again yields a function $J(\epsilon)$. From $dJ/d\epsilon = 0$ follow the Euler–Lagrange equations

$$\sum_{i=1}^{N} \frac{\partial}{\partial x_i} \frac{\partial F}{\partial \left(\partial y / \partial x_i \right)} - \frac{\partial F}{\partial y} = 0 \tag{12.32}$$

For the partial derivatives $\partial/\partial y$ and $\partial/\partial(\partial y/\partial x_i)$ the function F has to be considered in the form $F(y, \partial y/\partial x_i, x_i)$. For the derivative $\partial/\partial x_i$ then $\partial F/\partial(\partial y/\partial x_i)$ has to be considered as a function of the x_i alone.

A possible application of (12.32) is the deflection of a stretchable membrane (for example, a drum) which is clamped along a curve surrounding area B. Another example would be a soap skin in the gravitational field; gravity and surface tension compete with each other. In equilibrium, the sum of the potential energies is minimal.

The generalizations (12.26) and (12.31) can be combined. Then the functional contains several functions, each of which depends on several variables.

Higher derivatives

If higher derivatives occur, for example,

$$J = J[y] = \int_{x_1}^{x_2} dx\, F(y'', y', y, x) \tag{12.33}$$

then this results in correspondingly higher derivatives in the Euler–Lagrange equation. The boundary values $y(x_1)$, $y(x_2)$, $y'(x_1)$ and $y'(x_2)$ shall be fixed. Analogously to (12.16), we obtain:

$$\begin{aligned}\delta J &= \int_{x_1}^{x_2} dx \left(F_y\, \delta y + F_{y'}\, \delta y' + F_{y''}\, \delta y''\right) \\ &= \int_{x_1}^{x_2} dx \left(F_y - \frac{dF_{y'}}{dx} + \frac{d^2 F_{y''}}{dx^2}\right) \delta y = 0 \end{aligned} \tag{12.34}$$

Here, $F_{y''}\, \delta y''$ was converted to $(d^2 F_{y''}/dx^2)\, \delta y$ by a two-fold partial integration. From the arbitrariness of δy follows the Euler–Lagrange equation

$$F_y - \frac{dF_{y'}}{dx} + \frac{d^2 F_{y''}}{dx^2} = 0 \tag{12.35}$$

This is a fourth-order differential equation.

Variation at the boundary

We assume now that the boundary values are *not* fixed. Then in (12.14) the boundary terms $F_{y'}\, \eta$ remain, because $\eta(x_1)$ and $\eta(x_2)$ can now be arbitrary. In order that (12.14) becomes zero, $F_{y'}$ must vanish at the boundary:

$$F_{y'}\left(y(x_1),\, y'(x_1),\, x_1\right) = 0, \quad F_{y'}\left(y(x_2),\, y'(x_2),\, x_2\right) = 0 \tag{12.36}$$

Admitting a variation at the boundary results in two new boundary conditions. Therefore, one obtains in each case — boundary values given or not — the necessary number of boundary conditions for the determination of the integration constants.

Exercises

12.1 *Brachistochrone*

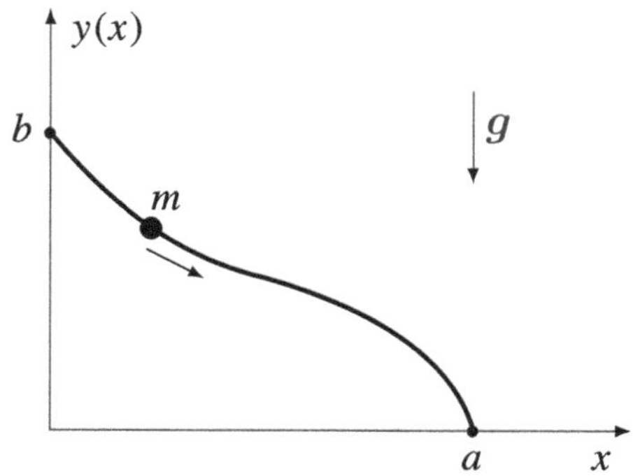

On which curve a frictionless gliding body in the gravitational field moves fastest from one point to another? The body is initially at rest. The vertical distance between the two points is b, the horizontal a. Sketch the solution for the cases $a/b < \pi/2$, $a/b = \pi/2$, and $a/b > \pi/2$, respectively. The solution curve is called *brachistochrone*.

Determine a first integral of the Euler–Lagrange equation. Show that the *cycloid* with the parameter representation $x(\tau) = A(\tau - \sin\tau)$ and $y(\tau) = b - A(1 - \cos\tau)$ satisfies the differential equation.

12.2 *Soap skin*

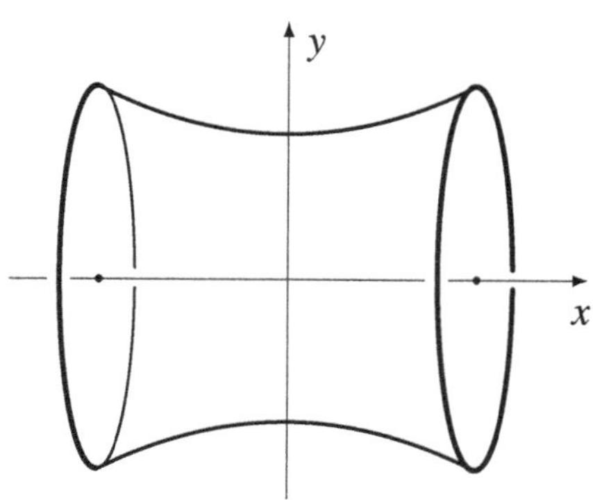

Between two parallel wire circles (radius R, distance D) a soap skin is stretched. The two circles are perpendicular to the line connecting the centers. Determine the shape of the soap skin. How does the skin behave when the wire rings are pulled apart slowly?

Chapter 13

Variation with Constraint

We study a variational problem in which the wanted function $y(x)$ must satisfy a constraint. We derive the Euler–Lagrange equations for isoperimetric and holonomic constraints. As examples, we consider the catenary line and a geodesic line.

Problem

Let a rope be suspended between two points in the gravitational field, Figure 13.1. This leads to the following question: Which curve describes the equilibrium position of the rope? The rope adjusts itself in such a way that the potential energy becomes minimal. We describe the position of the rope by the function $y(x)$. Then the equilibrium condition reads

$$J = U_{\rm pot} = \int_1^2 dm\, g\, y(x) = \rho\, g \int_{x_1}^{x_2} dx\; y\, \sqrt{1 + y'^2} = \text{minimal} \tag{13.1}$$

The mass elements dm are given by the (homogeneous) mass density ρ (mass per length) times the line element $ds = dx\,(1 + y'^2)^{1/2}$. The acceleration due to gravity is denoted by g. If any change $\delta y(x)$ of the position of the rope can make $U_{\rm pot}$ smaller, there are forces that act towards the displacement of y to $y + \delta y$. As a result $y(x)$ shifts until (13.1) holds. The solution curve is called brachistochrone.

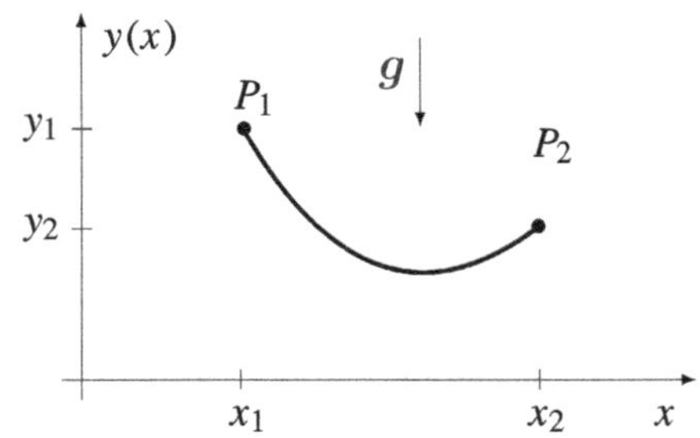

Figure 13.1 A rope of length L is suspended in the gravity field at points P_1 and P_2. The equilibrium position of the rope is determined by the condition that the potential energy is minimal.

Mathematically, this is again a variational problem, where the boundary values are not varied,

$$y(x_1) = y_1, \quad y(x_2) = y_2 \tag{13.2}$$

The possible curves $y(x)$ are subject to the additional restriction that the length L of the rope is fixed. So the $y(x)$ we are looking for must satisfy the extra constraint

$$K[y] = \int_{x_1}^{x_2} dx \sqrt{1 + y'^2} = L = \text{const.} \tag{13.3}$$

A similar problem is: Determine a curve of the length L between two points P_1 and P_2 in such a way that the area enclosed by the straight line $\overline{P_1 P_2}$ and the curve is maximal. Also, in this case, the constraint (13.3) must be fulfilled. From this problem comes the name *isoperimetric* (of equal circumference) for constraints of the form $K[y] = \text{const.}$

Besides that, *holonomous* auxiliary conditions can occur. As an example we consider curves in three-dimensional space, which can be represented by the functions $y = y(x)$ and $z = z(x)$. The curve shall connect the points

$$P_1 = \big(x_1, y(x_1), z(x_1)\big) \quad \text{and} \quad P_2 = \big(x_2, y(x_2), z(x_2)\big) \tag{13.4}$$

and lie on the surface

$$g(x, y, z) = 0 \tag{13.5}$$

Of course, P_1 and P_2 must also lie on the surface. We are looking for the curve which describes the shortest path between P_1 and P_2, i.e.

$$J[y, z] = \int_1^2 ds = \int_{x_1}^{x_2} dx \sqrt{1 + y'^2 + z'^2} = \text{minimal} \tag{13.6}$$

These shortest connections are called *geodesic lines* (of the surface).

The variation of (13.6) has to be performed under the constraint (13.5) This constraint is a much stronger restriction than an isoperimetric constraint: While (13.3) is one condition for the whole function, (13.5) is a condition for each x.

Lagrange Multiplier

We start with the problem of finding the extremum of a function of several variables under a constraint. To do this we consider the problem outlined in Figure 13.2:

$$f(x, y) = \text{minimal, where } g(x, y) = 0 \tag{13.7}$$

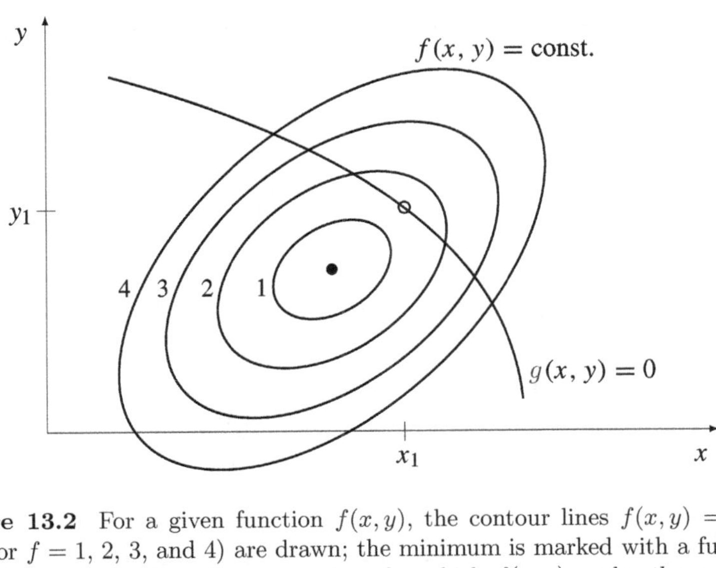

Figure 13.2 For a given function $f(x, y)$, the contour lines $f(x, y)$ = const. (e.g. for $f = 1$, 2, 3, and 4) are drawn; the minimum is marked with a full circle (•). We search for the values x_1 and y_1 for which $f(x, y)$ *under the constraint* $g(x, y) = 0$ becomes minimal. This minimum is marked by an open circle (○).

We assume that we can resolve the constraint $g(x, y) = 0$ to $y = y_g(x)$. Then

$$g\,(x, y_g(x)) = 0 \tag{13.8}$$

The minimum of f on the curve $y_g(x)$ can now be determined as the minimum of $h(x) = f(x, y_g(x))$:

$$\frac{d}{dx}\, f(x, y_g) = f_x(x, y_g) + f_y(x, y_g)\; y_g'(x) = 0 \tag{13.9}$$

This is an equation to determine the value x_1 we are looking for; the y value then follows from $y_1 = y_g(x_1)$.

With the method of the *Lagrange multipliers* we can avoid the explicit resolution of $g(x, y) = 0$ to $y = y_g(x)$. Instead of (13.9), we consider the following three equations for three unknowns x, y, and λ:

$$f_x(x, y) - \lambda\, g_x(x, y) = 0, \quad f_y(x, y) - \lambda\, g_y(x, y) = 0, \quad g(x, y) = 0 \tag{13.10}$$

We show that this is equivalent to (13.9). For this, we imagine that $g(x, y) = 0$ is resolved to $y = y_g(x)$ (but this step has not to be carried out actually). Then the constraint $g(x, y) = 0$ can be replaced by the equivalent condition

$$\bar{g}(x, y) = y - y_g(x) = 0 \tag{13.11}$$

We now write (13.10) with $\bar{g}$ instead of g and eliminate λ from the first two equations:

$$\left.\begin{aligned} f_x + \lambda\, y_g' &= 0 \\ f_y - \lambda\ \ &= 0 \end{aligned}\right\} \longrightarrow f_x(x, y_g) + f_y(x, y_g)\; y_g'(x) = 0 \tag{13.12}$$

Every solution of (13.10) is thus also a solution of (13.9); this is also true the other way round. The method (13.10) for determining the minimum of $f(x, y)$ under the constraint $g(x, y) = 0$ may be also formulated like this:

$$f^*(x, y) = f(x, y) - \lambda\, g(x, y) = \text{minimal} \quad \text{and} \quad g(x, y) = 0 \tag{13.13}$$

Practically this means:

- One searches the solution of $f^* =$ minimal *without regard to the constraint*. The solution $y = y(x, \lambda)$ contains then one parameter λ. This parameter is adjusted such that the constraint is fulfilled.

At first, it seems more cumbersome to consider three equations (13.10) instead of one equation (13.9). The advantage of (13.10) is that we avoid the explicit resolution of the constraint for y (such a resolution is feasible only for special functions).

Generalization: The presented method can be generalized to the problem

$$f(x_1, \ldots, x_N) = \text{minimal}, \quad \text{where } g_\alpha(x_1, \ldots, x_N) = 0 \tag{13.14}$$

with $\alpha = 1, \ldots, R$. One then obtains the $N + R$ equations

$$\begin{aligned} \frac{\partial}{\partial x_i}\left(f(x_1, \ldots, x_N) - \sum_{\alpha=1}^{R} \lambda_\alpha\, g_\alpha(x_1, \ldots, x_N)\right) = 0 \quad (i = 1, \ldots, N) \\ g_\alpha(x_1, \ldots, x_N) = 0 \quad (\alpha = 1, \ldots, R) \end{aligned} \tag{13.15}$$

for the $N + R$ wanted unknowns x_i and λ_α.

Isoperimetric Constraint

We formulate the catenary problem (13.1)–(13.3) in a more general form: We seek the function $y(x)$ for which

$$J = J[y] = \int_{x_1}^{x_2} dx\; F(y, y', x) = \text{minimal} \tag{13.16}$$

under the constraint that another functional K is equal to a given constant C,

$$K = K[y] = \int_{x_1}^{x_2} dx\; G(y, y', x) = C \tag{13.17}$$

Here, the boundary values are fixed again,

$$y(x_1) = y_1, \quad y(x_2) = y_2 \tag{13.18}$$

Let $y(x)$ be the function we are looking for. An arbitrary deviation $\delta y = \epsilon\eta(x)$ implies $K[y+\epsilon\eta] = C + \mathcal{O}(\epsilon)$ violating $K = C$ in general (for $\epsilon \neq 0$). Therefore, we consider two linearly independent functions $\eta_1(x)$ and $\eta_2(x)$, where η_1 and η_2 vanish at the boundary. Then, for $y + \epsilon_1\eta_1 + \epsilon_2\eta_2$ the constraint reads

$$K(\epsilon_1, \epsilon_2) = K[y + \epsilon_1\eta_1 + \epsilon_2\eta_2] = C \tag{13.19}$$

The condition $K(\epsilon_1, \epsilon_2) = C$ is a curve in the ϵ_1–ϵ_2 plane (Figure 13.3). A variation of the form $y + \epsilon_1\eta_1 + \epsilon_2\eta_2$ is compatible with the constraint if ϵ_1 and ϵ_2 lie on the curve $K(\epsilon_1, \epsilon_2) = C$.

The searched function y must obey the condition

$$J(\epsilon_1, \epsilon_2) = J[y + \epsilon_1\eta_1 + \epsilon_2\eta_2] = \text{minimal}, \quad \text{where } K(\epsilon_1, \epsilon_2) = C \tag{13.20}$$

The minimum must be at $\epsilon_1 = \epsilon_2 = 0$. According to (13.13), we can express this in the form

$$J(\epsilon_1, \epsilon_2) - \lambda K(\epsilon_1, \epsilon_2) = \text{minimal at } \epsilon_1 = \epsilon_2 = 0 \tag{13.21}$$

According to (13.13), the additional term would actually be $-\lambda[K(\epsilon_1, \epsilon_2) - C]$; however, the constant has no influence on the

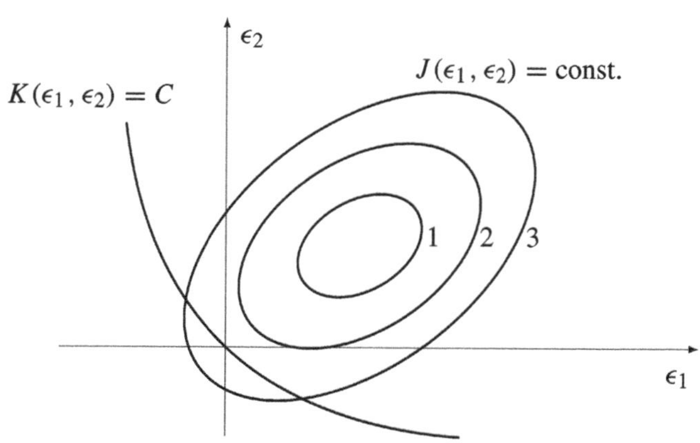

Figure 13.3 For $y \to y + \epsilon_1\eta_1 + \epsilon_2\eta_2$, the functional $J[y]$ reduces to a function $J(\epsilon_1, \epsilon_2)$; the sketch shows the contour lines $J = 1$, 2 and 3 of this function. The constraint $K[y] = C$ restricts the possible variation to the curve $K(\epsilon_1, \epsilon_2) = K[y+\epsilon_1\eta_1+\epsilon_2\eta_2] = C$. For the wanted $y(x)$ the minimum must be at $\epsilon_1 = \epsilon_2 = 0$.

position of the minimum. Now, (13.21) requires

$$\left(\frac{\partial (J - \lambda K)}{\partial \epsilon_i} \right)_{\epsilon_1 = \epsilon_2 = 0} = 0 \quad (i = 1, 2) \tag{13.22}$$

Since ϵ_1 and ϵ_2 are symmetric in J and K, respectively. both conditions $(i = 1, 2)$ are identical. The resulting condition is therefore

$$J^*[y] = \int_{x_1}^{x_2} dx\, F^*(y, y', x) = \text{minimal}, \quad \text{whereby } F^* = F - \lambda G \tag{13.23}$$

This yields the Euler–Lagrange equation

$$\frac{d}{dx} \frac{\partial F^*(y, y', x)}{\partial y'} = \frac{\partial F^*(y, y', x)}{\partial y} \tag{13.24}$$

or

$$\frac{d}{dx} \left(F_{y'} - \lambda G_{y'} \right) = F_y - \lambda G_y \tag{13.25}$$

The solution of this second-order differential equation contains two integration constants c_1 and c_2. Furthermore, it depends on λ, so $y = y(x, c_1, c_2, \lambda)$. The three parameters c_1, c_2, λ are given by the boundary conditions (13.18) and the constraint (13.17).

Generalization: The variational problem (13.16) may have several constraints of the kind (13.17):

$$K_i[y] = \int_{x_1}^{x_2} G_i(y, y', x)\, dx = C_i \quad (i = 1, \ldots, R) \tag{13.26}$$

Using (13.15), this leads to the Euler–Lagrange equation for

$$F^* = F(y, y', x) - \sum_{i=1}^{R} \lambda_i\, G_i(y, y', x) \tag{13.27}$$

The solution of the Euler–Lagrange equation depends on two integration constants and R parameters λ_i; these quantities are to be determined from the two boundary conditions and the R constraints.

Holonomic Constraint

We formulate the geodesic line problem (13.4)–(13.6) in general: We are looking for the functions $y_1(x)$ and $y_2(x)$ for which

$$J[y_1, y_2] = \int_{x_1}^{x_2} dx\; F(y_1, y_2, y_1', y_2', x) = \text{extremal} \tag{13.28}$$

Let the boundary values be fixed. The solution shall satisfy the holonomic constraint

$$g(y_1, y_2, x) = 0 \tag{13.29}$$

Geometrically $(x, y_1(x), y_2(x))$ represents a curve in the three-dimensional (x, y_1, y_2)-space, the condition (13.29) restricts this curve to a surface. Such a condition is much more restrictive than (13.17). While the integral condition (13.17) can be fulfilled by one parameter, (13.29) could specify one of the two functions we are looking for.

We solve the problem with the following strategy: The interval (x_1, x_2) is divided into I equal sections. Then (13.29) can be replaced by I isoperimetric constraints. For this, the solution is known from the previous section. In the limit $I \to \infty$, the substitution conditions become equivalent to (13.29).

The interval (x_1, x_2) is divided into I sections of length $d = (x_2 - x_1)/I$, which are defined by

$$\xi_i = x_1 + d\,(i-1) \quad (i = 1, 2, \ldots, I+1) \tag{13.30}$$

Then we define the auxiliary functions (Figure 13.4 on the left):

$$f_i(x) = \begin{cases} 1 & \text{for } \xi_i < x < \xi_{i+1} \\ 0 & \text{otherwise} \end{cases} \qquad (i = 1, 2, \ldots, I) \tag{13.31}$$

Now, I isoperimetric conditions follow immediately from (13.29), namely

$$K_i[y_1, y_2] = \int_{x_1}^{x_2} dx\; g(y_1, y_2, x)\, f_i(x) = 0 \quad (i = 1, 2, \ldots, I) \tag{13.32}$$

Conversely, (13.29) follows from these conditions for $I \to \infty$.

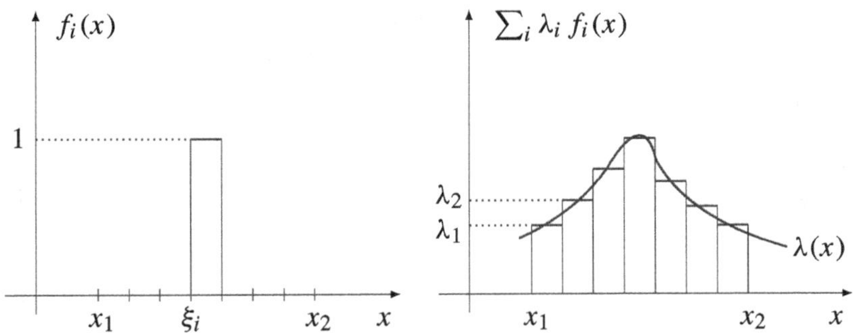

Figure 13.4 The auxiliary functions $f_i(x)$ are introduced (left) in order to replace the one holonomic constraint by I isoperimetric constraints. In the limit $I \to \infty$ the sum $\sum_i \lambda_i f_i(x)$ becomes the function $\lambda(x)$ (right).

In the form (13.32), the constraints can be taken into account according to (13.27). Thus, we have to solve the problem

$$J^*[y_1, y_2] = \int_{x_1}^{x_2} dx\; F^*(y_1, y_2, y_1', y_2', x) = \text{minimal} \tag{13.33}$$

with

$$F^* = F(y_1, y_2, y_1', y_2', x) - \sum_{i=1}^{I} \lambda_i\, g(y_1, y_2, x)\, f_i(x) \tag{13.34}$$

For $I \to \infty$, the sum $\sum_i \lambda_i f_i(x)$ becomes a continuous function $\lambda(x)$, so

$$\lim_{I\to\infty} \sum_{i=1}^{I} \lambda_i\, f_i(x) = \lambda(x) \tag{13.35}$$

This transition is graphically illustrated in Figure 13.4. Thereby F^* becomes

$$F^*(y_1, y_2, y_1', y_2', x) = F(y_1, y_2, y_1', y_2', x) - \lambda(x)\; g(y_1, y_2, x) \tag{13.36}$$

From (13.33) we get the Euler–Lagrange equations

$$\frac{d}{dx}\,\frac{\partial F}{\partial y_i'} = \frac{\partial F}{\partial y_i} - \lambda(x)\,\frac{\partial g}{\partial y_i} \qquad (i = 1, 2) \tag{13.37}$$

These two differential equations and $g(y_1, y_2, x) = 0$ determine the three functions $y_1(x)$, $y_2(x)$ and $\lambda(x)$. The integration constants are determined by the boundary conditions.

Generalization: We consider N functions $y_1(x), \ldots, y_N(x)$ and $R < N$ holonomic constraints $g_\alpha(y_1, \ldots, y_N, x) = 0$. The N Euler–Lagrange equations for

$$F^*(y_1, \ldots, y_N, y_1', \ldots, y_N', x) = F - \sum_{\alpha=1}^{R} \lambda_\alpha(x)\, g_\alpha(y_1, \ldots, y_N, x) \tag{13.38}$$

together with the R constraints determine the $N + R$ functions $y_i(x)$ and $\lambda_\alpha(x)$.

Applications

Catenary

We solve the problem sketched in Figure 13.1 for a rope or a chain in the gravitational field. The solution curve is called a *catenary*. For (13.1) and (13.3), the Euler–Lagrange equations (13.24) read

$$F^*(y, y') = F - \lambda G = \sqrt{1 + y'^2}\,(y - \lambda) \tag{13.39}$$

Here, λ is a parameter. The factor ρg from (13.1) is irrelevant here. Since F^* does not explicitly depend on x we use (12.24), i.e.

$$F^* - F^*_{y'}\, y' = \frac{y - \lambda}{\sqrt{1 + y'^2}} = a = \text{const.} \tag{13.40}$$

We resolve this to $y'^2 = (y - \lambda)^2/a^2 - 1$ and integrate to

$$y(x) = \lambda + a \cosh\left(\frac{x}{a} + b\right) \tag{13.41}$$

The parameters a, b, and λ of this solution are given by the conditions

$$y(x_1) = y_1, \quad y(x_2) = y_2, \quad \text{and} \quad \int_{x_1}^{x_2} dx\, \sqrt{1 + y'^2} = L \tag{13.42}$$

We specifically choose the suspension points

$$(x_1, y_1) = (-1, 0) \quad \text{and} \quad (x_2, y_2) = (1, 0) \tag{13.43}$$

Then we get $b = 0$ and $\lambda = -a \cosh(1/a)$. The remaining parameter a in

$$y(x) = a \cosh \frac{x}{a} - a \cosh \frac{1}{a} \tag{13.44}$$

is fixed by the length L. With $\sqrt{1 + y'^2} = \cosh(x/a)$ the constraint becomes

$$L = \int_{-1}^{1} dx \, \sqrt{1 + y'^2} = 2a \sinh \frac{1}{a} \tag{13.45}$$

For $L > 2$ there are two solutions, $a = a_0 > 0$ and $a = -a_0$. For the rope hanging downwards (Figure 13.1), the solution $a = a_0$ applies. For $a = -a_0$ the potential energy would be maximal.

The solution $a = -a_0$ describes the optimal line for an arch made of an incompressible material (a concrete shell, for example). This can be seen from the following: In the equilibrium solution for the rope the weight of each small section of the rope is compensated by tangential tension forces at both ends; in particular, there are no resulting lateral forces, i.e. no forces perpendicular to the catenary line. For a concrete arc, the weight can be compensated by compression forces alone; the lateral forces disappear.

Geodesic lines

We want to determine the geodesic lines on the surface

$$g(x, y, z) = 0 \tag{13.46}$$

Let the searched line be expressed in the parameter form $x(t)$, $y(t)$, $z(t)$. The meaning of t is not fixed; for a trajectory, t could be interpreted as a time coordinate. The parameter form is used for convenience.

The geodesic lines are the shortest paths between two points. So they are determined from

$$J[x,y,z] = \int_{t_1}^{t_2} dt\, \sqrt{\dot{x}^2 + \dot{y}^2 + \dot{z}^2} = \text{minimal} \tag{13.47}$$

under the constraint (13.46). According to the previous section, the Euler–Lagrange equations have to be solved for

$$F^*(x,y,z,\dot{x},\dot{y},\dot{z},t) = \sqrt{\dot{x}^2 + \dot{y}^2 + \dot{z}^2} - \lambda(t)\, g(x,y,z) \tag{13.48}$$

With the abbreviation $v = \sqrt{\dot{x}^2 + \dot{y}^2 + \dot{z}^2}$ the Euler–Lagrange equations read

$$\frac{d}{dt}\frac{\dot{x}}{v} = -\lambda\, g_x, \quad \frac{d}{dt}\frac{\dot{y}}{v} = -\lambda\, g_y, \quad \frac{d}{dt}\frac{\dot{z}}{v} = -\lambda\, g_z \tag{13.49}$$

For the evaluation, we use the path length $s = s(t)$ as the parameter (instead of t). With $ds = v\, dt$ we obtain

$$\frac{d}{dt} = v\,\frac{d}{ds}, \quad \dot{x} = \frac{dx(s)}{ds}\frac{ds}{dt} = v\, x'(s), \quad \frac{d}{dt}\frac{\dot{x}}{v} = v\, x''(s) \tag{13.50}$$

Thus, the Euler–Lagrange equations (13.49) become

$$x'' = \mu\, g_x, \quad y'' = \mu\, g_y, \quad z'' = \mu\, g_z \tag{13.51}$$

where the function $\lambda(t)$ has been replaced by the function $\mu(s) = -\lambda/v$.

The previous considerations are valid for any surface. Especially for the spherical surface

$$g(x,y,z) = x^2 + y^2 + z^2 - R^2 = 0 \tag{13.52}$$

(13.51) becomes

$$x'' = 2\mu x, \quad y'' = 2\mu y, \quad z'' = 2\mu z \tag{13.53}$$

The solution $x(s)$, $y(s)$, $z(s)$ must satisfy $g(x,y,z) = 0$; so we can already use this relation in the solution of the differential equation.

For this we differentiate $g(x(s), y(s), z(s)) = 0$ twice with respect to s,

$$0 = x'^2 + y'^2 + z'^2 + x\,x'' + y\,y'' + z\,z'' \tag{13.54}$$

Here, we substitute x'', y'', and z'' from (13.53). With (13.52) and $x'^2 + y'^2 + z'^2 = (dx^2 + dy^2 + dz^2)/ds^2 = 1$ we then get

$$\mu = -\frac{1}{2R^2} = \text{const.} \tag{13.55}$$

For an arbitrary surface, this step would yield μ in general as a function of (x, y, z, x', y', z'); also in this case one can eliminate μ from the Euler–Lagrange equations. We substitute (13.55) into (13.53),

$$x'' = -\frac{x}{R^2}, \quad y'' = -\frac{y}{R^2}, \quad z'' = -\frac{z}{R^2} \tag{13.56}$$

Each of these differential equations has the solution $A \sin(s/R) + B \cos(s/R)$; the constants are chosen in agreement with the constraint and the given boundary points. The resulting lines are great circles (circles with a circumference $2\pi R$) on the spherical surface. A special solution is the equator:

$$x(s) = R \sin\frac{s}{R} = R\sin\phi, \quad y(s) = R\cos\frac{s}{R} = R\cos\phi, \quad z = 0 \tag{13.57}$$

Exercises

13.1 *Occupation numbers for maximum entropy*

Determine the numbers $n_1, n_2, n_3, \ldots$ in such a way that the quantity

$$S(n_1, n_2, \ldots) = \sum_{i=1}^{\infty} \big((1 + n_i)\ln(1 + n_i) - n_i \ln(n_i)\big)$$

becomes extremal. The values for n_i shall be subject to the constraints

$$E = \sum_i \varepsilon_i\, n_i = \text{const.}, \quad N = \sum_i n_i = \text{const.}$$

The ε_i are given constants.

This problem arises in the treatment of an ideal Bose gas. Here, the ε_i are the energies of the discrete single-particle levels and the n_i are the number of particles in these levels. The thermal equilibrium is determined by the condition that the entropy S is maximal; thereby the total energy E and the number of particles N are fixed (closed system). The n_i which satisfy the extremal condition are the mean occupation numbers.

13.2 *Isoperimetric problem*

The ends of a rope (length L) are fixed at $(x_1, y_1) = (-d, 0)$ and $(x_2, y_2) = (d, 0)$ on the x-axis. For which position of the rope (in the x–y-plane) does the area between the rope and the x-axis become maximal?

13.3 *Geodesic lines on circular a cylinder*

Determine the geodesic lines on the surface of a circular cylinder. (i) Use a Lagrange multiplier in order to take into account the constraint "circular cylinder". (ii) Solve the problem once more by choosing suitable coordinates, such that the constraint is fulfilled automatically.

Chapter 14

Hamilton's Principle

Hamilton's principle is a variational principle whose Euler–Lagrange equations are the Lagrange equations of mechanics. Hamilton's principle is an elegant formulation of the basic laws of mechanics. This formulation has an exemplary character for other fields of physics.

Lagrange Equations of Second Kind

We assign to each trajectory $q(t)$ the *action functional*

$$S = S[q] = \int_{t_1}^{t_2} dt \; \mathcal{L}(q, \dot{q}, t) \tag{14.1}$$

This quantity S is called *action*. Here, q stands for $(q_1, \ldots, q_f)$ and $\dot{q}$ for $(\dot{q}_1, \ldots, \dot{q}_f)$. The Lagrange equations for $q_i(t)$ are equivalent to the requirement that the action functional is stationary:

$$\text{Hamilton's principle:} \quad \delta S[q] = 0 \tag{14.2}$$

In the variation, the endpoints $q(t_1)$ and $q(t_2)$ are fixed. The statement $\delta S[q] = 0$ is called *Hamilton's principle*, it is a condition for the trajectory $q(t)$. According to (12.17), this condition is equivalent to the Lagrangian equations of the second kind:

$$\delta S[q] = 0 \;\longleftrightarrow\; \frac{d}{dt}\frac{\partial \mathcal{L}}{\partial \dot{q}_i} = \frac{\partial \mathcal{L}}{\partial q_i} \tag{14.3}$$

Thus, the Lagrangian equations of mechanics are the Euler–Lagrange equations of the variational problem $\delta S = 0$.

Hamilton's principle states that the motion takes place in such a way, that the trajectories make the action stationary. It is not important whether the stationary point is an extremum or not. In specific applications, the solution of the Lagrangian equations, S is in general minimal (Exercise 14.1). Therefore, the Hamilton principle is also called the *principle of least action.*

The general solution of the Lagrange equations contains $2f$ integration constants since they are f differential equations of second order. These constants are determined either by the fixed boundary values of the variational problem or by the initial conditions of the physical problem. The two data sets

$$\begin{matrix} \text{boundary conditions:} & & \text{initial conditions:} \\ q_i(t_1),\ q_i(t_2) & \text{or} & q_i(t_1),\ \dot{q}_i(t_1) \end{matrix} \tag{14.4}$$

each specify $2f$ quantities; insofar they are equivalent.

Lagrange Equations of First Kind

For a system of N mass points we use $3N$ Cartesian coordinates $x = (x_1, \ldots, x_{3N})$. The internal and external forces shall be described by a potential. Then the Lagrange function is

$$\mathcal{L}(x, \dot{x}, t) = \frac{1}{2} \sum_{n=1}^{3N} m_n\, \dot{x}_n^2 - U(x_1, \ldots, x_{3N}, t) \tag{14.5}$$

The system shall be subject to the R constraints

$$g_\alpha(x, t) = g_\alpha(x_1, \ldots, x_{3N}, t) = 0 \quad (\alpha = 1, \ldots, R) \tag{14.6}$$

Then the Lagrangian equations of the first kind (7.14) apply, which can be expressed as

$$\frac{d}{dt}\frac{\partial \mathcal{L}}{\partial \dot{x}_n} = \frac{\partial \mathcal{L}}{\partial x_n} + \sum_{\alpha=1}^{R} \lambda_\alpha \frac{\partial g_\alpha}{\partial x_n} \tag{14.7}$$

We consider the variational problem

$$S = S[x] = \int_{t_1}^{t_2} dt\ \mathcal{L}(x, \dot{x}, t) = \text{stationary}, \quad \text{where } g_\alpha(x, t) = 0 \tag{14.8}$$

According to (13.38), this is solved by the Euler–Lagrange equations for

$$\mathcal{L}^*(x, \dot{x}, t) = \mathcal{L}(x, \dot{x}, t) + \sum_{\alpha=1}^{R} \lambda_\alpha(t)\, g_\alpha(x, t) \tag{14.9}$$

Compared to (13.38) the sign of λ is chosen differently here; this is a convention and has no influence on the result. The Euler–Lagrange equations for $\mathcal{L}^*$ are identical to (14.7). Therefore,

$$\delta \int_{t_1}^{t_2} dt\; \mathcal{L}^*(x, \dot{x}, t) = 0 \;\longleftrightarrow\; \text{Lagrange equations first kind} \tag{14.10}$$

In this way, the Hamilton principle can also be applied to systems with constraints. The physical constraints become constraints of the variational problem.

Notes on Hamilton's Principle

In Part I, we have given the basic laws of mechanics in form of the Newton's laws. In Part II, they were formulated as Lagrange equations; these equations are more general and (for more complex systems) much easier to formulate. In this chapter, we have introduced the Hamilton principle as a formulation which is equivalent to the Lagrangian equations. Another alternative is the Hamilton formalism (Part VII).

For specific problems, one first establishes the Lagrange function and then the equations of motion. The Hamilton principle is usually not explicitly written down, but with the Lagrangian equations, it is implicitly used. For general considerations (symmetries, building new theories) the Hamilton principle itself is a suitable starting point. It represents the basis of mechanics in the short and concise form

$$\delta \int_{t_1}^{t_2} dt\; \mathcal{L}(q, \dot{q}, t) = 0 \tag{14.11}$$

This is a single equation in contrast to the f Lagrange equations.

The statement (14.11) does not presuppose certain generalized coordinates. These coordinates have to be chosen such that all constraint conditions are fulfilled. Alternatively, constraints can be added as in (14.8)–(14.10).

In non-relativistic mechanics, $\mathcal{L}$ is the difference between kinetic and potential energy, and it has to be written as a function of suitable generalized coordinates. In contrast to kinetic or potential energy, $\mathcal{L}$ is not a physical quantity (not a measurable quantity). The Lagrangian function is rather a mathematical auxiliary quantity, which is defined in such a way, that from it the equations of motion follow. Due to the missing connection with a physical quantity, the condition of minimal (or stationary) action is not immediately obvious or plausible, as is the case for the condition $U_{\text{pot}} = \text{minimal}$ for the catenary or the soap skin.

The mathematically introduced quantity $\mathcal{L}(q, \dot{q}, t)$ is in general a particularly simple function of the variables in question. Often it is the simplest function compatible with the symmetries of the system. We explain this by the example of the Lagrangian function of a free particle. The degrees of freedom of a particle are described by the position vector $\boldsymbol{r}(t)$. If we restrict ourselves to first derivatives, then

$$\mathcal{L} = \mathcal{L}(\boldsymbol{r}, \dot{\boldsymbol{r}}, t) \tag{14.12}$$

is a general approach. The relevant symmetries imply further restrictions:

1. Due to the homogeneity of time, $\mathcal{L}$ cannot depend on t.
2. Due to the homogeneity of the space, $\mathcal{L}$ cannot depend on $\boldsymbol{r}$.
3. Due to the isotropy of the space, $\mathcal{L}$ can only depend on $\dot{\boldsymbol{r}}^2$.

The simplest possible choice is

$$\mathcal{L} = C\,\dot{\boldsymbol{r}}^2 \tag{14.13}$$

This expression already gives the correct equations of motion. The constant C is without influence on the equations of motion; it is commonly set equal to $m/2$.

The listed symmetry transformations are contained in the general Galilei transformation. It is noticeable, however, that we do not have the invariance to Galilei transformations in the narrower sense (spatial displacement by $\boldsymbol{v}t$). In fact, $\mathcal{L}$ is not invariant under this transformation:

$$\mathcal{L} = \frac{m}{2}\,\dot{\boldsymbol{r}}^2 \;\xrightarrow{\dot{\boldsymbol{r}} \to \dot{\boldsymbol{r}} + \boldsymbol{v}}\; \mathcal{L}^* = \frac{m}{2}\,\dot{\boldsymbol{r}}^2 + m\,\boldsymbol{v}\cdot\dot{\boldsymbol{r}} + \frac{m}{2}\,\boldsymbol{v}^2 \qquad (14.14)$$

Both Lagrange functions, $\mathcal{L}$ and $\mathcal{L}^*$, result, however, in the same equation of motion (see next section).

In other fields of physics, too, the simplest approach compatible with the symmetries of the system often leads to the correct Lagrange function. "Correct Lagrange function" means that fundamental laws of nature are reproduced. This simplicity is a great advantage of the Lagrangian formalism. In the development of new theories, the preferred approach starts from the Lagrange function (or Lagrange density).

The following chapter presents another advantage of Hamilton's principle. On the basis of this principle, the Noether theorem relates a general class of possible invariances to the corresponding conservation laws. Also, this discussion has an exemplary character for other areas of physics.

Gauge Transformation

The Lagrange function is a particularly simple starting point for the equations of motion. But there can be different functions, which lead to the same equations of motion. We consider such Lagrange functions as mutually equivalent to each other. Given the equivalence of $\delta \int dt\, \mathcal{L} = 0$ with the equations of motion, it is immediately clear, that a given Lagrange function $\mathcal{L}$ is equivalent to $\mathcal{L}^* = \text{const.} \cdot \mathcal{L}$ or to $\mathcal{L}^* = \mathcal{L} + \text{const.}$

An important class of equivalent Lagrangian functions results from the so-called *gauge transformations*. Thereby to $\mathcal{L}$ a total time derivative of an arbitrary function $f(q,t)$ is added

$$\mathcal{L}(q,\dot{q},t) \;\longrightarrow\; \mathcal{L}^*(q,\dot{q},t) = \mathcal{L}(q,\dot{q},t) + \frac{d}{dt}\,f(q,t) \qquad (14.15)$$

As before, q stands for $(q_1, \ldots, q_f)$. We now show that $\mathcal{L}$ and $\mathcal{L}^*$ lead to the same equations of motion. The action integral for $\mathcal{L}^*$ is

$$S^* = \int_{t_1}^{t_2} dt\, \mathcal{L}^* = \int_{t_1}^{t_2} dt\, \mathcal{L} + f(q(t_2), t_2) - f(q(t_1), t_1) \qquad (14.16)$$

When S is varied, the trajectories $q(t)$ are varied, i.e.

$$\delta S^* = \delta S + \sum_{i=1}^{f} \left(\frac{\partial f}{\partial q_i}\right)_{t_2} \delta q_i(t_2) - \sum_{i=1}^{f} \left(\frac{\partial f}{\partial q_i}\right)_{t_1} \delta q_i(t_1) = \delta S \qquad (14.17)$$

Since the boundary values are held fixed during the variation, $\delta q_i(t_1) = \delta q_i(t_2) = 0$ and $\delta S^* = \delta S$; the conditions $\delta S^* = 0$ and $\delta S = 0$ are identical. The Lagrangian functions $\mathcal{L}$ and $\mathcal{L}^*$ lead to the same equations of motion. This can also be shown directly by using the equations of motion for $\mathcal{L}$ and $\mathcal{L}^*$, Exercise 11.1.

The change of the Lagrange function under a Galilean transformation is such a gauge transformation because the additional term in (14.14) can be written as a total time derivative ($\boldsymbol{v}$ is here the constant velocity of the Galilean transformation):

$$\mathcal{L}^* - \mathcal{L} = \frac{d}{dt}\left(m\boldsymbol{v}\cdot\boldsymbol{r} + \frac{m}{2}\,\boldsymbol{v}^2 t\right) \qquad (14.18)$$

The gauge transformation of the Lagrange function takes its name from the gauge transformation of electrodynamics. The Lagrangian function of a particle in the electromagnetic field was given in (9.47):

$$\mathcal{L}(\boldsymbol{r}, \dot{\boldsymbol{r}}, t) = \frac{m}{2}\,\dot{\boldsymbol{r}}^2 - q\,\Phi(\boldsymbol{r}, t) + \frac{q}{c}\,\dot{\boldsymbol{r}}\cdot\boldsymbol{A}(\boldsymbol{r}, t) \qquad (14.19)$$

The physical fields $\boldsymbol{E}$ and $\boldsymbol{B}$,

$$\boldsymbol{E} = -\frac{1}{c}\frac{\partial \boldsymbol{A}}{\partial t} - \operatorname{grad}\Phi, \quad \boldsymbol{B} = \operatorname{curl}\boldsymbol{A} \qquad (14.20)$$

are invariant under the following *gauge transformation* of the potentials:

$$\boldsymbol{A} \to \boldsymbol{A} + \operatorname{grad}\Lambda(\boldsymbol{r}, t), \quad \Phi \to \Phi - \frac{1}{c}\frac{\partial \Lambda(\boldsymbol{r}, t)}{\partial t} \qquad (14.21)$$

Here, $\Lambda(\boldsymbol{r},t)$ is an arbitrary function. Just like the Lagrange function, the potentials are not physical quantities; like $\mathcal{L}$ they are introduced because many relations can be represented particularly simply with them. The transformation (14.21) is valid for the Lagrange function (14.19):

$$\mathcal{L} \longrightarrow \mathcal{L}^* = \mathcal{L} + \frac{q}{c}\,\frac{d\Lambda(\boldsymbol{r},t)}{dt} \tag{14.22}$$

Exercises

14.1 *Minimum action for particles in the gravitational field*

The Lagrangian function $\mathcal{L}(z,\dot z) = (m/2)\,\dot z^2 - mgz$ describes a particle in the gravitational field. Calculate the action integral

$$S = \int_0^{t_0} dt\ \mathcal{L}(z,\dot z) \quad \text{for } z(t) = -\frac{g}{2}\,t^2 + f(t)$$

Let the deviation $f(t)$ from the factual motion be a continuously differentiable function, which vanishes at the boundaries $f(0) = f(t_0) = 0$. Show that the action $S[f]$ takes its minimum value for $f(t) = 0$.

14.2 *Equivalence of Lagrange functions*

Which of the Lagrange functions

$$\mathcal{L}_1 = \frac{m}{2}\,\dot{\boldsymbol{r}}^2 + q\,\boldsymbol{E}_0\cdot\boldsymbol{r} \quad\text{or}\quad \mathcal{L}_2 = \frac{m}{2}\,\dot{\boldsymbol{r}}^2 - q\,\boldsymbol{E}_0\cdot\dot{\boldsymbol{r}}\,t \tag{14.23}$$

describes a charged particle in a constant, homogeneous electric field $\boldsymbol{E}_0$?

Chapter 15

Noether's Theorem

In Chapter 11, it was shown, that certain conservation laws follow from the space–time symmetries of inertial systems. On the basis of Hamilton's principle, the connection between symmetries and conservation laws can be formulated more generally: Any one-parameter set of transformations, under which the action is invariant, leads to a conserved quantity. This theorem was established by the mathematician Emmy Noether (1882–1935).

The connection, which we want to examine in general, may be sketched by a simple example: If in the functional

$$J[y] = \int dx\, F(y, y', x) \tag{15.1}$$

the integrand $F = F(y, y')$ does not explicitly depend on x, then F and thus J are invariant with respect to the transformation

$$x \to x^* = x + \epsilon \tag{15.2}$$

Then, according to (12.24), we get the conservation law

$$Q = F_{y'}\, y' - F = \text{const.} \tag{15.3}$$

This is a first-order differential equation for the sought-after function $y(x)$. It is a first integral of the Euler–Lagrange equation (or a conserved quantity or a constant of motion in a physical context).

The connection between invariance and conservation law is derived without specific reference to its physical interpretation. We start from Hamilton's principle:

$$\delta S[q] = \delta \int_{t_1}^{t_2} dt\ \mathcal{L}(q, \dot{q}, t) = 0 \tag{15.4}$$

In the argument of the Lagrange function

$$\mathcal{L} = \mathcal{L}(q, \dot{q}, t) = \mathcal{L}(q_1, \ldots, q_f, \dot{q}_1, \ldots, \dot{q}_f, t) \tag{15.5}$$

we use the abbreviations

$$q = (q_1(t), \ldots, q_f(t)) \quad \text{and} \quad \dot{q} = (\dot{q}_1(t), \ldots, \dot{q}_f(t)) \tag{15.6}$$

The $q_i(t)$ are arbitrary (generalized) coordinates $q = x = (x_1, \ldots, x_N)$.

We consider transformations of the coordinates and the time that depend on a continuous parameter ϵ:

$$\begin{aligned} q_i \to q_i^* &= \Psi_i(q, \dot{q}, t; \epsilon) = q_i + \epsilon\, \psi_i(q, \dot{q}, t) + \mathcal{O}(\epsilon^2) \\ t \to t^* &= \Phi(q, \dot{q}, t; \epsilon) = t + \epsilon\, \varphi(q, \dot{q}, t) + \mathcal{O}(\epsilon^2) \end{aligned} \tag{15.7}$$

This approach allows the transformation depends on coordinates, velocities, and time (via the functions $\psi_i(q, \dot{q}, t)$ and $\varphi(q, \dot{q}, t)$). The parameter value $\epsilon = 0$ corresponds to the identical transformation. We restrict ourselves to infinitesimal transformations, for example, an infinitesimal rotation around an axis instead of a finite rotation. The terms of the order ϵ^2 are omitted in the following; they have no influence on the result.

By the transformation (15.7) the functions $q_i(t)$ transform to new, different functions $q_i^*(t^*)$, where

$$q_i^*(t^*) = q_i(t) + \epsilon\, \psi_i(q(t), \dot{q}(t), t) \tag{15.8}$$

$$t^*(t) = t + \epsilon\, \varphi(q(t), \dot{q}(t), t) \tag{15.9}$$

For example, a time translation $t^* = t + \epsilon$ is of the form (15.7) with $\varphi = 1$ and $\psi_i = 0$, implying $q_i^*(t^*) = q_i(t)$. The functions $q_i^*(t^*)$ and $q_i(t)$ differ in a trivial way; for example, $q(t) = \sin(t)$ becomes $q^*(t^*) = \sin(t^* - \epsilon)$. The asterisk in q^* indicates that $q^*(t^*)$ is another function of the new variable t^*.

We now compare (i) the functional $S[q(t)]$ for the trajectory $q(t)$ and the boundary values t_1 and t_2 with (ii) the functional $S^* = S[q^*(t^*)]$, which is obtained for the transformed trajectory $q_i^*(t^*)$ and the corresponding boundary values t_1^* and t_2^*. If $S^* = S$, i.e. if

$$\int_{t_1^*}^{t_2^*} dt^* \, \mathcal{L}\left(q^*, \frac{dq^*}{dt^*}, t^*\right) = \int_{t_1}^{t_2} dt \, \mathcal{L}\left(q, \frac{dq}{dt}, t\right) \quad \text{(invariance)} \tag{15.10}$$

holds, then the functional is *invariant* under this transformation. Since the functional S determines the equations of motion the invariance $S^* = S$ is the mathematical expression for the symmetry of the system described by $\mathcal{L}$; this means that the symmetry with respect to the considered transformation. On the left side of (15.10) we first replace the integration variable t^* by t

$$\begin{aligned}\int_{t_1^*}^{t_2^*} dt^* \, \mathcal{L}\left(q^*, \frac{dq^*}{dt^*}, t^*\right) &= \int_{t_1}^{t_2} dt \, \mathcal{L}\left(q^*, \frac{dq^*}{dt^*}, t^*\right) \frac{dt^*}{dt} \\ &= \int_{t_1}^{t_2} dt \left[\mathcal{L}\left(q, \frac{dq}{dt}, t\right) + \epsilon \, \frac{d}{d\epsilon}\left(\mathcal{L}\left(q^*, \frac{dq^*}{dt^*}, t^*\right) \frac{dt^*}{dt}\right)_{\epsilon=0}\right]\end{aligned} \tag{15.11}$$

Here, the integrand $\mathcal{L}(q^*, dq^*/dt^*, t^*)\, dt^*/dt$ was taken as a function $f(\epsilon)$ and expanded according to $f(\epsilon) = f(0) + f'(0)\,\epsilon + \mathcal{O}(\epsilon^2)$. Necessary and sufficient for the invariance condition (15.10) is thus

$$\frac{d}{d\epsilon}\left[\mathcal{L}\left(q^*, \frac{dq^*}{dt^*}, t^*\right) \frac{dt^*}{dt}\right]_{\epsilon=0} = 0 \quad \text{(invariance condition)} \tag{15.12}$$

The index $\epsilon = 0$ at the bracket means that the function differentiated with respect to ϵ has to be taken at this point. From the condition (15.12) we derive a conservation law of the form $Q = Q(q, \dot{q}, t) =$ const. We do this using the Euler–Lagrange equations

$$\frac{d}{dt}\frac{\partial \mathcal{L}}{\partial \dot{q}_i} = \frac{\partial \mathcal{L}}{\partial q_i} \tag{15.13}$$

Therefore, the statements to be derived are valid only for the actual trajectory curve. From (15.8) and (15.9) we obtain

$$\frac{dt^*}{dt} = 1 + \epsilon\,\frac{d\,\varphi(q,\dot{q},t)}{dt} = 1 + \epsilon\,\frac{d\varphi}{dt} \tag{15.14}$$

and

$$\frac{dq_i^*}{dt^*} = \frac{dq_i^*}{dt}\,\frac{dt}{dt^*} = \left(\dot{q}_i + \epsilon\,\frac{d\psi_i}{dt}\right)\left(1 - \epsilon\,\frac{d\varphi}{dt}\right) = \dot{q}_i + \epsilon\,\frac{d\psi_i}{dt} - \epsilon\,\dot{q}_i\,\frac{d\varphi}{dt} \tag{15.15}$$

As agreed above, the terms $\mathcal{O}(\epsilon^2)$ are omitted. We evaluate (15.12)

$$\begin{aligned}
&\frac{d}{d\epsilon}\left[\mathcal{L}\left(q_i + \epsilon\,\psi_i,\ \dot{q}_i + \epsilon\,\frac{d\psi_i}{dt} - \epsilon\,\dot{q}_i\,\frac{d\varphi}{dt},\ t + \epsilon\,\varphi\right)\left(1 + \epsilon\,\frac{d\varphi}{dt}\right)\right]_{\epsilon=0} \\
&= \sum_{i=1}^{f}\frac{\partial\mathcal{L}}{\partial q_i}\,\psi_i + \sum_{i=1}^{f}\frac{\partial\mathcal{L}}{\partial\dot{q}_i}\,\frac{d\psi_i}{dt} - \sum_{i=1}^{f}\frac{\partial\mathcal{L}}{\partial\dot{q}_i}\,\dot{q}_i\,\frac{d\varphi}{dt} + \frac{\partial\mathcal{L}}{\partial t}\,\varphi + \mathcal{L}\,\frac{d\varphi}{dt} \\
&= \frac{d}{dt}\sum_{i=1}^{f}\frac{\partial\mathcal{L}}{\partial\dot{q}_i}\,\psi_i + \left(\mathcal{L} - \sum_{i=1}^{f}\frac{\partial\mathcal{L}}{\partial\dot{q}_i}\,\dot{q}_i\right)\frac{d\varphi}{dt} + \varphi\,\frac{\partial\mathcal{L}}{\partial t} = 0
\end{aligned} \tag{15.16}$$

Since the result is to be taken at $\epsilon = 0$, there appear now again the non-transformed quantities in the argument of $\mathcal{L} = \mathcal{L}(q,\dot{q},t)$. By (9.35),

$$\frac{d}{dt}\left(\mathcal{L} - \sum_{i=1}^{f}\frac{\partial\mathcal{L}}{\partial\dot{q}_i}\,\dot{q}_i\right) = \frac{\partial\mathcal{L}}{\partial t} \tag{15.17}$$

we obtain from (15.16)

$$\begin{aligned}
&\frac{d}{d\epsilon}\left[\mathcal{L}\left(q^*,\ \frac{dq^*}{dt^*},\ t^*\right)\frac{dt^*}{dt}\right]_{\epsilon=0} \\
&\quad = \frac{d}{dt}\left[\sum_{i=1}^{f}\frac{\partial\mathcal{L}}{\partial\dot{q}_i}\,\psi_i + \left(\mathcal{L} - \sum_{i=1}^{f}\frac{\partial\mathcal{L}}{\partial\dot{q}_i}\,\dot{q}_i\right)\varphi\right] = 0
\end{aligned} \tag{15.18}$$

The invariance condition (15.12) implies that the left-hand side disappears. Thus, the *Noether theorem* states: The invariance of the functional $S[q]$ with respect to a one-parameter transformation (15.7) implies the conservation law

$$Q = Q(q, \dot{q}, t) = \sum_{i=1}^{f} \frac{\partial \mathcal{L}}{\partial \dot{q}_i}\,\psi_i + \left(\mathcal{L} - \sum_{i=1}^{f} \frac{\partial \mathcal{L}}{\partial \dot{q}_i}\,\dot{q}_i\right)\varphi = \text{const.} \quad (15.19)$$

This is a first-order differential equation; it is the first integral of the equations of motion. It is valid for the actual trajectories. It may also be called a constant of motion, and physically it means that Q is a conserved quantity.

The procedure for applying 65 Noether's theorem is as follows: First, one writes down the functions ψ_i and φ in (15.7) for the transformation under consideration. Then one checks by evaluation of (15.12), whether the symmetry holds with respect to these transformations. This depends on the Lagrange function $\mathcal{L}$ of the system under investigation. If the invariance condition is satisfied, one determines the conserved quantity Q from $\mathcal{L}$, ψ_i, and φ.

We summarize: The Noether theorem connects a symmetry (consisting of the invariance under a one-parameter transformation) with a conservation law:

$$\begin{array}{ccc} \text{symmetry:} & \overset{\text{Noether}}{\longrightarrow} & \text{conserved quantity:} \\ S^* = S \text{ or } (15.12) & & Q = Q(q, \dot{q}, t) = \text{const.} \end{array} \quad (15.20)$$

We consider two simple examples:

1. $\mathcal{L}$ does not depend explicitly on t, so $\mathcal{L} = \mathcal{L}(q_i, \dot{q}_i)$. We, therefore, consider the transformation

$$\begin{aligned} q_i^* &= q_i, & \psi_i &= 0 \\ t^* &= t + \epsilon, & \varphi &= 1 \end{aligned} \quad (15.21)$$

From this we get

$$q_i^*(t^*) = q_i(t), \quad \frac{dt^*}{dt} = 1 \quad \text{and} \quad \frac{dq_i^*}{dt^*} = \frac{dq_i}{dt} \quad (15.22)$$

We evaluate the invariance condition (15.12):

$$\frac{d}{d\epsilon}\left[\mathcal{L}\left(q_i^*, \frac{dq_i^*}{dt^*}\right)\frac{dt^*}{dt}\right]_{\epsilon=0} = \frac{d}{d\epsilon}\,\mathcal{L}(q_i, \dot{q}_i) = 0 \quad (15.23)$$

Since the invariance condition is satisfied, it follows

$$Q = \mathcal{L} - \sum_{i=1}^{f} \frac{\partial \mathcal{L}}{\partial \dot{q}_i}\, \dot{q}_i = \text{const.} \tag{15.24}$$

For $\mathcal{L} = \sum m_i \dot{x}_i^2/2 - U(x)$ (with Cartesian coordinates $q_i = x_i$), $-Q$ is equal to the energy E of the system.

2. $\mathcal{L} = \mathcal{L}(q_1, \ldots, q_{k-1}, q_{k+1}, \ldots, q_f, \dot{q}_1, \ldots, \dot{q}_f, t)$ does not depend explicitly on a certain coordinate q_k. Therefore, we consider the transformation

$$\begin{aligned} q_i^* &= q_i + \epsilon\, \delta_{ik}, \qquad & \psi_i &= \delta_{ik} \\ t^* &= t, & \varphi &= 0 \end{aligned} \tag{15.25}$$

Using this and

$$\frac{dq_k^*}{dt^*} = \frac{dq_k}{dt} \tag{15.26}$$

we evaluate the invariance condition (15.12)

$$\frac{d}{d\epsilon} \left[\mathcal{L}\left(q^*, \frac{dq_i^*}{dt^*}, t^*\right) \frac{dt^*}{dt} \right] = \frac{d}{d\epsilon}\, \mathcal{L}(q, \dot{q}, t) = 0 \tag{15.27}$$

Here, $\partial \mathcal{L}/\partial q_k = 0$ was taken into account. Since the invariance condition is satisfied, the following holds:

$$Q = \frac{\partial \mathcal{L}}{\partial \dot{q}_k} = p_k = \text{const.} \tag{15.28}$$

This is the generalized momentum associated with the coordinate q_k.

In this way, we obtain all the statements known from Chapters 9 and 11. The Noether theorem is of particular advantage for more general Lagrangian functions or densities, where it is not a priori clear which is the conservation laws. Furthermore, after a simple extension (in the following section), we can also treat symmetries in point mechanics (spherical oscillator, Kepler problem), which do not belong to the symmetries considered in Chapters 9 and 11.

Extended Noether Theorem

We started in (15.10) from the invariance condition $S^* = S$. Since the equations of motion are equivalent to $\delta S = 0$, the weaker condition $\delta S^* = \delta S$ is also sufficient for the symmetry:

$$\delta \int_{t_1^*}^{t_2^*} dt^* \, \mathcal{L}\left(q^*, \frac{dq^*}{dt^*}, t^*\right) = \delta \int_{t_1}^{t_2} dt \, \mathcal{L}\left(q, \frac{dq}{dt}, t\right) \quad \text{(invariance)} \tag{15.29}$$

Now, we know from (14.15)–(14.17) that a total time derivative of a function $f(q,t)$ can be added to $\mathcal{L}$ without changing δS. We may therefore replace the invariance condition (15.12) with the weaker condition

$$\frac{d}{d\epsilon}\left[\mathcal{L}\left(q^*, \frac{dq^*}{dt^*}, t^*\right)\frac{dt^*}{dt}\right]_{\epsilon=0} = \frac{d}{dt}\, f(q,t) \quad \text{(invariance condition)} \tag{15.30}$$

Here, $f(q,t) = f(q_1, \ldots, q_f, t)$ can be any function of the coordinates and time.

In the above derivation, we replace (15.12) by (15.30). Then the term df/dt occurs on the right-hand side of (15.16) and (15.18), and (15.19) becomes the *extended Noether theorem*: If $\mathcal{L}$ under the transformation (15.7) satisfies the invariance condition (15.30), then we obtain the conservation law in the form

$$Q = \sum_{i=1}^{f} \frac{\partial \mathcal{L}}{\partial \dot{q}_i}\, \psi_i + \left(\mathcal{L} - \sum_{i=1}^{f} \frac{\partial \mathcal{L}}{\partial \dot{q}_i}\, \dot{q}_i\right)\varphi - f(q,t) = \text{const.} \tag{15.31}$$

This Q is a conserved quantity for the actual solution.

Galilei invariance

The invariance under a Galilean transformation is called *Galilean invariance*. For a system of mass points, we derive the associated conservation law.

For this purpose, we investigate the special Galilean transformation

$$x^* = x + \epsilon t, \quad y^* = y, \quad z^* = z, \quad t^* = t \tag{15.32}$$

for a closed system of N mass points,

$$\mathcal{L} = \frac{1}{2}\sum_{\nu=1}^{N} m_\nu\, \dot{\boldsymbol{r}}_\nu^2 - \sum_{\nu=2}^{N}\sum_{\mu=1}^{\nu-1} U_{\nu\mu}(|\boldsymbol{r}_\nu - \boldsymbol{r}_\mu|) \tag{15.33}$$

In order to apply the above formulas, we express the N position vectors $\boldsymbol{r}_\nu$ by $3N$ Cartesian coordinates x_n:

$$x_n = x_{3\nu+j-3} = \boldsymbol{r}_\nu \cdot \boldsymbol{e}_j \tag{15.34}$$

The particle index ν runs from 1 to N, the index j from 1 to 3, and n from 1 to $3N$. The transformation (15.32) has to be applied to all x-coordinates, i.e.

$$\begin{aligned} x^*_{3\nu-2} &= x_{3\nu-2} + \epsilon\, t, & \psi_{3\nu-2} &= t \\ x^*_{3\nu-1} &= x_{3\nu-1}, & \psi_{3\nu-1} &= 0 \\ x^*_{3\nu} &= x_{3\nu}, & \psi_{3\nu} &= 0 \\ t^* &= t, & \varphi &= 0 \end{aligned} \tag{15.35}$$

When evaluating the invariance condition, we use

$$\frac{dx^*_{3\nu-2}}{dt^*} = \dot{x}_{3\nu-2} + \epsilon, \qquad \frac{dx^*_{3\nu-1}}{dt^*} = \dot{x}_{3\nu-1}, \qquad \frac{dx^*_{3\nu}}{dt^*} = \dot{x}_{3\nu} \tag{15.36}$$

and

$$|\boldsymbol{r}^*_\nu - \boldsymbol{r}^*_\mu| = |\boldsymbol{r}_\nu - \boldsymbol{r}_\mu|, \qquad \frac{dt^*}{dt} = 1 \tag{15.37}$$

Thus, (15.30) becomes

$$\begin{aligned} &\frac{d}{d\epsilon}\left[\mathcal{L}\left(\ldots, |\boldsymbol{r}^*_\nu - \boldsymbol{r}^*_\mu|, \ldots, \frac{dx^*_{3\nu-2}}{dt^*}, \frac{dx^*_{3\nu-1}}{dt^*}, \frac{dx^*_{3\nu}}{dt^*}, \ldots\right)\frac{dt^*}{dt}\right]_{\epsilon=0} \\ &= \left[\frac{d}{d\epsilon}\,\mathcal{L}\left(\ldots, |\boldsymbol{r}_\nu - \boldsymbol{r}_\mu|, \ldots, \dot{x}_{3\nu-2} + \epsilon,\, \dot{x}_{3\nu-1},\, \dot{x}_{3\nu}, \ldots\right)\right]_{\epsilon=0} \\ &= \sum_{\nu=1}^{N} \frac{\partial \mathcal{L}}{\partial \dot{x}_{3\nu-2}} = \sum_{\nu=1}^{N} m_\nu\, \dot{x}_{3\nu-2} = \frac{d}{dt}\sum_{\nu=1}^{N} m_\nu \boldsymbol{r}_\nu \cdot \boldsymbol{e}_1 = \frac{d}{dt}\, M\boldsymbol{R}\cdot \boldsymbol{e}_1 \end{aligned} \tag{15.38}$$

In the last step, we have introduced the center-of-mass vector $\boldsymbol{R} = \sum_\nu m_\nu \boldsymbol{r}_\nu / M$. We see that the invariance condition (15.30) holds for the function

$$f = \sum_{\nu=1}^{N} m_\nu \boldsymbol{r}_\nu \cdot \boldsymbol{e}_1 = M\boldsymbol{R} \cdot \boldsymbol{e}_1 \tag{15.39}$$

Then from (15.31) follows the conservation law:

$$Q = \sum_{\nu=1}^{N} \frac{\partial \mathcal{L}}{\partial \dot{x}_{3\nu-2}} \psi_{3\nu-2} - f(x,t) = M(\dot{\boldsymbol{R}}t - \boldsymbol{R}) \cdot \boldsymbol{e}_1 = \text{const.} \tag{15.40}$$

This symmetry applies in the same way to a special Galilean transformation in y- or z-direction. Therefore,

$$\dot{\boldsymbol{R}}t - \boldsymbol{R} = \text{const.} \tag{15.41}$$

These are three conservation quantities. They can be solved to

$$\boldsymbol{R}(t) = \boldsymbol{A}t + \boldsymbol{B} \tag{15.42}$$

The same result is also obtained from the translational invariance of (15.33), see (11.10) to (11.13), and Exercise 15.3.

Closed systems are invariant under the 10-parameter group of Galilean transformations. As already shown in Chapter 11, this results in 10 conservation laws, namely the energy E, the momentum $\boldsymbol{P}$, the angular momentum $\boldsymbol{L}$ and the quantity $\boldsymbol{R} - \dot{\boldsymbol{R}}t$.

Other examples[1] to which the extended Noether theorem may be applied are the harmonic oscillator (SU(3)-symmetry, Exercise 15.4) and the Kepler problem (Lenz's vector, Exercise 17.1).

Exercises

15.1 *Symmetry of the potential* $U = \alpha/r^2$

Show that the action for a particle in the potential $U(\boldsymbol{r}) = \alpha/r^2$ is invariant under the one-parameter transformation

$$\boldsymbol{r}^* = \lambda\,\boldsymbol{r}, \quad t^* = \lambda^2 t$$

Derive the associated conservation law and simplify it with the help of the energy conservation law.

---

[1] *American Journal of Physics* **45** (1977) 336 and **39** (1971) 502.

15.2 *Particles in the homogeneous electric field*

A charged particle in the constant, homogeneous electric field can be described by the Lagrange function

$$\mathcal{L}(\boldsymbol{r}, \dot{\boldsymbol{r}}) = \frac{m}{2}\, \dot{\boldsymbol{r}}^2 + q\, \boldsymbol{E}_0 \cdot \boldsymbol{r}$$

The system is invariant under spatial translations. Derive the conservation quantity associated with this symmetry.

15.3 *Translational invariance in the many-particle system*

The Lagrange function of a closed system of N mass points is given by

$$\mathcal{L} = \frac{1}{2} \sum_{\nu=1}^{N} m_\nu\, \dot{\boldsymbol{r}}_\nu^2 - \sum_{\nu=2}^{N} \sum_{\mu=1}^{\nu-1} U_{\nu\mu}(|\boldsymbol{r}_\nu - \boldsymbol{r}_\mu|) \tag{15.43}$$

Show the invariance under the translation $\boldsymbol{r}_\nu^* = \boldsymbol{r}_\nu + \boldsymbol{\epsilon}$, and determine the corresponding conserved quantity.

15.4 *Conservation laws of the spherical oscillator*

In appropriate units, the Lagrange function of the spherical harmonic oscillator reads

$$\mathcal{L} = \frac{1}{2} \sum_{i=1}^{3} \left(\dot{x}_i^2 - x_i^2\right)$$

Show that the transformations

$$x_i^* = x_i + \frac{\epsilon}{2} \left(\delta_{ik}\, \dot{x}_l + \delta_{il}\, \dot{x}_k\right), \quad t^* = t \tag{15.44}$$

leaves the variation of the action invariant (and thus the equations of motion).

In (15.44), k and l can take the values 1, 2, and 3; there are therefore, nine possible transformations. Determine the associated nine conservation laws Q_{kl}. Show that this includes the energy and angular momentum conservation.

Hint: Show that

$$\ell_i^2 = \frac{1}{2} \sum_{k,l=1}^{3} \epsilon_{ikl}^2 \left(Q_{kk} Q_{ll} - Q_{kl}^2 \right)$$

is valid for the angular momentum components ℓ_i. Here, ϵ_{ikl} is the Levi-Civita symbol (has nothing to do with the ϵ in (15.44)).

PART IV
Central Potential

Chapter 16

Two-Body Problem

We investigate the motion of two bodies in a closed system. This includes the Kepler problem, the classical treatment of the hydrogen atom, the Rutherford scattering, and the dynamics of a diatomic molecule (vibrations, rotations).

The two bodies are treated as mass points (positions $\boldsymbol{r}_i$, masses m_i, $i = 1, 2$). The conservative forces are described by a potential of the form $U(|\boldsymbol{r}_1 - \boldsymbol{r}_2|)$; other forces do not occur in the closed system. Thus, the Lagrangian function reads

$$\mathcal{L}(\boldsymbol{r}_1, \boldsymbol{r}_2, \dot{\boldsymbol{r}}_1, \dot{\boldsymbol{r}}_2) = \frac{m_1}{2}\,\dot{\boldsymbol{r}}_1^{\,2} + \frac{m_2}{2}\,\dot{\boldsymbol{r}}_2^{\,2} - U(|\boldsymbol{r}_1 - \boldsymbol{r}_2|) \tag{16.1}$$

The Lagrange equations represent a coupled system of six differential equations of second order. The symmetries of the problem,

1. Symmetry with respect to spatial translations.
2. Symmetry with respect to rotations.
3. Symmetry with respect to time translations.

Lead to conservation laws, which allow the stepwise simplification of the problem:

1. Conservation of the center-of-mass momentum: reduction to a one-body problem.
2. Conservation of angular momentum: reduction to a radial equation.
3. Conservation of energy: reduction to a differential equation of first order.

The solution of the remaining differential equation can be written down as an integral. For the evaluation of this integral, the central potential $U(r)$ must be specified.

The procedure outlined here is a prime example for the exploitation of symmetries for the solution of physical problems. The symmetries do not depend on whether the system is treated classically or quantum mechanically. Therefore, they are used in a similar way for the quantum mechanical treatment of the hydrogen atom.

Reduction to One-Body Problem

The symmetries appear more clearly in the Lagrangian function if one introduces suitable generalized coordinates. For the translational invariance these are the center-of-mass vector and the relative vector, for the rotational invariance spherical or cylindrical coordinates.

By

$$\boldsymbol{R} = \frac{m_1 \boldsymbol{r}_1 + m_2 \boldsymbol{r}_2}{m_1 + m_2}, \qquad \boldsymbol{r} = \boldsymbol{r}_1 - \boldsymbol{r}_2 \tag{16.2}$$

we introduce the center-of-mass vector $\boldsymbol{R}$ and the relative vector $\boldsymbol{r}$. We solve (16.2) for $\boldsymbol{r}_1$ and $\boldsymbol{r}_2$:

$$\boldsymbol{r}_1 = \boldsymbol{R} + \frac{m_2}{m_1 + m_2}\,\boldsymbol{r}, \qquad \boldsymbol{r}_2 = \boldsymbol{R} - \frac{m_1}{m_1 + m_2}\,\boldsymbol{r} \tag{16.3}$$

For (16.1), this is a transformation to new generalized coordinates. We substitute (16.3) into (16.1) and obtain

$$\mathcal{L} = \mathcal{L}(\boldsymbol{r}, \dot{\boldsymbol{r}}, \dot{\boldsymbol{R}}) = \frac{M}{2}\,\dot{\boldsymbol{R}}^2 + \frac{\mu}{2}\,\dot{\boldsymbol{r}}^2 - U(|\boldsymbol{r}|) \tag{16.4}$$

Here, we have introduced the total mass M and the *reduced mass* μ,

$$M = m_1 + m_2, \qquad \mu = \frac{m_1 m_2}{m_1 + m_2} \tag{16.5}$$

Due to the translational symmetry of the system, $\boldsymbol{R}$ is a cyclic coordinate, and the corresponding generalized momentum (center-of-mass momentum) is constant:

$$\frac{\partial \mathcal{L}}{\partial \dot{\boldsymbol{R}}} = M\dot{\boldsymbol{R}} = \text{const.}, \quad \text{hence } \boldsymbol{R}(t) = \boldsymbol{A}\,t + \boldsymbol{B} \tag{16.6}$$

The center of mass moves along a straight line with constant velocity. The solution $\boldsymbol{R}(t)$ does not enter into the equation of motion for $\boldsymbol{r}(t)$; the equations of motion are *decoupled*. This is generally true if the Lagrange function does not contain any coupling terms, i.e. if it is of the form

$$\mathcal{L}(\boldsymbol{r}, \boldsymbol{R}, \dot{\boldsymbol{r}}, \dot{\boldsymbol{R}}) = \mathcal{L}_1(\boldsymbol{R}, \dot{\boldsymbol{R}}) + \mathcal{L}_2(\boldsymbol{r}, \dot{\boldsymbol{r}}) \tag{16.7}$$

Due to this decoupling of the equations of motion, we can restrict the discussion of $\mathcal{L}_2$ only. Thus, we have only to solve the *one-body problem* with

$$\mathcal{L}_2 = \mathcal{L}(\boldsymbol{r}, \dot{\boldsymbol{r}}) = \frac{\mu}{2}\, \dot{\boldsymbol{r}}^2 - U(r) \quad (r = |\boldsymbol{r}|) \tag{16.8}$$

In the following, we denote $\mathcal{L}_2$ again by $\mathcal{L}$. The Lagrangian function (16.8) describes a fictitious particle of mass μ moving in a central potential $U(r)$. For $m_1 \ll m_2$ (for example, $m_{\text{planet}} \ll m_{\text{sun}}$) $\mu \approx m_1$ holds; and the motion of the fictitious particle is approximately equal to that of the body 1.

We now introduce cylindrical coordinates, $\boldsymbol{r} := (\rho, \varphi, z)$. With (10.26) and $r^2 = \rho^2 + z^2$ we obtain

$$\mathcal{L}(\rho, z, \dot{\rho}, \dot{\varphi}, \dot{z}) = \frac{\mu}{2}\left(\dot{\rho}^2 + \rho^2\, \dot{\varphi}^2 + \dot{z}^2\right) - U\left(\sqrt{\rho^2 + z^2}\right) \tag{16.9}$$

Thus, the Lagrangian equations become

$$\mu\, \ddot{\rho} = \mu\, \rho\, \dot{\varphi}^2 - \frac{\partial U}{\partial \rho} \tag{16.10}$$

$$\frac{d}{dt}\left(\mu \rho^2\, \dot{\varphi}\right) = 0 \tag{16.11}$$

$$\mu\, \ddot{z} = -\frac{\partial U}{\partial z} \tag{16.12}$$

It is convenient not to carry out the time derivative in the equation for $\varphi(t)$.

Reduction to the Radial Equation

According to (3.7), the angular momentum for a central force is conserved:

$$\boldsymbol{\ell} = \mu\, \boldsymbol{r} \times \dot{\boldsymbol{r}} = \text{const.} \tag{16.13}$$

This conservation quantity can also be derived as in Chapter 11 from the isotropy of the space, or from the Noether theorem as a consequence of the invariance of the Lagrangian function under rotations.

We place the z-axis of our reference frame in the $\boldsymbol{\ell}$ direction:

$$\boldsymbol{\ell} = \ell\, \boldsymbol{e}_z \quad \text{(choice of IS)} \tag{16.14}$$

This specification is possible only because the direction of $\boldsymbol{\ell}$ is constant; otherwise, the so specified reference frame would not be an inertial frame. From $\boldsymbol{\ell} = \mu\, \boldsymbol{r} \times \dot{\boldsymbol{r}} \parallel \boldsymbol{e}_z$ follows $\boldsymbol{r} \perp \boldsymbol{e}_z$ and thus

$$z(t) = 0 \tag{16.15}$$

The entire trajectory lies in the x–y plane. We put $z = 0$ into the equations of motion.[1] Equation (16.12) is then trivially fulfilled. The equation of motion (16.11) may be solved to

$$\mu\, \rho(t)^2\, \dot{\varphi}(t) = \ell = \text{const.} \tag{16.16}$$

Due to the rotational symmetry about the z-axis, φ is a cyclic coordinate in (16.9). The corresponding constant of motion is the z-component of the angular momentum. Thus, we have used the angular momentum conservation twice: First the constancy of the direction (16.14) and then the constancy of the magnitude.

In the equation of motion (16.10), we put $z = 0$ and $\dot{\varphi} = \ell/(\mu\rho^2)$:

$$\mu\, \ddot{\rho} = \frac{\ell^2}{\mu\, \rho^3} - \frac{dU(\rho)}{d\rho} \quad \text{(radial equation)} \tag{16.17}$$

The first term on the right-hand side of this *radial equation* can be interpreted as a centrifugal force. In the second term, we see now a total derivative because U depends on ρ only.

---

[1] We would get the correct result, too, if we insert $z(t) = 0$ in the Lagrange function (16.9) itself. In principle, however, it is inadmissible to use a partial solution such as $z(t) = 0$ or a conserved quantity in $\mathcal{L}$, because $\mathcal{L}$ is not a physical quantity. Example: For $\mathcal{L} = T - U$ the conservation of energy shall hold, $T + U = E_0 = \text{const}$. The substitution of $T = E_0 - U$ into $\mathcal{L}$ would result in nonsense, namely $\mathcal{L} = E_0 - 2U$. For $\mathcal{L}$ the functional dependence of the coordinates and velocities is decisive. In contrast to this, the equations of motion and the conservation laws can always be inserted into each other.

Conservation of Energy

The Lagrange function (16.9) does not depend explicitly on time t; the system is symmetric under time translations. From $\partial \mathcal{L}/\partial t = 0$ and (9.36) follows:

$$E = \frac{\partial \mathcal{L}}{\partial \dot{\rho}}\,\dot{\rho} + \frac{\partial \mathcal{L}}{\partial \dot{\varphi}}\,\dot{\varphi} + \frac{\partial \mathcal{L}}{\partial \dot{z}}\,\dot{z} - \mathcal{L}$$
$$= \frac{\mu}{2}\left(\dot{\rho}^2 + \rho^2\dot{\varphi}^2 + \dot{z}^2\right) + U\big(\sqrt{\rho^2 + z^2}\big) = \text{const.} \qquad (16.18)$$

This conserved quantity (or constant of motion) equals the energy $E = T + U$ of the one-body problem. With $z = 0$ and $\dot{\varphi} = \ell/(\mu\rho^2)$ we obtain

$$E = \frac{\mu}{2}\,\dot{\rho}^2 + \frac{\ell^2}{2\mu\rho^2} + U(\rho) \qquad (16.19)$$

This is a first-order differential equation for $\rho(t)$. Substituting the solution $\rho(t)$ into (16.16), we may obtain $\varphi(t)$ by a further integration. The full solution of the two-body problem is thus of the following form:

$$\left.\begin{array}{l} \boldsymbol{R}(t) = \boldsymbol{A}\,t + \boldsymbol{B} \\[1ex] \left.\begin{array}{l} \rho(t) \ \text{ from (16.19)} \\ \varphi(t) \ \text{ from (16.16)} \\ z(t) = 0 \end{array}\right\} \rightarrow \boldsymbol{r}(t) \end{array}\right\} \stackrel{(16.3)}{\longrightarrow} \boldsymbol{r}_1(t),\, \boldsymbol{r}_2(t) \qquad (16.20)$$

The general solution of the six second-order differential equations for $\boldsymbol{r}_1(t)$ and $\boldsymbol{r}_2(t)$ must contain 12 integration constants. The center-of-mass motion contains six integration constants $\boldsymbol{A}$ and $\boldsymbol{B}$. The choice $z(t) = 0$ implies the determination of two integration constants. The conservation laws ℓ and E represent one integration constant each. Two further constants result from the integration of (16.16) and (16.19). All 12 integration constants may be fixed by the initial conditions. The two integration constants of the z-movement do not occur explicitly, because we have restricted ourselves a priori to $z(t) = 0$.

We recall the path towards the solution: The translational invariance implies the conservation of the center-of-mass momentum;

thus the motion of the center of mass can be calculated trivially. The rotational invariance implies the conservation of the angular momentum of the relative motion; thus the motion is confined to a plane and the angular motion is determined by $\ell = \mu \rho^2 \dot{\varphi}$. The time shift invariance implies the conservation of energy; thus one obtains a differential equation of first order. With $\ell = \mu \rho^2 \dot{\varphi}$ it can be written as a differential equation for the radial motion $\rho(t)$ alone, (16.17).

Solution

We present the solution for $\rho(t)$ and for the trajectory in the form of integrals. To do this we write (16.19) as

$$\dot{\rho} = \frac{d\rho}{dt} = \sqrt{\frac{2}{\mu}\left(E - U(\rho) - \frac{\ell^2}{2\mu\rho^2}\right)} \tag{16.21}$$

As the initial condition, we assume $\rho(t_0) = \rho_0$; the time point t_0 is arbitrary. We integrate (16.21) from t_0 to t:

$$t = t_0 + \int_{\rho_0}^{\rho} d\rho' \sqrt{\frac{\mu/2}{E - U(\rho') - \ell^2/(2\mu\rho'^2)}} \tag{16.22}$$

Executing this integral yields a function $t = t(\rho)$, from which $\rho = \rho(t)$ follows. Whether and how (analytically, numerically) the integral can be solved depends on the potential $U(\rho)$.

We can substitute the $\rho(t)$ determined by (16.22) into (16.16) and determine $\varphi(t)$ by further integration. However, one can also first determine the orbital curve $\rho = \rho(\varphi)$ without reference to the time; together with $\rho = \rho(t)$ this also completely fixes the solution. From (16.21) and (16.16) we obtain

$$\frac{d\varphi}{d\rho} = \frac{d\varphi}{dt}\,\frac{dt}{d\rho} = \frac{\ell}{\mu\rho^2}\Bigg/\sqrt{\frac{2}{\mu}\left(E - U(\rho) - \frac{\ell^2}{2\mu\rho^2}\right)} \tag{16.23}$$

A further integration yields the trajectory $\varphi = \varphi(\rho)$:

$$\varphi = \varphi_0 + \int_{\rho_0}^{\rho} d\rho' \frac{\ell/\rho'^2}{\sqrt{2\mu\,(E - U(\rho')) - \ell^2/\rho'^2}} \tag{16.24}$$

Whether the integrals (16.21)–(16.24) can be solved analytically depends on the functional form of the potential $U(\rho)$. In the following, we will investigate the motion for various potential functions.

Discussion of Motion

Even without explicitly solving the integrals of the last section, qualitative statements about the motion can be made on the basis of (16.19) and (16.16). Specifically, we discuss the potential

$$U(\rho) = \begin{cases} -\alpha/\rho \\ \alpha\,\rho^2 \\ \alpha\,(\rho-\rho_0)^2 \\ -\alpha/\rho^3 \end{cases} \tag{16.25}$$

Since $z = 0$, the arguments $r = \sqrt{\rho^2 + z^2}$ and ρ are interchangeable:

$$U(r) = U(\rho) \tag{16.26}$$

In particular, the first case in (16.25) refers to

$$U(r) = \begin{cases} -\dfrac{G m_1 m_2}{r} & \text{gravitational potential} \\ \dfrac{q_1\, q_2}{r} & \text{Coulomb potential} \end{cases} \tag{16.27}$$

The gravitational potential occurs in the Kepler problem, e.g., for the planetary motion in the solar system. The Coulomb potential describes the interaction in the hydrogen atom (proton plus electron), or the interaction between an α-particle and a nucleus (Rutherford scattering). For (16.27) the integrals (16.22) – (16.24) are elementarily solvable. This is also true for the oscillator, i.e. the second and third case in (16.25). The third case describes approximately the rotations and vibrations of a diatomic molecule; here ρ_0 is the equilibrium distance of the two atoms. The last case $U = -\alpha/\rho^3$ results in elliptic integrals, which cannot be solved analytically.

We discuss the time-dependence of the motion graphically. For this we refer to the examples (16.25); however, this discussion is easily transferable to other potentials. We start from

$$E = \frac{\mu}{2}\,\dot{\rho}^2 + U_{\text{eff}}(\rho) \tag{16.28}$$

with the *effective potential*

$$U_{\text{eff}}(\rho) = U(\rho) + \frac{\ell^2}{2\,\mu\,\rho^2} \tag{16.29}$$

One now plots the graph of $U_{\text{eff}}(\rho)$, Figures 16.1–16.3. Then the distance of the graph to the horizontal line E = const. equals the kinetic energy $\mu\dot{\rho}^2/2$ (Figure 16.1). Depending on the sign of $\dot{\rho}$ the particle runs towards the center or outwards. Hereby, the motion is restricted to regions for which $E - U_{\text{eff}} \geq 0$ holds. The direction of the ρ-motion changes at the turning points which are given by

$$E = U_{\text{eff}}(\rho_i) \quad \text{(turning points)} \tag{16.30}$$

In Figures 16.1–16.3, we consider possible motion types for different potentials. This results in the following types of motion, which are marked in the figures by (i), (ii) or (iii):

(i) *Scattering*: A particle comes from infinity and approaches the point ρ_0 (point of closest approach). The repulsion at ρ_0 is caused by the centrifugal barrier for $\ell \neq 0$. At ρ_0, the direction of the ρ motion reverses, and the particle flies away from the center ($\rho \to \infty$). Such a process is called scattering. The constant angular momentum $\ell = \mu\,\rho^2\,\dot{\varphi}$ together with $\rho(t)$ determines the simultaneous angular motion. Therefore, the particle is deflected by a certain angle during the scattering. Such a trajectory curve may occur for the potentials $U = -\alpha/\rho$ and $U = -\alpha/\rho^3$ but not for the oscillator potential.

(ii) *Bound motion*: It occurs when a physically accessible region ($E - U_{\text{eff}} > 0$) is confined by the two solutions of (16.30), ρ_{min} and ρ_{max}. In particular, the planetary ellipses are such bound solutions (Figure 16.2).

(iii) *Fall to the center*: For $\ell = 0$ and an attractive, non-singular potential, the particle passes through the center. If the potential is attractive and singular (e.g. $U(\rho) = -\alpha/\rho$), then the velocity at $\rho = 0$ becomes infinite and the model is unrealistic.

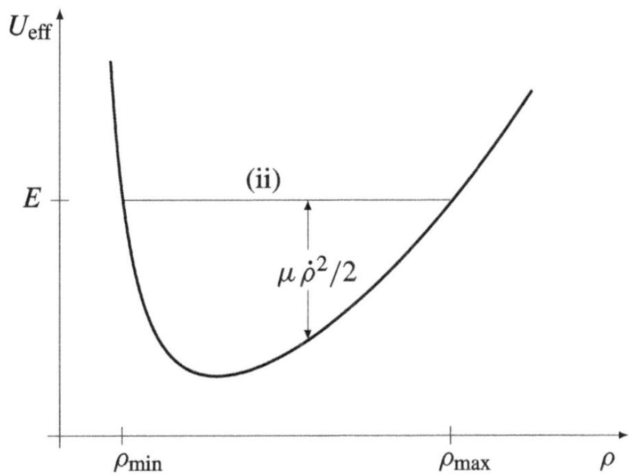

Figure 16.1 The effective potential $U_{\text{eff}} = \alpha\rho^2 + \ell^2/2\mu\rho^2$ for $\alpha > 0$ and $\ell \neq 0$ as a function of ρ. The kinetic energy $\mu\,\dot\rho^2/2$ for the radial motion equals the vertical distance between U_{eff} and the horizontal line $E = \text{const}$. For this potential only a bound motion (ii) is possible; the particle oscillates between the turning points ρ_{min} and ρ_{max}. In the ρ–φ-plane, the trajectory is a closed ellipse whose center coincides with the center of the potential.

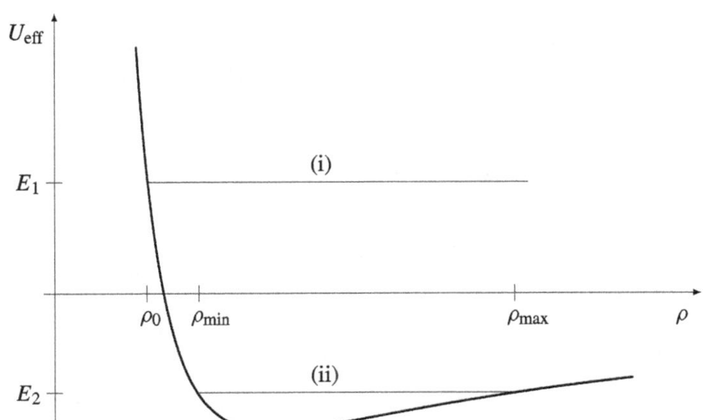

Figure 16.2 The effective potential $U_{\text{eff}} = -\alpha/\rho + \ell^2/2\mu\rho^2$ for $\alpha > 0$ and $\ell \neq 0$ as a function of ρ. For $E = E_2 < 0$ the motion is bound, for $E = E_1 > 0$, on the other hand, it is unbound. The bound motion (ii) oscillates between the turning points ρ_{min} and ρ_{max}. In the ρ–φ-plane, the associated orbit is an ellipse whose one focal point coincides with the center of the potential. In the case of unbound motion (i), the particle approaches a minimum distance ρ_0 and then moves away again; in the ρ–φ-plane, the orbit is a hyperbola.

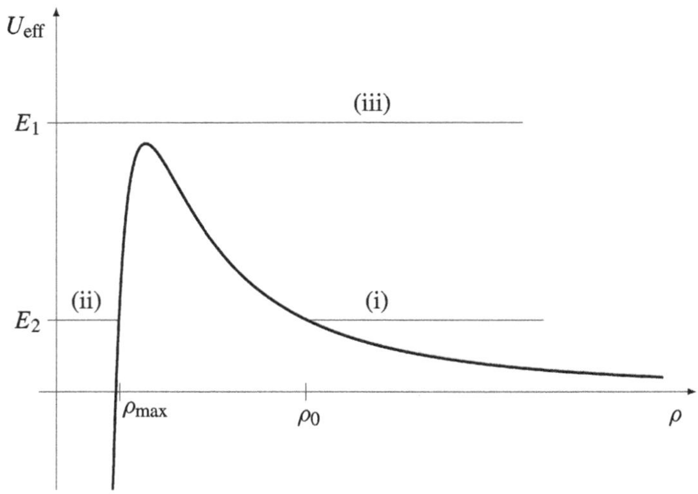

Figure 16.3 The effective potential $U_{\rm eff} = -\alpha/\rho^3 + \ell^2/2\mu\rho^2$ for $\alpha > 0$ and $\ell \neq 0$ as a function of ρ. The unbound motion (i) is a scattering with the turning point ρ_0. The unbound motion (iii) passes through the center; however, because of $U(0) = -\infty$ this is not a realistic model. The bound motion (ii) runs also through the center, and it is limited to the range $\rho \le \rho_{\rm max}$.

If an attractive potential $U(\rho)$ for $\rho \to 0$ grows stronger than $1/\rho^2$, then the particle can also reach the center for $\ell \neq 0$. This case is sketched for $U = -\alpha/\rho^3$ in Figure 16.3.

We discuss the bound motion between $\rho_{\rm min}$ and $\rho_{\rm max}$ in a bit more detail. In the ρ–φ-plane, such motions are confined by the circles $\rho = \rho_{\rm min}$ and $\rho = \rho_{\rm max}$, Figure 16.4. For $\rho = \rho_{\rm min}$ or $\rho = \rho_{\rm max}$, the radial velocity vanishes $\dot\rho = 0$, whereas $\dot\varphi = \ell/(\mu\rho^2)$ always has the same value. From this and from the isotropy of the potential follows that the shape of the trajectory curve between two turning points (at the outer or the inner circle) is always the same. The whole path is thus determined by such a partial loop. Between two temporally successive points at $\rho = \rho_{\rm max}$ (or $\rho_{\rm min}$) there is an angle change $\Delta\varphi$ which we can calculate with (16.24):

$$\Delta\varphi = 2\int_{\rho_{\rm min}}^{\rho_{\rm max}} d\rho\, \frac{\ell/\rho^2}{\sqrt{2\mu\,[E - U_{\rm eff}(\rho)]}} \qquad (16.31)$$

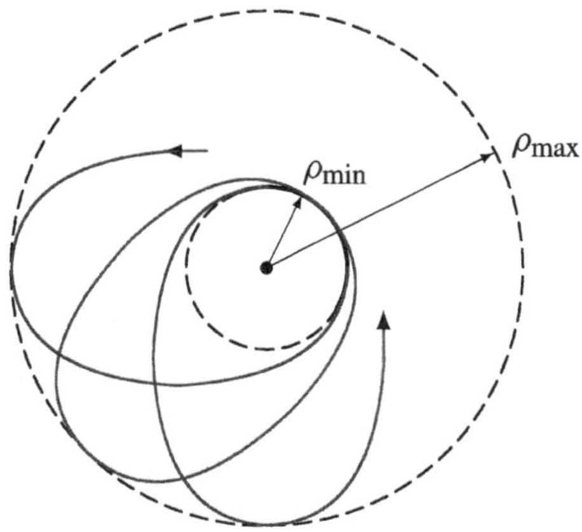

Figure 16.4 A sketch of a possible bound orbit in a central field. The motion takes place in the ρ–φ-plane between $\rho = \rho_{min}$ and $\rho = \rho_{max}$. The parts between ρ_{min} and ρ_{max} always have the same shape. For special potentials (gravitational and oscillator potential) a closed orbit (ellipse) results.

For the $1/\rho$ potential, we get $\Delta\varphi = 2\pi$, i.e. a closed orbit. As we will see in the following chapter, this is an ellipse with a focal point at $\rho = 0$. It is also possible that after n revolutions $n\,\Delta\varphi = 2m\pi$ holds, that is the orbit closes only after several loops. Especially for the oscillator $2\,\Delta\varphi = 2\pi$ holds; the trajectory is an ellipse with $\rho = \rho_{min}$ as an inscribed circle. In general, however, we have $\Delta\varphi \neq 2\pi\,(m/n)$, and the orbit is not closed. Then the particle comes arbitrarily close to any point between the circles in Figure 16.4 if only the motion is followed long enough.

Exercises

16.1 *On the choice of generalized coordinates*

Set up the Lagrange function for two particles (masses m_1 and m_2) with the potential $U(|\boldsymbol{r}_1 - \boldsymbol{r}_2|)$, using the generalized coordinates $\boldsymbol{\varrho} = \boldsymbol{r}_1 + \boldsymbol{r}_2$ and $\boldsymbol{r} = \boldsymbol{r}_1 - \boldsymbol{r}_2$. Show that $\boldsymbol{\varrho}$ is a cyclic coordinate. Why is this choice of coordinates less favorable than the choice $\boldsymbol{R} = (m_1\,\boldsymbol{r}_1 + m_2\,\boldsymbol{r}_2)/M$ with $M = m_1 + m_2$ and $\boldsymbol{r} = \boldsymbol{r}_1 - \boldsymbol{r}_2$?

16.2 *Substitution of conservation laws into the Lagrange function*

The Lagrange function

$$\mathcal{L}(\dot\rho, \dot\varphi, \rho) = \frac{\mu}{2}\left(\dot\rho^2 + \rho^2\,\dot\varphi^2\right) - U(\rho)$$

describes the motion in a central potential. Herein, one could eliminate the variable $\dot\varphi$ by using $\ell = \mu\,\rho^2\,\dot\varphi$. This leads to a new function $\mathcal{L}'(\dot\rho, \rho)$. Is $\mathcal{L}'$ a correct Lagrange function?

16.3 *Trajectories in the spherical oscillator potential*

A particle moves in the oscillator potential

$$U(r) = \alpha\, r^2 \quad \text{with } \alpha = \mu\omega^2/2 > 0 \tag{16.32}$$

First, discuss the motion qualitatively using a sketch of the effective potential. Then calculate the trajectories $\rho = \rho(\varphi)$. What is the shape of the trajectories? Determine ρ_{min} and ρ_{max} and sketch a trajectory curve. Compare these results with the solution of the equations of motion $\mu\,\ddot{\boldsymbol{r}} = -\operatorname{grad} U(r)$ in Cartesian coordinates.

Chapter 17

Kepler Problem

We solve the equations of motion for the $1/r$ potential. This solution describes approximately the motion of a planet in the field of the sun (Kepler problem).

Both, the Newtonian gravitational potential and the electrostatic Coulomb potential are of the form

$$U(r) = -\frac{\alpha}{r} \tag{17.1}$$

where

$$\alpha = \begin{cases} G\, m_1 m_2 & \text{gravitational potential} \\ -q_1 q_2 & \text{Coulomb potential} \end{cases} \tag{17.2}$$

Figure 16.2 shows the effective potential

$$U_{\text{eff}}(\rho) = -\frac{\alpha}{\rho} + \frac{\ell^2}{2\mu\rho^2} \tag{17.3}$$

for $\alpha > 0$. We can use $r = \rho$ because the trajectory lies ρ-φ-plane.

With (16.24) we calculate the trajectory curve

$$\varphi = \int d\rho\, \frac{\ell/\rho^2}{\sqrt{2\mu E + 2\mu\alpha/\rho - \ell^2/\rho^2}} = \arccos \frac{\ell/\rho - \mu\alpha/\ell}{\sqrt{2\mu E + \mu^2\alpha^2/\ell^2}} + \text{const.} \tag{17.4}$$

We choose the reference direction of φ so that the constant vanishes. Introducing the *parameter*[1]

$$p = \frac{\ell^2}{\mu\alpha} \tag{17.5}$$

and the *eccentricity*

$$\varepsilon = \sqrt{1 + \frac{2E\ell^2}{\mu\alpha^2}} \tag{17.6}$$

(17.4) becomes

$$\frac{p}{\rho} = 1 + \varepsilon \cos\varphi \tag{17.7}$$

This equation describes conic sections:

$$\begin{aligned} &\varepsilon > 1 \quad E > 0 \quad \text{hyperbola} \\ &\varepsilon = 1 \quad E = 0 \quad \text{parabola} \\ &\varepsilon < 1 \quad E < 0 \quad \text{ellipse} \end{aligned} \tag{17.8}$$

The circle ($\varepsilon = 0$) is a special ellipse.

We first consider the bound motion ($E < 0$, $\varepsilon < 1$) in the attractive potential ($\alpha > 0$, $p > 0$). For this purpose, we introduce the positive quantities

$$a = \frac{p}{1-\varepsilon^2} \quad \text{and} \quad b = \frac{p}{\sqrt{1-\varepsilon^2}} \tag{17.9}$$

With $\rho = \sqrt{x^2+y^2}$ and $\cos\varphi = x/\sqrt{x^2+y^2}$, equation (17.7) becomes

$$\frac{(x + a\,\varepsilon)^2}{a^2} + \frac{y^2}{b^2} = 1 \tag{17.10}$$

This is the equation of an ellipse with semi-axes a and b. The lengths a, b, and p are shown in Figure 17.1.

[1]The parameter p is also called *latus rectum* according to its geometrical meaning, Figure 17.1.

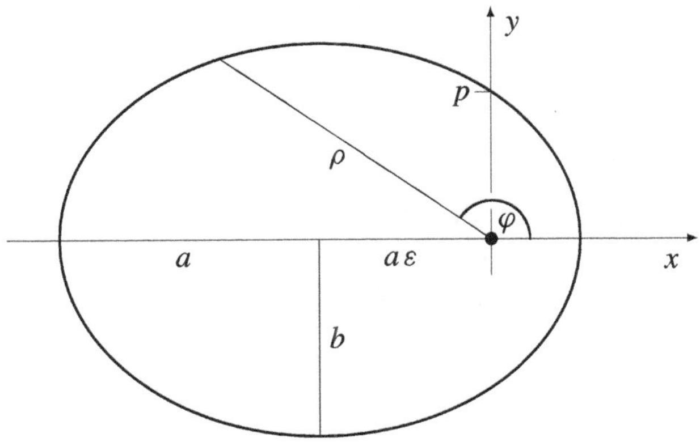

Figure 17.1 The ellipse is an orbital curve in the attractive $1/r$-potential; here a focal point (•) of the ellipse lies at the center ($\rho = 0$ or $x = y = 0$) of the potential. The shown ellipse has the axial ratio $a/b = 4/3$ and the eccentricity $\varepsilon = \sqrt{7}/4$. At $y = p$, the y axis intersects the ellipse; the parameter p is also called latus rectum. The distance $a\varepsilon$ between the center and the focal point is proportional to the eccentricity ε.

Kepler's Laws of Planetary Motion

We discuss the elliptic solution for a system consisting of the sun and a planet. We denote the coordinates with $\boldsymbol{r}_2 = \boldsymbol{r}_{\text{sun}} = \boldsymbol{r}_\odot$ and $\boldsymbol{r}_1 = \boldsymbol{r}_{\text{planet}} = \boldsymbol{r}_\text{P}$. As reference frame, we take the center-of-mass system (CS) with $\boldsymbol{R} = 0$; because of (16.6) this is an inertial frame. Then according to (16.3),

$$\boldsymbol{r}_\odot = \boldsymbol{r}_2 = \boldsymbol{R} - \frac{m_1}{M}\,\boldsymbol{r} \overset{\text{CS}}{=} -\frac{m_\text{P}}{M}\,\boldsymbol{r} \approx 0 \tag{17.11}$$

$$\boldsymbol{r}_\text{P} = \boldsymbol{r}_1 = \boldsymbol{R} + \frac{m_2}{M}\,\boldsymbol{r} \overset{\text{CS}}{=} \frac{m_\odot}{M}\,\boldsymbol{r} \approx \boldsymbol{r} \tag{17.12}$$

Both bodies traverse elliptical paths in opposite directions, where the common center of gravity lies in one of the foci of the ellipse. The approximations given in the last expression are valid for $m_\text{P} \ll m_\odot$ and yield Kepler's first law: "The planetary orbits are ellipses with the sun in a focal point". By this approximation terms of the order $m_\text{P}/m_\odot$ are neglected; for the heaviest planet, the Jupiter, they are of the size 10^{-3}.

The second Kepler law states that the area dA swept out by the driving beam per time dt is constant. From (16.16) follows:

$$\frac{dA}{dt} = \frac{\rho^2}{2}\frac{d\varphi}{dt} = \frac{\ell}{2\mu} = \text{const.} \tag{17.13}$$

This equal area law had already been illustrated in Figure 3.1.

The third Kepler law states that the square of the orbital time T is proportional to the third power of the major semi-axis a:

$$T^2 = \text{const.} \cdot a^3 \tag{17.14}$$

According to this law, the different planets $1, 2, \ldots$ show the same ratio $T_1^2/a_1^3 = T_2^2/a_2^3 = \cdots$. To derive this, we set the area $A = \pi\, a\, b$ of the ellipse equal to the area that follows from (17.13) for a full revolution:

$$A = \int_0^T dt\, \frac{dA}{dt} = T\, \frac{\ell}{2\mu} = \pi\, a\, b \tag{17.15}$$

From (17.5), (17.6), and (17.9) follows:

$$b = \frac{p}{\sqrt{1-\varepsilon^2}} = \frac{\ell}{\sqrt{2\mu|E|}} = \sqrt{\frac{a\,\ell^2}{\mu\,\alpha}} \tag{17.16}$$

We put this into (17.15):

$$T^2 = 4\pi^2\, \frac{\mu}{\alpha}\, a^3 \tag{17.17}$$

The factor μ/α is approximately the same for all planets:

$$\frac{\mu}{\alpha} = \frac{m_\text{P}\, m_\odot}{m_\text{P} + m_\odot}\, \frac{1}{G\, m_\text{P}\, m_\odot} = \frac{1}{G\,(m_\text{P} + m_\odot)} \approx \frac{1}{G\, m_\odot} = \text{const.} \tag{17.18}$$

In this approximation, (17.14) holds, i.e. Kepler's third law.

Realistic planetary orbits

The actual planetary orbits show deviations from ellipses due to various effects. The most important ones are:

1. Gravitational forces of the planets among themselves.
2. Relativistic effects.
3. Quadrupole moment of the sun.

All these effects constitute small disturbances. The first effect is small because the masses of the other planets are much smaller than the mass of the sun. The largest planet is Jupiter; for its mass, we get the ratio $m_J/M_\odot \approx 10^{-3}$. Roughly speaking we expect relative corrections of the order 10^{-3}. In detail, however, the different distances between all masses (sun, planets, moons) had to be taken into account.

For a first classification of the magnitude of relativistic effects, we estimate the orbital velocity v_E of the earth. In $v_E = 2\pi r/T$, we set $T = 1\,\text{year}$ for the orbital period and $r = 8 \cdot 60\,c\text{s}$ for the Sun–Earth distance (corresponding to 8 light minutes). This results in $v_E/c \approx 10^{-4}$. Relativistic effects are therefore rather small. In a consistent relativistic theory, the gravitational field itself must be treated relativistically, too.[2]

A quadrupole moment of the sun is to be expected due to its self-rotation and associated oblateness. This results in a gravitational quadrupole field, which has a $1/r^3$-dependence. Due to this radial dependence, it will mainly influence the sun nearest planets. For example, it contributes to the perihelion precession of Mercury, the sun's nearest planet.

All these effects lead to a deviation from the closed elliptical path. After each orbit, the line from the center to the closest point to the sun, the perihelion of the orbit, is slightly turned (Exercise 17.4). This perihelion precession is for Mercury of the order of one second of arc per orbit; it is caused to about 9/10 by the other planets and to 1/10 by relativistic effects. The influence of the quadrupole moment (oblateness) of the sun is again much smaller.

Deterministic Many-Body Problem

Already in the 19th century, it was possible to determine the influence of the other planets on the individual planetary orbits (especially on the perihelion) in a perturbative approach. Today such problems are treated numerically on a computer; this holds, of course, also for

[2]See T. Fließbach, *Allgemeine Relativitätstheorie*, 7th edition, Springer Spektrum 2016.

the calculation of the trajectory of a space shuttle to the moon. We sketch the principle of such a numerical solution.

The equations of motion for N mass points (sun, planets, moons) are

$$m_\nu\, \ddot{\boldsymbol{r}}_\nu(t) = -\mathrm{grad}_\nu \sum_{\mu\neq\nu}^{N} U_{\nu\mu}(|\boldsymbol{r}_\nu - \boldsymbol{r}_\mu|), \quad U_{\nu\mu}(r) = -\frac{G\, m_\nu\, m_\mu}{r} \tag{17.19}$$

We now assume that at a given time t the positions $\boldsymbol{r}_\nu(t)$ and velocities $\dot{\boldsymbol{r}}_\nu(t)$ are known. Then we calculate from (17.19) the accelerations $\ddot{\boldsymbol{r}}_\nu(t)$. The positions and velocities at the slightly later time $t + \Delta t$ are then

$$\begin{aligned} \dot{\boldsymbol{r}}_\nu(t+\Delta t) &= \dot{\boldsymbol{r}}_\nu(t) + \ddot{\boldsymbol{r}}_\nu(t)\; \Delta t \\ \boldsymbol{r}_\nu(t+\Delta t) &= \boldsymbol{r}_\nu(t) + \dot{\boldsymbol{r}}_\nu(t)\; \Delta t \end{aligned} \tag{17.20}$$

For this step from t to $t + \Delta t$ one writes a computer program. Then one lets the computer start with the known initial conditions $\boldsymbol{r}_\nu(0)$, $\dot{\boldsymbol{r}}_\nu(0)$, and let it perform successively the steps from $t = 0$ over $t = \Delta t$, $t = 2\Delta t$, $t = 3\Delta t, \ldots$, until, for example, $t = 10^6\, \Delta t$. The numerical convergence (to the true solution) can be checked by doubling the number of steps at half the step size. This describes the principle of the numerical solution; for the practical application, there are efficient algorithms for the integration of differential equations. In this numerical treatment, given external forces may be taken into account, too.

The orbits $\boldsymbol{r}_\nu(t)$ of all (relevant) bodies in the solar system determine the gravitational potential:

$$\Phi(\boldsymbol{r}, t) = -\sum_\nu \frac{G m_\nu}{|\boldsymbol{r} - \boldsymbol{r}_\nu(t)|} \tag{17.21}$$

This potential determines the force $\boldsymbol{F} = -m\, \mathrm{grad}\, \Phi$ on an additional small body like a comet or a spaceship. If the additional body is small enough, its contribution to (17.21) can be neglected, i.e. its own gravitational field is not taken into account.

The N-particle system considered in (17.19) and (17.20) is *deterministic*: For given initial conditions, the equations of motion define the system state (defined by $\boldsymbol{r}_\nu(t)$ and $\dot{\boldsymbol{r}}_\nu(t)$) at each time t. This follows from the construction (17.20) of the solution. The statement "deterministic" applies in general to mechanical systems; the prerequisite for the applicability of (17.20) is only that the right-hand sides of the equations of motion (17.19) are known as a function of the coordinates, the velocities, and the time.

Regular or chaotic motion

We consider a lottery machine with 49 balls. The rotation of the machine implies external forces, and balls scatter at each other (and also the rotational motion of the lotto balls has to be taken into account). But all this does not change the principle, that (17.20) describes the motion deterministically. This deterministic system behaves in fact, however *chaotically*; its course of motion is not really predictable. Such a scenario may be called *deterministic chaos*.

The study of chaotic dynamics is a broad field.[3] We confine ourselves here to some qualitative remarks on the question of whether the motion of a mechanical many-particle system is regular (solar system?) or chaotic (lotto, weather).

The characteristic of chaotic behavior is that small differences in the initial conditions can quickly become large (Figure 17.2). For two equal lotto apparatuses with very similar initial conditions, the difference between the two systems grows exponentially. The initial conditions can only be determined with a certain accuracy. Therefore, the motion can be predicted only for a limited period of time (in the case of the lottery, perhaps over a few collisions; in the case of the weather, over a few days). In addition, there are always small external disturbances, for example, a person coughing in the lottery room or a butterfly flapping its wings as a disturbance in the atmosphere. They might lead to an extremely small change of the system state. But an even extremely small disturbance may result in a completely different course of events.

[3]For an introduction, see, for example, G. L. Baker and J. P. Gollub, *Chaotic dynamics*, Cambridge University Press, 1990.

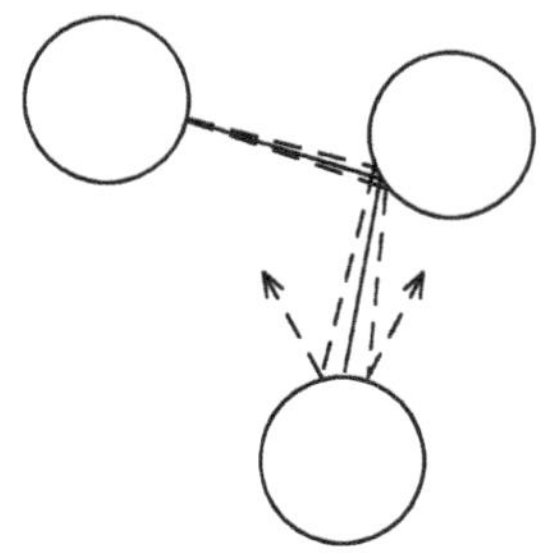

Figure 17.2 The circles are billiard balls, lottery balls, or the atoms of a gas. The next two collisions of the left upper sphere are sketched schematically. The figure illustrates that small differences in the initial conditions can very quickly become large and then lead to completely different trajectories. For a few collisions the course is still predictable, for many collisions, however, it becomes chaotic.

In contrast to the lottery machine, our solar system seems to execute a completely regular motion. In fact, however, there are chaotic deviations from the regular orbits on a small scale. The question arises whether these deviations might become large in the course of time.

A completely regular motion is obtained for so-called *integrable* systems. These are systems, which can be completely integrated (like the two-body problem). This is the case, if the number of the independent conservation laws is equal to the number f of degrees of freedom. In the two-body problem $f = 6$, and the six independent conservation laws are the center-of-mass momentum (3 components), the angular momentum (size and direction), and the energy. Already the three-body problem is a non-integrable system.

In our solar system, one can observe chaotic behavior on a small scale: At one point of a planetary orbit, one puts a (fictitious) plane perpendicular to the orbit. For each pass of the trajectory one marks the respective point of penetration. Due to the disturbances by the other planets (and other disturbances), the penetration points in this *Poincaré section* show a chaotic pattern.

So, the solar system shows chaotic behavior in the small, but a regular behavior in the large. This system is almost integrable: The subsystem sun-planet (no other planets etc.) is integrable (the idealized Kepler problem treated here). But this subsystem is subject to small, non-integrable perturbations. In the foreseeable future (say in the next 10^4 years), no deviation from the now observed regularity is to be expected. However, it remains open whether the small perturbations in the distant future (say in 10^8 years) will lead to a substantially different course of motion.[4]

Hyperbolic Orbit

Depending on the energy range, the orbit equation (17.7) describes ellipses or hyperbolas. The limiting case of the parabola corresponds to a point on the energy scale; it will not be considered in detail in the following.

We first consider the unconstrained motion ($E > 0$, $\varepsilon > 1$) in the attractive potential ($\alpha > 0$, $p > 0$). For this the curve (17.7) is a hyperbola; it is sketched in Figure 17.3 on the left. This could be the

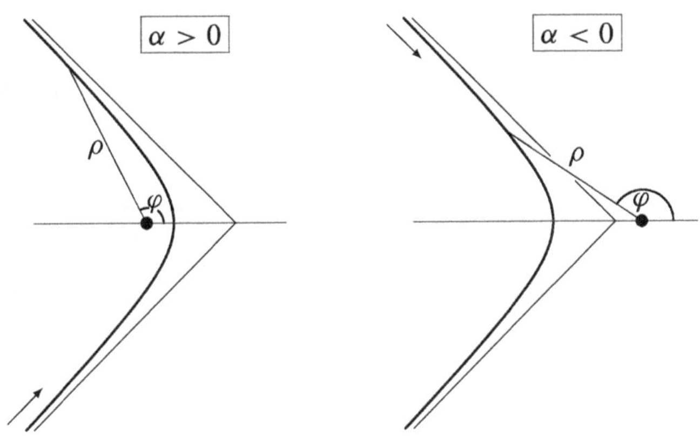

Figure 17.3 For positive energies, the orbital curves in the $1/r$ potential are hyperbolas whose focal point (•) coincides with the center of the potential. In the left part, the scattering at an attractive potential is shown, e.g. for a comet in the gravitational field of the sun. The right part is valid for a repulsive potential, e.g. for the scattering of an alpha particle at an atomic nucleus. The direction indicated by the arrow corresponds to $\ell = \mu\,\rho^2\,\dot{\varphi} > 0$.

---

[4]The KAM-theorem (by Kolmogorov, Arnold and Moser) establishes the conditions under which the motion of a nearly-integrable system remains regular.

orbit of a comet in the field of the sun. The possible angular range is limited by $\cos\varphi \geq -1/\varepsilon$. By $\cos\varphi_\infty = -1/\varepsilon$ the directions of the asymptotes of the hyperbola are given.

We now consider a repulsive potential ($\alpha < 0$), such as a Coulomb potential between two positive charges. Then $U_{\text{eff}}(\rho)$ is positive and monotonically decreasing. There is only an unbound motion ($E > 0$, $\varepsilon > 1$). For $\alpha < 0$ the p defined in (17.5) is negative. We write the equation (17.7) therefore in the form

$$\frac{|p|}{\rho} = -1 - \varepsilon\cos\varphi \tag{17.22}$$

This describes a hyperbola; it is sketched in Figure 17.3 on the right. Since the left-hand side of (17.22) is positive, the angles are restricted by $\cos\varphi \leq -1/\varepsilon$ to a range around $\varphi = \pi$. The boundary angles ($\cos\varphi_\infty = -1/\varepsilon$) define the asymptotes of the hyperbola. The hyperbola could be the trajectory of an alpha particle in the Coulomb field of an atomic nucleus. This Rutherford scattering will be discussed in more detail in Chapter 18.

In both cases shown in Figure 17.3, the particle approaches the center to the minimum distance r_0, which is defined by $U_{\text{eff}}(r_0) = E$. After that, it moves away again ($r \to \infty$ for $t \to \infty$). The angle φ changes mainly in the range $r \sim r_0$ (we assume $\ell \neq 0$). The particle is deflected by an angle depending on E and ℓ.

The discussed hyperbolic trajectories are valid for the relative vector $\boldsymbol{r}$, i.e. for the motion of the (fictitious) particle of mass μ in the potential $U(r)$. Then (17.11) and (17.12) determine the orbits of the two individual mass points.

Time Dependence of the Path

We still have to calculate the solution for $\rho(t)$, which determines the time dependence of the trajectory. We consider the attractive potential $U = -\alpha/\rho$ with $\alpha > 0$. From (16.21) follows:

$$t = \sqrt{\frac{\mu}{2|E|}} \int \frac{\rho\, d\rho}{\sqrt{-\rho^2 + \alpha\rho/|E| - \ell^2/(2\mu|E|)}} \tag{17.23}$$

Here, we insert a from (17.16) and ε from (17.6):

$$t=\sqrt{\frac{\mu}{2|E|}}\int\frac{\rho\, d\rho}{\sqrt{a^2\varepsilon^2-(\rho-a)^2}} \tag{17.24}$$

The solution $t(\rho)=\cdots\sqrt{a^2\varepsilon^2-(\rho-a)^2}+\cdots\arcsin(\cdots\rho+\cdots)$ of this integral cannot be resolved elementarily to $\rho=\rho(t)$. We give instead a parameter representation of the solution. To do this we substitute

$$\rho-a=-a\,\varepsilon\cos\xi \tag{17.25}$$

in the integral (17.24):

$$t=\sqrt{\frac{\mu a^3}{\alpha}}\int d\xi\,\big(1-\varepsilon\cos\xi\big)=\sqrt{\frac{\mu a^3}{\alpha}}\,\big(\xi-\varepsilon\sin\xi\big)+t_0 \tag{17.26}$$

We arbitrarily set $t_0=0$. This gives us the following parameter representation of the time dependence:

$$\rho=a\big(1-\varepsilon\cos\xi\big),\quad t=\sqrt{\frac{\mu a^3}{\alpha}}\,\big(\xi-\varepsilon\sin\xi\big)\quad(E<0) \tag{17.27}$$

For $E<0$ we have $\varepsilon<1$, so that ξ grows monotonically with t. And ρ oscillates between $\rho_{\min}=a(1-\varepsilon)$ and $\rho_{\max}=a(1+\varepsilon)$.

For $E>0$ we get some sign changes (relative to $E<0$). Then the analogous calculation leads to

$$\rho=a\big(\varepsilon\cosh\xi-1\big),\quad t=\sqrt{\frac{\mu a^3}{\alpha}}\,\big(\varepsilon\sinh\xi-\xi\big)\quad(E>0) \tag{17.28}$$

For the repulsive potential $U=-\alpha/\rho$ with $\alpha<0$ one obtains instead $\rho=a\,(\varepsilon\cosh\xi+1)$ and $t=\cdots(\varepsilon\sinh\xi+\xi)$.

Exercises

17.1 *Lenz's vector*

For the trajectory $\boldsymbol{r}(t)$ in the potential $U=-\alpha/r$ we define the *Lenz vector* $\boldsymbol{\Lambda}$ by

$$\boldsymbol{\Lambda}=\frac{\mu}{\alpha}\,\dot{\boldsymbol{r}}\times(\boldsymbol{r}\times\dot{\boldsymbol{r}})-\frac{\boldsymbol{r}}{r}=\frac{\boldsymbol{p}\times\boldsymbol{\ell}}{\alpha\,\mu}-\frac{\boldsymbol{r}}{r} \tag{17.29}$$

Show:

(a) $\boldsymbol{\Lambda}$ is a constant of motion, $d\boldsymbol{\Lambda}/dt = 0$.
(b) $|\boldsymbol{\Lambda}|$ is equal to the eccentricity ε.
(c) $\boldsymbol{\Lambda}$ points to the center-nearest orbital point (perihelion), so $\boldsymbol{\Lambda} \parallel \boldsymbol{r}_{\min}$.
(d) The evaluation of $\boldsymbol{\Lambda} \cdot \boldsymbol{r}$ yields the orbital curve $p/\rho = 1 + \varepsilon \cos \varphi$.

17.2 *Kepler orbits in Cartesian coordinates*

Consider the Kepler problem $U = -\alpha/r$ in Cartesian coordinates with the initial conditions $\boldsymbol{r}(0) := (s, 0, 0)$ and $\dot{\boldsymbol{r}}(0) := (0, v_0, 0)$. Derive the trajectories

$$y^2 = \lambda(\lambda - 2)\, x^2 - 2\lambda(\lambda - 1)\, s\, x + \lambda^2 s^2, \quad \lambda = \frac{\mu s v_0^2}{\alpha} \qquad (17.30)$$

from the conservation laws for the angular momentum and the Lenz vector (Exercise 17.1). Which parameter values lead to a parabola, circle, straight line, ellipse, or hyperbola?

17.3 *Earth satellite on a circular orbit*

What is the effective potential $U_{\text{eff}}(\rho)$ for the orbit of an earth satellite with mass m? Use the conservation of the energy $E = m\,\dot{\rho}^2/2 + U_{\text{eff}}(\rho)$. Determine the radius ρ_0 of the orbit as a function of the orbital frequency ω of the satellite. What is the radius for a geostationary orbit?

17.4 *Perihelion precession*

In the gravitational potential $U_0(\rho) = -\alpha/\rho$ of the sun, a planet moves on an elliptic orbit ($\alpha = G m_{\odot} m_{\text{P}}$, $\mu \approx m_{\text{P}}$). Small perturbations δU in the potential $U = U_0 + \delta U$ usually lead to a *perihelion precession*: After each revolution the direction of the perihelion (closest point to the sun) changes by the angle $\delta\varphi$.

On the way from perihelion to perihelion the angle changes by

$$\Delta\varphi = -2\sqrt{2\mu}\, \frac{d}{d\ell} \int_{\rho_{\min}(\ell)}^{\rho_{\max}(\ell)} d\rho\, \sqrt{E - U(\rho) - \ell^2/(2\mu\rho^2)} \qquad (17.31)$$

First check the validity of this relation. Show $\varDelta\varphi = 2\pi$ for $\delta U = 0$ (closed elliptic orbit). Now, calculate the perihelion precession $\delta\varphi$ to first order in δU. The result is

$$\delta\varphi = \varDelta\varphi - 2\pi = 2\mu\,\frac{d}{d\ell}\left[\frac{1}{\ell}\int_0^{\pi} d\varphi\;\rho^2\,\delta U(\rho)\right] \tag{17.32}$$

where $\rho = \rho(\varphi)$ is the unperturbed solution. Evaluate the result for the perturbation potentials $\delta U = \gamma/\rho^3$ and $\delta U = \beta/\rho^2$.

Chapter 18

Scattering

A beam of particles is directed toward the matter. The particles scattered by the matter are then counted by a detector. The ratio of the scattered particle current to the incident current density defines the cross-section. The measured cross-section allows conclusions about the structure of the matter. We calculate the cross-section for the case that the interaction between a projectile and a target particle is described by a potential.

The unbound motion in the two-body problem is — after the reduction to a one-particle problem — like this: A (fictitious) particle of mass μ runs from infinity towards the potential center, reaches a minimal distance, and runs away again towards infinity (like in Figure 17.3). Though this process, the particle is deflected by a certain angle. This can be used to describe the collision of a billiard ball with another, or the deflection of a comet in the gravitational field of the sun, or the scattering of an alpha particle in the Coulomb field of an atomic nucleus.

Of particular practical interest is the following situation: A particle beam is running towards a collection of other, generally stationary particles (such as a piece of matter). If each of the incoming *projectiles* interacts with maximally one particle of the *target*, then this process can be treated as a two-body problem. The solution of this two-body problem (which according to Chapter 16 can be reduced to a one-body problem) determines the cross-section of the scattering experiment. For the Rutherford scattering, i.e. for the scattering of alpha particles at atomic nuclei, we will explicitly calculate this cross-section.

Definition of the Cross-Section

We consider a beam of incident particles that are scattered at a potential (representing one target particle). We assume that the incident beam has a homogeneous *particle current density* across the beam width. The current density j is defined by

$$j = \frac{\text{number of incident particles}}{\text{time} \cdot \text{area}} = \frac{\Delta N_0}{\Delta t \, \Delta A} \tag{18.1}$$

Far away from the scattering center, all particles in the beam have the same velocity $\boldsymbol{v}_\infty$ and therefore the same energy

$$E = \frac{\mu}{2} \, v_\infty^2 \tag{18.2}$$

We consider a beam that is directed towards one single target particle, Figure 18.1. The interaction between a projectile particle and the target particle is described by a central potential. The projectile particles should not influence each other. The considered process is then a *superposition* of the independent scatterings of the individual incoming particles at the potential.

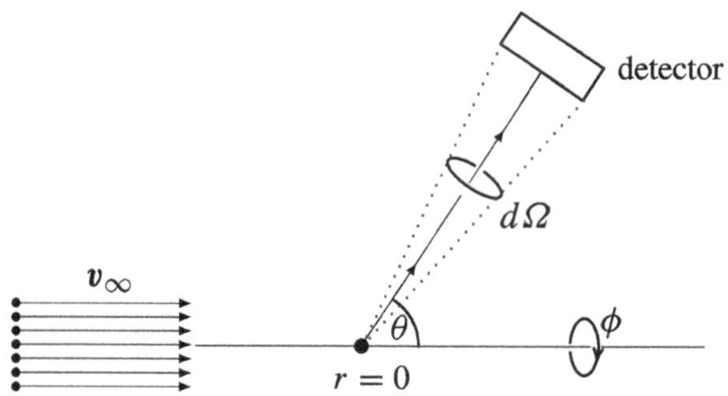

Figure 18.1 A particle beam travels towards a scattering center at $r = 0$. The asymptotic velocity $\boldsymbol{v}_\infty$ defines the beam direction. A detector counts all particles in the solid angle $d\Omega$. Due to the cylindrical symmetry around the beam axis, the number of scattered particles depends on θ but not on ϕ.

A picked-out particle is deflected by an angle θ (Figure 18.1). The problem is cylindrically symmetric with respect to rotations about the beam axis passing through the scattering center; therefore the number of scattered particles does not depend on the angle ϕ. The problem, however, is not spherically symmetric (like the potential), because the incident beam has a distinguished direction.

A detector set up at a distance R counts all particles which hit it. Let the distance R be so large and the detector area $dA_{\rm D}$ be so small that thereby a small solid angle $d\Omega$ is defined:

$$d\Omega = d\phi \, d\cos\theta = d\phi \, \sin\theta \, d\theta = \frac{dA_{\rm D}}{R^2} \tag{18.3}$$

Let all differentials be positive here.

During a time interval Δt, the detector counts a certain number $\Delta N(\theta)$ of scattered particles. Thus, it measures the flux

$$\text{particle flux in } d\Omega = \frac{\Delta N(\theta)}{\Delta t} \tag{18.4}$$

This particle flux is proportional to the solid angle $d\Omega$ and to the incident current density j. The ratio of (18.4) to j and $d\Omega$ is called *differential cross-section* $d\sigma/d\Omega$:

$$\begin{aligned}\frac{d\sigma}{d\Omega} &= \frac{\Delta N(\theta)}{j \, d\Omega \, \Delta t} = \frac{\text{particle flux per } d\Omega}{\text{incident particle current density}} \\ &= \frac{\text{number of scattered particles per time and per } d\Omega}{\text{number of incident particles per time and area}}\end{aligned} \tag{18.5}$$

If we sum this over all directions, we get the *total cross-section* σ:

$$\sigma = \int d\Omega \, \frac{d\sigma}{d\Omega} = \int_0^{2\pi} d\phi \int_0^{\pi} d\theta \, \sin\theta \, \frac{d\sigma}{d\Omega} \tag{18.6}$$

Since $d\sigma/d\Omega$ does not depend on ϕ, we get $\int d\phi = 2\pi$. From (18.5) and (18.6) it follows that the total number of scattered particles per time is equal to σj. Therefore, σ is the area at which the incident particle current density j is effectively scattered. For the scattering of a nucleon from a heavy atomic nucleus (with a radius $R \sim 6 \cdot 10^{-15}$ m) we therefore expect the order of magnitude $\sigma \sim \pi R^2 \approx 10^{-28}\,\text{m}^2$, for the scattering of two atoms ($R \sim 2\,\text{Å}$) from each other, $\sigma \sim 10^{-19}\,\text{m}^2$.

The actual measurement is always performed on a target with many scattering centers. In an experiment, a thin foil might be placed into the incident beam. The number M of scattering centers is then equal to the number of target particles in the beam area. The foil is chosen so thin that each projectile particle is scattered only by one target particle and then leaves the foil without further interaction. If these conditions are fulfilled, then the scattering rate measured in the detector must be divided by M to relate it to a single scattering center.

The beam width L (about some millimeters) leads to an indeterminacy of the size L/R in the angle; one does not know from which region of the beam a particle detected in the detector comes. Also, the finite size of the detector and the finite thickness of the target foil result in an angular uncertainty. These uncertainties can be reduced by a sufficiently large distance R of the detector.

The definitions (18.1)–(18.6) are valid independently of whether the scattering is treated classically (as here) or quantum mechanically.

Calculation of the Cross-Section

In the last section, the cross-section was reduced to the measurable quantities j, $d\Omega$, and $\Delta N(\theta)/\Delta t$; thus, it is itself a measurable quantity. Under the discussed conditions (no interaction of the incident particles with each other, collision of each projectile particle with at most one target particle), the theoretical treatment of the scattering can be reduced to the two-body scattering problem. On the basis of the solution of the two-body problem (Chapters 16 and 17) we calculate the cross-section.

All incident particles have the same energy E. For a central potential, the scattering angle θ can then only depend on the distance s of the scattered particle from the beam axis (Figure 18.2):

$$\theta = \theta(s) \quad \text{or} \quad s = s(\theta). \tag{18.7}$$

The *impact parameter* s determines the angular momentum ℓ with respect to the scattering center:

$$\ell = \mu\, v_\infty\, s = \sqrt{2\mu E}\, s \tag{18.8}$$

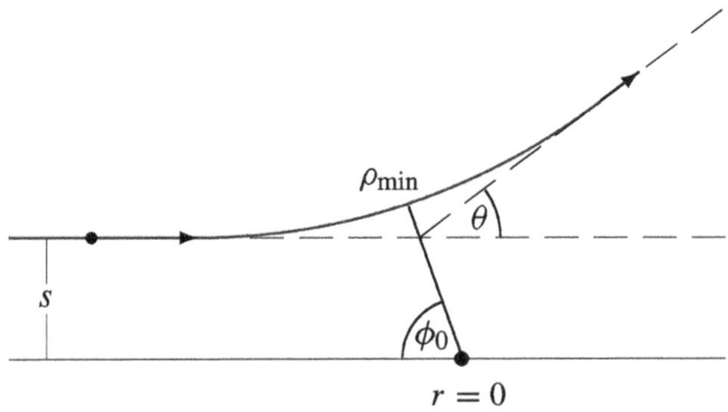

Figure 18.2 Asymptotically each projectile particle has a certain distance s from the beam axis, which passes through the scattering center (center of the target particle). The scattering angle θ is a function of this impact parameter s.

In the incident beam, particles with different values of s occur. All particles with an impact parameter between s and $s+ds$ are scattered in an angular range between θ and $\theta + d\theta$; i.e. $ds = (ds/d\theta)\, d\theta$. Since no particles are lost, the incident current between s and $s + ds$ must be equal to the outgoing current between θ and $\theta + d\theta$:

$$j\ 2\pi\, s\, \big|ds\big| = \left| \int_0^{2\pi} d\phi \int_\theta^{\theta+d\theta} d\theta'\ \sin\theta'\ j\ \frac{d\sigma}{d\Omega} \right| = j\ 2\pi\ \sin\theta\ \frac{d\sigma}{d\Omega}\ \big|d\theta\big| \tag{18.9}$$

In general, the scattering angle decreases as the impact parameter increases, so $ds/d\theta < 0$. In the balance equation (18.9), both sides must be positive. Therefore, the absolute values of ds and $d\theta$ are taken, and the other factors are always positive (j and $d\sigma/d\theta$ by definition, and $\sin\theta$ because of $0 \le \theta \le \pi$). The balance equation (18.9) yields then

$$\frac{d\sigma}{d\Omega} = \frac{s(\theta)}{\sin\theta} \left| \frac{ds(\theta)}{d\theta} \right| \tag{18.10}$$

Thus, the determination of $d\sigma/d\Omega$ is based on that of $s = s(\theta)$.

The position vector of the scattered particle changes between $\rho = \infty$ and $\rho_{\min}$ by the angle ϕ_0 (Figure 18.2). For the positive angle ϕ_0, we obtain from (16.24)

$$\phi_0 = \int_{\rho_{\rm min}}^{\infty} d\rho \, \frac{\ell/\rho^2}{\sqrt{2\mu(E-U(\rho)) - \ell^2/\rho^2}} \tag{18.11}$$

As can be seen in Figure 18.2, the angular change ϕ_0 occurs twice, namely during the approach of the projectile to $\rho_{\rm min}$, and again during the flight away until $\rho = \infty$; the trajectory of the projectile is symmetric with respect to $\rho_{\rm min}$. The scattering angle θ is equal to

$$\theta(s) = \pi - 2\,\phi_0(s) = \pi - 2\int_{\rho_{\rm min}}^{\infty} d\rho \, \frac{s/\rho^2}{\sqrt{1 - s^2/\rho^2 - U(\rho)/E}} \tag{18.12}$$

For ℓ, (18.8) was used. Equation (18.12) determines $\theta(s)$ and thus also $s(\theta)$. From this the cross-section (18.10) can be calculated.

The main contributions to the integral (18.12) come from the range $\rho \sim \rho_{\rm min}$ because for larger distances the potential is so small that the scattering angle hardly changes. It therefore does not matter that the particles counted by the detector are at a finite distance (for example, some decimeters) and not at infinity (upper integral limit).

For the evaluation of (18.12) we still need the minimum distance $\rho_{\rm min}$. It is given by $d\rho/d\varphi = 0$, which according to (16.23) holds for $E = U + \ell^2/2\mu\rho^2$. By (18.8) this condition becomes

$$1 = \frac{s^2}{\rho_{\rm min}^2} + \frac{U(\rho_{\rm min})}{E} \tag{18.13}$$

Thus, at the lower integral limit, the integrand in (18.12) is singular; the integral however exists.

Rutherford Cross-Section

We apply our results to the scattering of two atomic nuclei, whose interaction is given by the Coulomb potential,

$$U(r) = -\frac{\alpha}{r} = \frac{Z_1 Z_2\, e^2}{r} \tag{18.14}$$

Here, Z_1 and Z_2 are the nuclear charge numbers. Specifically, we consider alpha particles ($Z_1 = 2$) with an energy of a few MeV, which are scattered by a gold foil ($Z_2 = 79$). Alpha particles may be obtained from natural radioactivity in an energy range $E \approx 4 \cdots 8\,\text{MeV}$.

The maximum approach of the projectile occurs for a central collision ($s = 0$). In this case, the angular momentum vanishes ($\ell = \mu\, s\, v_\infty = 0$) and (18.13) becomes $U(\rho_{\text{min}}) = E$. For the considered Coulomb potential we get then

$$\rho_{\text{min}} = \frac{Z_1 Z_2\, e^2}{E} \approx 30 \cdots 60\,\text{fm} \quad (\ell = 0) \tag{18.15}$$

The numerical value is obtained for $Z_1 = 2$, $Z_2 = 79$, and $E \approx 4 \cdots 8\,\text{MeV}$; in nuclear physics one uses the length unit $1\,\text{fm} = 1\ \text{fermi} = 10^{-15}\,\text{m}$.

For $\ell \neq 0$, the minimum distances are even larger than those in (18.15). Therefore, the alpha particles do not feel anything of the strong interaction. The strong interaction takes effect only when the surfaces of the two nuclei have approached about 1 fm; for this, we needed a distance $\rho \approx 10\,\text{fm}$. In the considered energy range, there is no correction of the Coulomb potential due to the strong interaction.

Besides the atomic nuclei, the gold foil contains also electrons (of the gold atoms). Since the mass of the alpha particles is 10^4 times larger than that of electrons, the deflection of the alpha particles by the electrons is negligible.

The trajectories of a particle in the potential (18.14) are according to (17.22) hyperbolas

$$\frac{|p|}{\rho} = -1 - \varepsilon \cos\varphi \tag{18.16}$$

with $|p| = \ell^2/\mu|\alpha|$, where $|\alpha| = Z_1 Z_2\, e^2$. We substitute (18.8) into the eccentricity (17.6):

$$\varepsilon = \sqrt{1 + \frac{2E\ell^2}{\mu\,\alpha^2}} = \sqrt{1 + \left(\frac{2Es}{\alpha}\right)^2} > 1 \tag{18.17}$$

The hyperbola (18.16) is sketched on the right in Figure 17.3. The minimum distance ρ_{min} is obtained at the angle $\varphi_{\text{min}} = \pi$, while $\rho = \infty$ leads to $\cos\varphi_\infty = -1/\varepsilon$. We need the (positive) angular difference ϕ_0 from Figure 18.2, i.e. $\phi_0 = |\varphi_{\min} - \varphi_\infty| = |\pi - \varphi_\infty|$. From

$$\cos\phi_0 = \cos(\pi - \varphi_\infty) = -\cos\varphi_\infty = \frac{1}{\varepsilon} = \frac{1}{\sqrt{1+(2Es/\alpha)^2}} \tag{18.18}$$

follows

$$\tan\phi_0 = \frac{2Es}{|\alpha|} \quad (0 \le \phi_0 \le \pi/2) \tag{18.19}$$

By (18.12), $\phi_0 = (\pi - \theta)/2$ holds. We substitute this into (18.19) and get

$$s = s(\theta) = \frac{|\alpha|}{2E}\cot\frac{\theta}{2} \tag{18.20}$$

This result can also be obtained by evaluating the integral (18.12). From (18.20) follows

$$\theta \to \begin{cases} 0 & \text{for } s \to \infty \\ \pi & \text{for } s \to 0 \end{cases} \tag{18.21}$$

In Figure 18.2, the reader may sketch some trajectories for different values of the impact parameter s. Using (18.20) we compute the differential cross-section (18.10):

$$\frac{d\sigma}{d\Omega} = \left(\frac{Z_1 Z_2 e^2}{4E}\right)^2 \frac{1}{\sin^4(\theta/2)} \quad \text{Rutherford's formula} \tag{18.22}$$

This is the famous *Rutherford cross-section.* In the year 1913, Rutherford could verify this angular dependence experimentally. Thus, it was clear that the scattering potential for the alpha particles indeed has the Coulomb form (18.14) and that the charge of the atomic nucleus is concentrated within the range $\rho < \rho_{\min}$. This confirmed the Rutherford atomic model (1911), which assumed a very small atomic nucleus in the center of an atom.

In quantum mechanics, the motion of the alpha particles is described by waves. In the considered energy range, the wavelengths are of the same size as the nucleus; therefore it cannot be expected that (18.22) is more than a rough estimate. For the Coulomb potential, however, the quantum mechanical cross-section happens to coincide with that calculated in mechanics.

Just like σ, $d\sigma/d\Omega$ has the dimension of an area. From $d\sigma/d\Omega \sim L_{\rm C}^2$ we read off the characteristic length scale of the Coulomb scattering:

$$L_{\rm C} = \frac{Z_1 Z_2 e^2}{E} \tag{18.23}$$

According to (18.15), we obtain for this $L_{\rm C} \sim 30 \cdots 60\,\mathrm{fm}$. Particles with an impact parameter $s \lesssim L_{\rm C}$ are deflected by a considerable angle; for $s \gg L_{\rm C}$, the scattering angles are small.

The total cross-section is not — as one might expect — of the size $L_{\rm C}^2$. Rather, it diverges:

$$\sigma = \int d\Omega\, \frac{d\sigma}{d\Omega} = 2\pi \int_0^{\pi} d\theta\, \sin\theta\, \frac{d\sigma}{d\Omega} \overset{(18.22)}{\longrightarrow} \infty \tag{18.24}$$

This is due to the fact that the Rutherford differential cross-section diverges at $\theta = 0$. Experimentally, the differential cross-section shows for small θ increase according to (18.22). For $\theta = 0$ itself, $d\sigma/d\Omega$ cannot be measured since this is the beam direction. The divergence at $\theta = 0$ is due to the fact that also particles with a large impact parameter are scattered, although by decreasing angles. In this sense, the Coulomb potential has an infinite range. In the actual experiment, the Coulomb potential (18.14) is effectively shielded by the atomic shell. Therefore, particles with a parameter larger than the atomic radius do not experience any deflection; the cross-section is in fact finite.

Transformation to Laboratory System

After the reduction from the two-body to the one-body problem (Chapter 16), the relative motion $\boldsymbol{r}(t)$ is discussed as if a particle of mass μ moves in the potential $U(r)$. In this picture, we have derived the cross-section. We relate this cross-section now to the actually underlying two-body problem.

As a reference system, we first consider the *center-of-mass system* (CS), in which the center of the masses is at rest, $\boldsymbol{R} = 0$ (right part of Figure 18.3). According to (16.3), $\boldsymbol{r}_1(t) = m_2\,\boldsymbol{r}(t)/M$ and $\boldsymbol{r}_2(t) = -m_1\,\boldsymbol{r}(t)/M$; thus the motion of the individual mass points

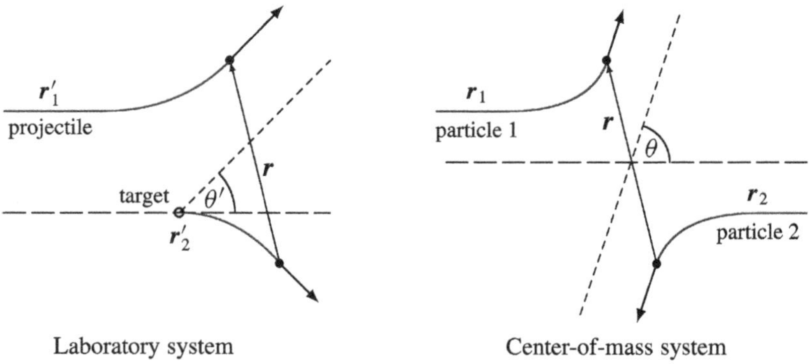

Figure 18.3 The scattering of two particles in the laboratory system (LS, left) and in the center-of-mass system (CS, right). The scattering angle of the projectile in LS is denoted by θ', that in CS by θ. The angles are depicted at the origins of the respective reference systems. The relative vector $\boldsymbol{r}(t)$ is the same in both systems; it turns around by the angle $\pi-\theta$ in the course of the scattering.

is directly related to $\boldsymbol{r}(t)$. Here, $\pi - \theta$ is the angle of rotation of the relative vector $\boldsymbol{r}$, and θ is the scattering angle.

The situation in the experiment is usually different: For example, shooting alpha particles onto a gold foil, the target particle is at rest before the scattering. This reference frame is called the *laboratory system* (LS); it is sketched in the left part of Figure 18.3. During scattering, momentum is transferred to the target particle (gold nucleus) so that it moves after the scattering.

Both systems, CS and LS, are supposed to be inertial systems that are connected by a Galilean transformation (with a constant relative velocity). The occurring quantities of CS and LS are listed in Table 18.1. The LS quantities are marked by a dash.

Since $\boldsymbol{r}_1(t) = m_2\,\boldsymbol{r}(t)/M$, the angle θ is not only the scattering angle in the reduced one-body problem (i.e. for $\boldsymbol{r}$) but also the scattering angle of the projectiles in CS (i.e. for $\boldsymbol{r}_1$). Moreover, since each particle (mass μ) in the reduced one-body problem corresponds to one projectile particle (mass m_1), $d\sigma/d\Omega$ from (18.5) and (18.10) is the differential cross-section for the projectiles in CS. In LS, the cross-section is denoted by $d\sigma'/d\Omega'$; it describes the scattering of projectiles in LS. We will establish the relation between $d\sigma/d\Omega$ and $d\sigma'/d\Omega'$.

Table 18.1 In this table, we display various quantities that occur in the description of the scattering process. The CS and LS expressions for various quantities are related to each other. The expressions on the left follow from (16.3) and $\boldsymbol{R} = 0$. The expressions on the right result from the expression for $\boldsymbol{R}'$ and $\boldsymbol{r}$, for which we use $\dot{\boldsymbol{r}}'_2(-\infty) = 0$ and $\dot{\boldsymbol{R}}' = \text{const.}$

CS	Quantity/relation	LS
$\boldsymbol{r}_1,\ \boldsymbol{r}_2$	Position vectors	$\boldsymbol{r}'_1,\ \boldsymbol{r}'_2$
$\boldsymbol{v}_1,\ \boldsymbol{v}_2$	Velocities	$\boldsymbol{v}'_1,\ \boldsymbol{v}'_2$
θ	Scattering angle	θ'
$\sigma,\ \dfrac{d\sigma}{d\Omega}$	Cross-section	$\sigma',\ \dfrac{d\sigma'}{d\Omega'}$
$\boldsymbol{R} = \dfrac{m_1\boldsymbol{r}_1 + m_2\boldsymbol{r}_2}{M}$	Center-of-mass vector	$\boldsymbol{R}' = \dfrac{m_1\boldsymbol{r}'_1 + m_2\boldsymbol{r}'_2}{M}$
$\boldsymbol{r} = \boldsymbol{r}_1 - \boldsymbol{r}_2$	Relative vector	$\boldsymbol{r} = \boldsymbol{r}'_1 - \boldsymbol{r}'_2$
$\boldsymbol{R}(t) = 0$	Precondition	$\boldsymbol{r}'_2(-\infty) = 0$ $\dot{\boldsymbol{r}}'_2(-\infty) = 0$
$\boldsymbol{r}_1(t) = \dfrac{m_2}{M}\,\boldsymbol{r}(t)$ $\boldsymbol{r}_2(t) = -\dfrac{m_1}{M}\,\boldsymbol{r}(t)$	Conclusions	$\dot{\boldsymbol{R}}'(t) = \dfrac{m_1}{M}\,\dot{\boldsymbol{r}}'_1(-\infty)$ $\dot{\boldsymbol{r}}'_1(-\infty) = \dot{\boldsymbol{r}}(-\infty)$

Since the center of mass rests at the origin of CS, $\boldsymbol{R}'$ is the vector from the origin of LS to that of CS. Thus, the transformation for the position vectors of the two particles read

$$\boldsymbol{r}'_i(t) = \boldsymbol{R}'(t) + \boldsymbol{r}_i(t) \tag{18.25}$$

Due to the conservation of the center-of-mass momentum ($\dot{\boldsymbol{R}}' =$ const.), this is a Galilean transformation. We use the notation $\boldsymbol{v}_i = \dot{\boldsymbol{r}}_i$ and $\boldsymbol{v}'_i = \dot{\boldsymbol{r}}'_i$. From (18.25) follows:

$$\boldsymbol{v}'_1(t) = \dot{\boldsymbol{R}}' + \boldsymbol{v}_1(t) \tag{18.26}$$

Figure 18.4 depicts this relationship for $t = -\infty$ (before the collision) and for $t = \infty$ (after the collision). Let the time scale be chosen such that the closest approach is reached at $t = 0$. By "$t = \pm\infty$"

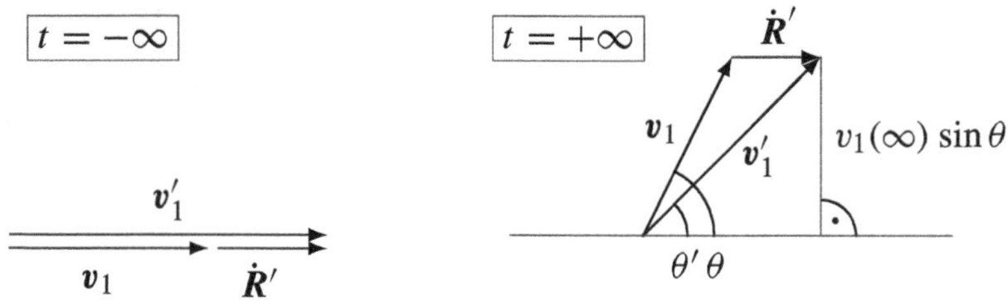

Figure 18.4 Sketch of the vector relation $\boldsymbol{v}_1'(t) = \dot{\boldsymbol{R}} + \boldsymbol{v}_1(t)$ for the velocities of the projectile in LS and CS. The left part shows the vectors before the scattering ($t = -\infty$), the right part after the scattering ($t = +\infty$).

we mean practically "far away" from the scattering center, for example, $\rho = 10^6\,\rho_{\text{min}}$. From Figure 18.4 we read off the relationship between θ' and θ:

$$\tan\theta' = \frac{v_1(\infty)\,\sin\theta}{v_1(\infty)\,\cos\theta + \dot{R}'} \tag{18.27}$$

From the lower part of Table 18.1, we find

$$v_1(\infty) = \dot{r}_1(\infty) = \frac{m_2}{M}\,\dot{r}(\infty) \quad \text{and} \quad \dot{R}' = \frac{m_1}{M}\,\dot{r}(-\infty) \tag{18.28}$$

From the energy conservation follows $2E = \mu\,\dot{r}(\infty)^2 = \mu\,\dot{r}(-\infty)^2$, thus $\dot{r}(\infty) = \dot{r}(-\infty)$. The last two equations yield

$$\tan\theta' = \frac{\sin\theta}{\cos\theta + m_1/m_2} \tag{18.29}$$

This is the relation between the scattering angles in LS and CS. For $m_1 \ll m_2$ we get $\theta' \approx \theta$; in this case, the target remains approximately at rest during the scattering.

We now determine the relation between $d\sigma'/d\Omega'$ and $d\sigma/d\Omega$. The Galilean transformation between LS and CS does not change the time intervals. Therefore, the number of scattered particles per time in the associated angular ranges $d\theta$ and $d\theta'$ must be equal:

$$j\,\frac{d\sigma}{d\Omega}\,2\pi\,\sin\theta\,d\theta = j'\,\frac{d\sigma'}{d\Omega'}\,2\pi\,\sin\theta'\,d\theta' \tag{18.30}$$

The current density j' of the incoming beam in LS equals the particle density $\Delta N_1/\Delta V$ multiplied by the velocity $v_1'(-\infty)$. We show $j' = j$:

$$j' = \frac{\Delta N_1}{\Delta V}\, v_1'(-\infty) = \frac{\Delta N_1}{\Delta V}\, \dot{r}(-\infty) = \frac{\Delta N_0}{\Delta V}\, \dot{r}(-\infty) = j \qquad (18.31)$$

For this, we used $v_1'(-\infty) = \dot{r}(-\infty)$ from the last row in Table 18.1. Moreover, $\Delta N_1 = \Delta N_0$ is valid with ΔN_0 from (18.1), because each of the ΔN_1 projectile particles corresponds to one of the ΔN_0 particles in the reduced one-body problem. The volume element ΔV is unchanged in the Galilean transformation. With $j = j'$ it follows from (18.30).

$$\frac{d\sigma'}{d\Omega'} = \frac{d\sigma}{d\Omega}\,\frac{\sin\theta}{\sin\theta'}\,\frac{d\theta}{d\theta'} = \frac{d\sigma}{d\Omega}\,\frac{d\cos\theta}{d\cos\theta'} \qquad (18.32)$$

Substituting on the right $\theta = \theta(\theta')$ from (18.29) we get $d\sigma'/d\Omega'$ as a function of θ'. In general, this leads to complex expressions. We, therefore, consider the special cases $m_1 \ll m_2$ and $m_1 = m_2$. In the first case, we get

$$\theta' \approx \theta, \quad \frac{d\sigma'}{d\Omega'} \approx \frac{d\sigma}{d\Omega} \quad m_1 \ll m_2 \qquad (18.33)$$

This applies approximately to the scattering of alpha particles from gold nuclei in the target foil; the corrections to this are in the percentage range.

For $m_1 = m_2$, (18.29) gives

$$\tan\theta' = \frac{\sin\theta}{\cos\theta + 1} = \tan\frac{\theta}{2} \quad (m_1 = m_2) \qquad (18.34)$$

Thus, we obtain the simple relation

$$\theta' = \frac{\theta}{2} \quad \left(0 \le \theta' \le \pi/2, \quad m_1 = m_2\right) \qquad (18.35)$$

The range for the angle θ' follows from $0 \le \theta \le \pi$. Specifically, $\theta' \le \pi/2$ means that a billiard ball in a carom (collision, scattering) with another billiard ball is deflected by at most $\pi/2$ (without rotational effects, like, for example, backspin). From $\theta' = \theta/2$ and

(18.32) we obtain for the differential cross-section in LS

$$\frac{d\sigma'}{d\Omega'} = 4\,\cos\theta'\,\frac{d\sigma(2\theta')}{d\Omega} \qquad (m_1 = m_2) \tag{18.36}$$

Thus, the Rutherford cross-section for the scattering of two equally heavy particles in the laboratory system reads

$$\frac{d\sigma'}{d\Omega'} = \left(\frac{Z_1 Z_2\, e^2}{2E}\right)^2 \frac{\cos\theta'}{\sin^4\theta'} \qquad (m_1 = m_2) \tag{18.37}$$

The energy $E = \mu\,\dot{r}(-\infty)^2/2$ can be expressed by the initial energy of the projectiles $E_1(-\infty) = m_1\,v_1'(-\infty)^2/2$; this yields $E = E_1/2$.

Energy Balance

Since $\partial\mathcal{L}/\partial t = 0$ for (16.1), the total energy of the system is conserved; this is true for any inertial system, i.e. in CS as well as in LS. Due to the conservation of energy, the scattering is called *elastic*. Inelastic scattering, on the other hand, means that energy from the relative motion is transferred to inner degrees of freedom (like, for example, an excitation of the target nucleus).

In LS, the energies of the projectile before and after the collision are of particular interest; the energies of the target particles after the collision are usually not measured. Since the projectile transfers momentum to the target, it loses part of its kinetic energy. We want to calculate the energy loss of the projectile. For this we square $\boldsymbol{v}_1 = \boldsymbol{v}_1' - \dot{\boldsymbol{R}}'$ and insert $t = \infty$:

$$v_1(\infty)^2 = v_1'(\infty)^2 + \dot{R}'^2 - 2\,v_1'(\infty)\,\dot{R}'\,\cos\theta' \tag{18.38}$$

From Table 18.1, we take $\dot{R}' = m_1\,v_1'(-\infty)/M$ and

$$v_1(\infty) = \frac{m_2}{M}\,\dot{r}(\infty) = \frac{m_2}{M}\,\dot{r}(-\infty) = \frac{m_2}{M}\,v_1'(-\infty) \tag{18.39}$$

Thus, (18.38) becomes

$$\frac{m_2^2}{M^2}\,v_1'(-\infty)^2 = v_1'(\infty)^2 + \frac{m_1^2}{M^2}\,v_1'(-\infty)^2 - 2\,v_1'(\infty)\,\frac{m_1}{M}\,v_1'(-\infty)\,\cos\theta' \tag{18.40}$$

The ratio of the velocities of the projectile after and before the scattering we denote by x,

$$x = \frac{v_1'(\infty)}{v_1'(-\infty)} \tag{18.41}$$

For this, (18.40) leads to the quadratic equation

$$x^2 - \frac{2\,m_1}{M}\, x\cos\theta' + \frac{m_1^2 - m_2^2}{M^2} = 0 \tag{18.42}$$

The value of x determines the change $\Delta E_1'/E_1'$ of the energy of the projectile:

$$\frac{\Delta E_1'}{E_1'} = \frac{E_1'(\infty) - E_1'(-\infty)}{E_1'(-\infty)} = x^2 - 1 = \begin{cases} -\sin^2\theta' & (m_1 = m_2) \\ \mathcal{O}(m_1/m_2) & (m_1 \ll m_2) \\ \mathcal{O}(m_2/m_1) & (m_1 \gg m_2) \end{cases} \tag{18.43}$$

The energy loss is small if the mass of a projectile particle is much larger or much smaller than the mass of the target particle. The energy loss is maximal for particles with the same mass. For this reason, a moderator used to slow down neutrons in a reactor is built from the lightest possible elements. For example, deuterium is used (in practice, this means heavy water); hydrogen (with $m_1 \approx m_2$) itself is less suitable because it easily captures neutrons.

As another example, we consider the carom (collision, scattering) of two billiard balls ($m_1 = m_2$, without effects of rotational or rolling motion). For a grazing collision ($\theta' \ll 1$), the energy loss of the played ball is small. For $\theta' \to \pi/2$ (this corresponds to $s \to 0$, i.e. to a near central collision) the loss approaches 100%. For a central carom, the cue ball transfers then all its energy and stops after the collision. For a near central carom, the cue ball loses most of its energy and moves on slowly with θ' slightly below $\pi/2$.

Our considerations for the transformation between CS and LS make no reference to the nature of the interaction process. Therefore, they remain valid also in quantum mechanics. However, the equations must be modified for relativistic velocities, because for this case the Galilean transformation is no longer valid.

Exercises

18.1 *Rutherford scattering*

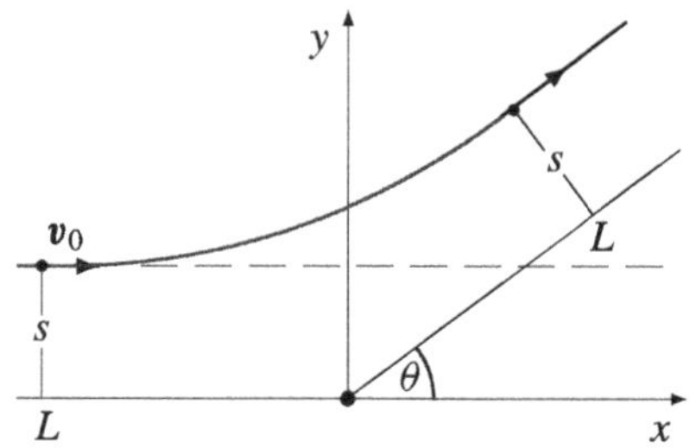

Calculate the impact parameter $s(\theta)$ as a function of the scattering angle θ for the Rutherford scattering (potential $U(r) = -\alpha/r$). For this purpose, consider the Lenz vector (Exercise 17.1) for the times $t \to -\infty$ and $t \to +\infty$, and use the constancy of this vector.

18.2 *Scattering cross-section for* $U(r) = \alpha/r^2$

Calculate the differential cross-section $d\sigma/d\Omega$ for the repulsive potential $U(r) = \alpha/r^2$ with $\alpha > 0$. Does the total scattering cross-section σ exist?

18.3 *Scattering of hard spheres*

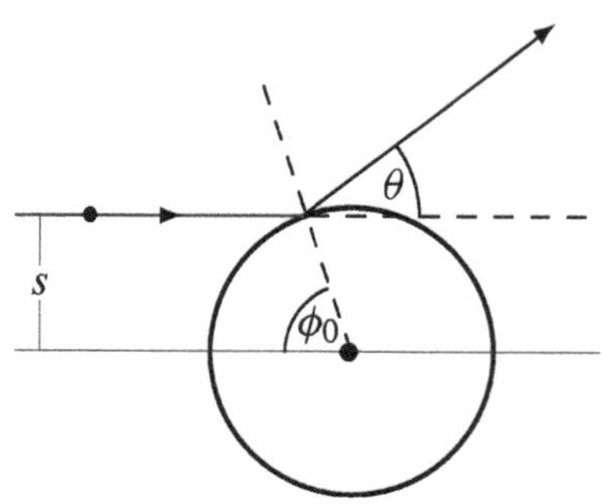

A particle scatters at the potential

$$U(r) = \begin{cases} \infty & (r \le R) \\ 0 & (r > R) \end{cases} \tag{18.44}$$

Determine the scattering angle $\theta(s)$ as a function of the impact parameter s. Check the result by a geometric consideration. Calculate $d\sigma/d\Omega$ and σ.

The interaction between two billiard balls (mass m, radius a) can be described by the potential (18.44) with $R = 2a$. Determine the scattering cross-section $d\sigma'/d\Omega'$ for the scattering of two billiard balls in the laboratory system.

PART V

Rigid Body

Chapter 19

Kinematics

In Part V, we treat the motion of a rigid body. As generalized coordinates for the rotational motion, we introduce the Euler angles (this chapter). The inertia of the body during the rotational motion is described by the inertia tensor (Chapter 20). In this context, we discuss the tensor concept in more detail (Chapter 21). The equations of motion themselves are given as Euler's equations (Chapter 22) or as Lagrangian equations (Chapter 23).

A *rigid body* is a system of mass points m_ν whose distances $|\boldsymbol{r}_{\nu\mu}| = |\boldsymbol{r}_\nu - \boldsymbol{r}_\mu|$ are fixed. The rigid body can be used as a model for a piece of ordinary, solid matter, such as a stone or a ballpoint pen. In this model, the internal degrees of freedom (such as vibrations) of the body are neglected.

We first consider $N = 3$ mass points. Then the $3N = 9$ Cartesian coordinates are restricted by $R = 3$ constraints, namely, the fixed values of $|\boldsymbol{r}_{12}|$, $|\boldsymbol{r}_{23}|$ and $|\boldsymbol{r}_{31}|$. Thus, the system has $f = 3N - R = 6$ degrees of freedom. The position of each further mass point is fixed by three distances from already existing mass points. Thus, each further mass point results in three additional coordinates *and* three additional constraints. Therefore, independent of N we stay at 6 degrees of freedom:

$$f = 3N - R = 6 \quad \text{(rigid body)} \tag{19.1}$$

The following alternative reasoning leads to the same result: The position of a rigid body can be determined by specifying its center of mass (three coordinates) and its orientation relative to the axes of an inertial system (three angles). The six degrees of freedom can be divided into three degrees of freedom of translation and three of rotation. The motion of the rigid body is thus described by the time course of six suitable coordinates.

The *kinematics* deals with the mere description of this motion. The *dynamics* of rigid bodies on the other hand is described by the physical laws according to which the motion proceeds. Since the rotation is comparatively complex, kinematics has its own separate chapter here; the dynamics will be investigated later in Chapters 22 and 23.

A rigid body for which one point is held fixed is called a *gyroscope* or gyro or top. A gyroscope has only three rotational degrees of freedom. One also talks about (non-supported) gyros, if the translational motion is decoupled from the rotational motion.

Angular Velocity

To describe the motion of a rigid body, we introduce a *body-fixed* Cartesian coordinate system, which we denote by BS in Part V. Beside BS we consider also an inertial system IS, in which Newton's laws are valid. In contrast to the body-fixed BS, this IS is also called a *space-fixed* system of reference. We use the following notation for the Cartesian coordinates and the corresponding ortho-normalized basis vectors of IS and BS:

$$\begin{array}{llll} \text{space-fixed IS:} & x,\ y,\ z & \text{and} & \boldsymbol{e}_x,\ \boldsymbol{e}_y,\ \boldsymbol{e}_z \\ \text{body-fixed BS:} & x_1,\ x_2,\ x_3 & \text{and} & \boldsymbol{e}_1,\ \boldsymbol{e}_2,\ \boldsymbol{e}_3 \end{array} \tag{19.2}$$

Since BS is fixed to the body, the basis vectors $\boldsymbol{e}_i(t)$ are in general time-dependent. Then BS is not an inertial system.

The body-fixed BS is not uniquely defined. We may choose a body-fixed BS' with a different origin $0'$ and a different orientation of the axes (with constant angles relative to BS). This freedom of choice will be used later to simplify the equations of motion.

The origin 0 of the body-fixed system BS has in IS the position vector $\boldsymbol{r}_0(t)$, Figure 19.1. Its velocity is

$$\boldsymbol{v}_0(t) = \frac{d\boldsymbol{r}_0}{dt} \tag{19.3}$$

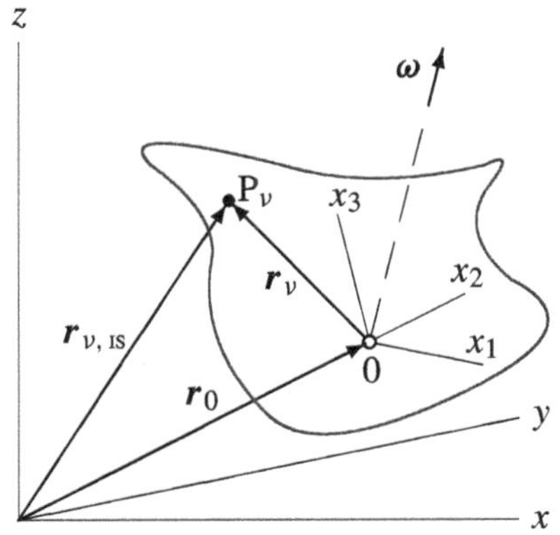

Figure 19.1 A body-fixed system BS (with Cartesian coordinates x_1, x_2, and x_3) is fixed to the rigid body (irregular contour). The motion of the rigid body can then be described by the position and orientation of BS relative to an inertial system IS (with x, y and z). The vector $\boldsymbol{r}_0$ determines the position of the origin 0 of BS. The orientation of the axes of BS can be defined by three suitable angles.

The time derivative refers (as always, unless otherwise stated) to the inertial system). The body-fixed BS may rotate with the *angular velocity*

$$\boldsymbol{\omega}(t) = \frac{d\boldsymbol{\varphi}}{dt} \tag{19.4}$$

relative to the IS. Here, $d\boldsymbol{\varphi}$ denotes a rotation by the angle $|d\boldsymbol{\varphi}|$ around an axis in the direction of $d\boldsymbol{\varphi}$ (Figure 6.1). The angular velocity will in general be time-dependent.

We now consider an arbitrary point P_ν of the rigid body. Let its position vector in IS be $\boldsymbol{r}_{\nu,\,\text{IS}}$, Figure 19.1. We denote the vector from the origin 0 of BS to P_ν by

$$\boldsymbol{r}_\nu = \boldsymbol{r}_{\nu,\,\text{IS}} - \boldsymbol{r}_0 \tag{19.5}$$

The velocity of the point P_ν in IS is then

$$\boldsymbol{v}_{\nu,\,\text{IS}} = \frac{d}{dt}\,\boldsymbol{r}_{\nu,\,\text{IS}} = \frac{d}{dt}\left(\boldsymbol{r}_0 + \boldsymbol{r}_\nu\right) = \boldsymbol{v}_0 + \frac{d\boldsymbol{r}_\nu}{dt} \tag{19.6}$$

The time derivative refers to the change of the vector $\boldsymbol{r}_\nu$ with respect to the inertial system (IS). Using (6.11), the time derivative can be expressed as follows:

$$\frac{d\boldsymbol{r}_\nu}{dt} = \left(\frac{d\boldsymbol{r}_\nu}{dt}\right)_{\text{IS}} = \left(\frac{d\boldsymbol{r}_\nu}{dt}\right)_{\text{BS}} + \boldsymbol{\omega} \times \boldsymbol{r}_\nu = \boldsymbol{\omega} \times \boldsymbol{r}_\nu \tag{19.7}$$

Relative to the body-fixed BS, $\boldsymbol{r}_\nu$ is time-independent, because 0 and P_ν are points of the rigid body; from this follows the last step in (19.7). The relation (19.7) is also valid if $\boldsymbol{\omega}$ is time-dependent. From (19.6) and (19.7) we obtain

$$\boldsymbol{v}_{\nu,\,\text{IS}} = \boldsymbol{v}_0 + \boldsymbol{\omega} \times \boldsymbol{r}_\nu \tag{19.8}$$

We now choose another point $0'$ of the rigid body as the origin of the body-fixed system, see Figure 19.2. Let the vector between 0 and $0'$ be

$$\boldsymbol{a} = \boldsymbol{r}_0 - \boldsymbol{r}_{0'} \tag{19.9}$$

Then, the vector from $0'$ to the point P_ν becomes

$$\boldsymbol{r}'_\nu = \boldsymbol{r}_\nu + \boldsymbol{a} \tag{19.10}$$

The same reasoning that led to (19.8) yields

$$\boldsymbol{v}_{\nu,\,\text{IS}} = \boldsymbol{v}_{0'} + \boldsymbol{\omega}' \times \boldsymbol{r}_{\nu'} \tag{19.11}$$

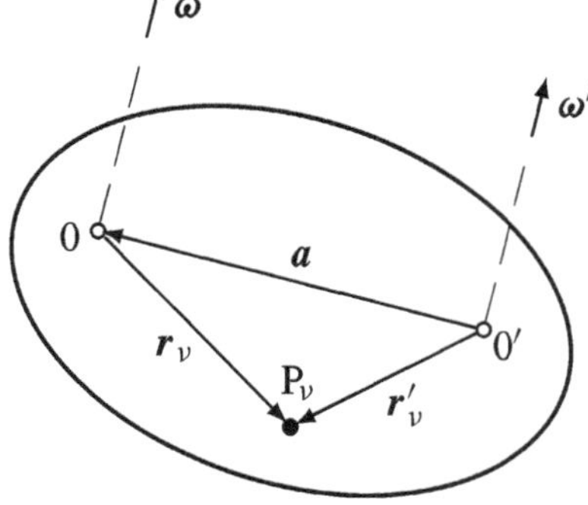

Figure 19.2 As the origin of the body-fixed BS any point of the rigid body (elliptic contour) may be chosen. This choice has no influence on the angular velocity; the translational velocity changes, however, according to (19.14).

The velocity $\boldsymbol{v}_{\nu,\,\text{IS}}$ of the point P_ν in IS is independent of the choice of BS. The comparison of (19.8) and (19.11) shows

$$\boldsymbol{v}_0 + \boldsymbol{\omega} \times \boldsymbol{r}_\nu = \boldsymbol{v}_{0'} + \boldsymbol{\omega}' \times \boldsymbol{r}_\nu + \boldsymbol{\omega}' \times \boldsymbol{a} \tag{19.12}$$

This equation is valid for any point of the rigid body, i.e. for any vector $\boldsymbol{r}_\nu$. From this follows:

$$\boldsymbol{\omega}' = \boldsymbol{\omega} \tag{19.13}$$

and

$$\boldsymbol{v}_{0'} = \boldsymbol{v}_0 - \boldsymbol{\omega} \times \boldsymbol{a} \tag{19.14}$$

Thus, the angular velocity $\boldsymbol{\omega}$ does not depend on the specific choice for BS. Therefore, the angular velocity $\boldsymbol{\omega}$ characterizes the rotation of the rigid body. The translational velocity $\boldsymbol{v}_0$ depends, however, on the choice of the origin of BS. The choice of this origin is a question of convenience and will be made later.

We show that angular velocities can be added like vectors. For this purpose, we consider two infinitesimal rotations $d\boldsymbol{\varphi}_a = \boldsymbol{\omega}_a\, dt$ and $d\boldsymbol{\varphi}_b = \boldsymbol{\omega}_b\, dt$. The subsequent rotations of the vector $\boldsymbol{r}$ yield the changes $d\boldsymbol{r}_a$ and $d\boldsymbol{r}_b$:

$$d\boldsymbol{r}_a = d\boldsymbol{\varphi}_a \times \boldsymbol{r}, \quad d\boldsymbol{r}_b = d\boldsymbol{\varphi}_b \times (\boldsymbol{r} + d\boldsymbol{r}_a) = d\boldsymbol{\varphi}_b \times \boldsymbol{r} \tag{19.15}$$

For infinitesimal rotations, the quadratic terms can be omitted. Thus,

$$d\boldsymbol{r} = d\boldsymbol{r}_a + d\boldsymbol{r}_b = (d\boldsymbol{\varphi}_a + d\boldsymbol{\varphi}_b) \times \boldsymbol{r} = (\boldsymbol{\omega}_a + \boldsymbol{\omega}_b) \times \boldsymbol{r}\, dt \tag{19.16}$$

From this, we see that the sequence of any two infinitesimal rotations can be interchanged. Equation (19.16) holds for any vector. Therefore, angular velocities may be added

$$\boldsymbol{\omega} = \boldsymbol{\omega}_a + \boldsymbol{\omega}_b \tag{19.17}$$

Finite rotations, on the other hand, do not interchange in general; in the representation of rotations by matrices (Chapter 21) this corresponds to the non-commutability of the matrix multiplication.

Euler Angles

To specify the position of the rigid body at any time, we need suitable generalized coordinates. For this we use three Cartesian coordinates for $\boldsymbol{r}_0$, and the three *Euler angles* ϕ, ψ and θ, which specify the direction of the axes of BS relative to IS.

To define the Euler angles, consider Figure 19.3. The x–y plane and the x_1–x_2 plane intersect in the nodal line N; the unit vector $\boldsymbol{e}_N$ is assigned to the N-direction. The Euler angles are then defined as follows:

$$\begin{aligned} &\phi: \quad \text{angle between the } x\text{-axis and } N \\ &\psi: \quad \text{angle between } N \text{ and the } x_1\text{-axis} \\ &\theta: \quad \text{angle between the } z\text{-axis and } x_3\text{-axis} \end{aligned} \tag{19.18}$$

These three Euler angles define the orientation of the rigid body. We will use them later as generalized coordinates in the Lagrange function (Chapter 23).

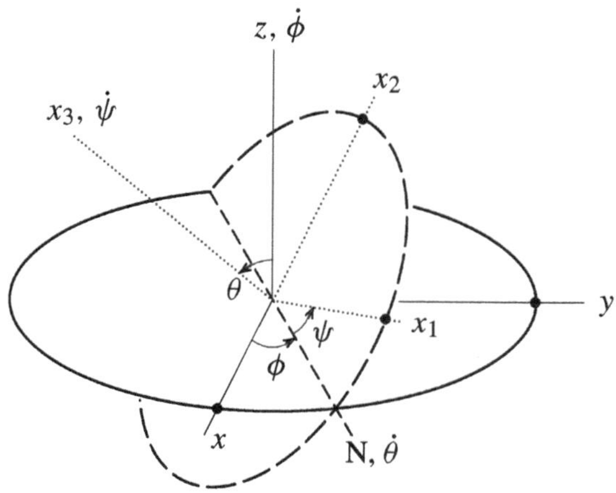

Figure 19.3 By the three Euler angles ϕ, ψ, and θ the position of the x_i-axes of the body-fixed system relative to the space-fixed IS (with x, y, and z) is specified. The angular velocities $\dot{\phi}$, $\dot{\psi}$ and $\dot{\theta}$ are denoted at the corresponding axis of rotation.

We now establish the relationship between the angular velocity and the Euler angles. For this purpose, we consider the three special rotations, for which two Euler angles are held fixed:

$$d\phi = d\psi = 0 \qquad \boldsymbol{\omega}_\theta = \dot{\theta}\, \boldsymbol{e}_N \tag{19.19}$$

$$d\psi = d\theta = 0 \qquad \boldsymbol{\omega}_\phi = \dot{\phi}\, \boldsymbol{e}_z \tag{19.20}$$

$$d\theta = d\phi = 0 \qquad \boldsymbol{\omega}_\psi = \dot{\psi}\, \boldsymbol{e}_3 \tag{19.21}$$

In the first case, the nodal line N is the axis of rotation. In the second case, the rotation is about the z-axis. In the third case, the rigid body rotates around its x_3-axis; the nodal line is constant. The reader may visualize these rotations in Figure 19.3.

We express the unit vectors $\boldsymbol{e}_N$ and $\boldsymbol{e}_z$ by those of the body-fixed BS:

$$\boldsymbol{e}_N = \cos\psi\, \boldsymbol{e}_1 - \sin\psi\, \boldsymbol{e}_2 \tag{19.22}$$

$$\boldsymbol{e}_z = \sin\theta\, \sin\psi\, \boldsymbol{e}_1 + \sin\theta \cos\psi\, \boldsymbol{e}_2 + \cos\theta\, \boldsymbol{e}_3 \tag{19.23}$$

Arbitrary infinitesimal rotations correspond to independent changes $d\phi$, $d\psi$, and $d\theta$ of Euler angles. Since the order of infinitesimal rotations is arbitrary, the following applies

$$d\boldsymbol{\varphi} = \boldsymbol{\omega}\, dt = (\boldsymbol{\omega}_\theta + \boldsymbol{\omega}_\phi + \boldsymbol{\omega}_\psi)\, dt \tag{19.24}$$

and thus

$$\boldsymbol{\omega} = \boldsymbol{\omega}_\theta + \boldsymbol{\omega}_\phi + \boldsymbol{\omega}_\psi = \dot{\theta}\, \boldsymbol{e}_N + \dot{\phi}\, \boldsymbol{e}_z + \dot{\psi}\, \boldsymbol{e}_3 \tag{19.25}$$

The vector $\boldsymbol{\omega}$ can — like any other vector — be represented by its components in the body-fixed BS or in the space-fixed IS:

$$\boldsymbol{\omega} = \begin{cases} \omega_1\, \boldsymbol{e}_1 + \omega_2\, \boldsymbol{e}_2 + \omega_3\, \boldsymbol{e}_3 := (\omega_1,\, \omega_2,\, \omega_3) \\ \omega_x\, \boldsymbol{e}_x + \omega_y\, \boldsymbol{e}_y + \omega_z\, \boldsymbol{e}_z := (\omega_x,\, \omega_y,\, \omega_z) \end{cases} \tag{19.26}$$

The components of $\boldsymbol{\omega}$ are obtained by projection onto the respective basis vectors. In applications, we will preferably use the components

ω_1, ω_2 and ω_3 in the body-fixed BS:

$$\begin{aligned}
\omega_1 &= \boldsymbol{\omega} \cdot \boldsymbol{e}_1 = \dot{\phi}\ \sin\theta\ \sin\psi + \dot{\theta}\ \cos\psi \\
\omega_2 &= \boldsymbol{\omega} \cdot \boldsymbol{e}_2 = \dot{\phi}\ \sin\theta\ \cos\psi - \dot{\theta}\ \sin\psi \\
\omega_3 &= \boldsymbol{\omega} \cdot \boldsymbol{e}_3 = \dot{\phi}\ \cos\theta\ + \dot{\psi}
\end{aligned} \tag{19.27}$$

The right-hand side follows from (19.25) and (19.22, 19.23). Hereby, the angular velocities are given by the Euler angles and their time derivatives.

Chapter 20

Inertia Tensor

We determine the kinetic energy of the rotational motion of a rigid body. This energy is a quadratic form in the angular velocity. The coefficients Θ_{ik} of this quadratic form define the inertia tensor.

Kinetic Energy

Let the rigid body consist of N mass points m_ν, $\nu = 1, \ldots, N$. The kinetic energy T is then the sum of the kinetic energies of the individual mass points:

$$2T = \sum_{\nu=1}^{N} m_\nu \, \boldsymbol{v}_{\nu,\,\text{IS}}^2 \overset{(19.8)}{=} \sum_{\nu=1}^{N} m_\nu \, \boldsymbol{v}_0^2 + 2 \sum_{\nu=1}^{N} m_\nu \, (\boldsymbol{\omega} \times \boldsymbol{r}_\nu) \cdot \boldsymbol{v}_0 + \sum_{\nu=1}^{N} m_\nu \, (\boldsymbol{\omega} \times \boldsymbol{r}_\nu)^2 \tag{20.1}$$

Here, $\boldsymbol{r}_\nu$ is the vector from the origin 0 of the body-fixed BS to the mass point m_ν (Figure 19.1). By a suitable choice of the origin, the linear term in $\boldsymbol{\omega}$ can be made to vanish:

$$\sum_{\nu=1}^{N} m_\nu \, (\boldsymbol{\omega} \times \boldsymbol{r}_\nu) \cdot \boldsymbol{v}_0 = (\boldsymbol{v}_0 \times \boldsymbol{\omega}) \cdot \sum_{\nu=1}^{N} m_\nu \, \boldsymbol{r}_\nu = \begin{cases} 0 \ \text{ for } \boldsymbol{v}_0 = 0 \\ 0 \ \text{ for } \sum_\nu m_\nu \, \boldsymbol{r}_\nu = 0 \end{cases} \tag{20.2}$$

The first case assumes that the origin of BS is at rest; this is true for a gyroscope for which the origin 0 is the supporting point. The second case assumes that the origin 0 is equal to the center of mass of the rigid body; then the center-of-mass vector $\sum m_\nu \boldsymbol{r}_\nu / M$ vanishes. In the following, we will always assume one of these two cases. Using (20.2), the kinetic energy (20.1) becomes a sum of the kinetic energies of the translation ($T_{\rm trans}$) and of the rotation ($T_{\rm rot}$):

$$T = \frac{M}{2}\,\boldsymbol{v}_0^2 + \frac{1}{2}\sum_{\nu=1}^{N} m_\nu\,(\boldsymbol{\omega}\times\boldsymbol{r}_\nu)^2 = T_{\rm trans} + T_{\rm rot} \tag{20.3}$$

We evaluate $T_{\rm rot}$ in the body-fixed BS with the Cartesian coordinates x_1, x_2, and x_3. The components of the vectors $\boldsymbol{\omega}$ and $\boldsymbol{r}_\nu$ in BS are denoted according to

$$\boldsymbol{\omega} := (\omega_1,\, \omega_2,\, \omega_3) \quad \text{and} \quad \boldsymbol{r}_\nu := (x_1^\nu,\, x_2^\nu,\, x_3^\nu) \tag{20.4}$$

By

$$\begin{aligned}(\boldsymbol{\omega}\times\boldsymbol{r})^2 &= \omega^2\, r^2 - (\boldsymbol{\omega}\cdot\boldsymbol{r})^2 = \sum_k^3 \omega_k^2\, r^2 - \sum_{i,k=1}^{3} \omega_i\, x_i\, \omega_k\, x_k \\ &= \sum_{i,k=1}^{3} \left(r^2\,\delta_{ik} - x_i\, x_k\right)\omega_i\,\omega_k \end{aligned} \tag{20.5}$$

$T_{\rm rot}$ becomes

$$T_{\rm rot} = \frac{1}{2}\sum_{\nu=1}^{N} m_\nu\,(\boldsymbol{\omega}\times\boldsymbol{r}_\nu)^2 = \frac{1}{2}\sum_{i,k=1}^{3} \Theta_{ik}\,\omega_i\,\omega_k \tag{20.6}$$

Here, we have introduced the twofold indexed quantity

$$\Theta_{ik} = \sum_{\nu=1}^{N} m_\nu \left(r_\nu^2\,\delta_{ik} - x_i^\nu\, x_k^\nu\right) \quad \text{inertia tensor} \tag{20.7}$$

This quantity is called the *inertia tensor*. Together with the $\boldsymbol{r}_\nu$ the inertia tensor Θ_{ik} is *time-independent*.

The term *tensor* refers to the transformational behavior of the Θ_{ik} under orthogonal transformations. Tensors will be formally defined and discussed in some detail in the following chapter.

Mass density

We calculate the inertia tensor for a continuous mass distribution with

$$\varrho(\boldsymbol{r}) = \text{mass density} = \frac{\text{mass}}{\text{volume}} = \frac{\Delta m}{\Delta V} \tag{20.8}$$

Here, ΔV is a small volume at $\boldsymbol{r}$ in which the mass Δm is located. The volume ΔV is chosen large enough to contain many atoms; we refer here to ordinary matter (solid, liquid, gas). On the other hand, ΔV should be small enough so that macroscopic inhomogeneities are not averaged out. For a solid, for example, $(\Delta V)^{1/3} \approx 10^{-8}$ m could be chosen; then about 10^4 to 10^5 atoms are contained in ΔV.

A body with a time-independent, continuous mass density $\varrho(\boldsymbol{r})$ in the body-fixed system can be treated as a rigid body. For this purpose, we think of the volume of the body as decomposed into N partial volumes ΔV_ν located at $\boldsymbol{r}_\nu$. The mass elements

$$m_\nu = \varrho(\boldsymbol{r}_\nu)\, \Delta V_\nu \tag{20.9}$$

of each partial volume are thought to be replaced by a point mass m_ν at $\boldsymbol{r}_\nu$. Then we have a rigid body as introduced in the previous chapter. The error by the substitution with point masses goes to zero in the limit $N \to \infty$ (and $\Delta V_\nu \to 0$):

$$\sum_{\nu=1}^{N} m_\nu \,\ldots = \sum_{\nu=1}^{N} \varrho(\boldsymbol{r}_\nu)\, \Delta V_\nu \,\ldots \stackrel{N\to\infty}{\longrightarrow} \int d^3r\; \varrho(\boldsymbol{r}) \,\ldots \tag{20.10}$$

The integration can go over the whole space because outside the body, $\varrho(\boldsymbol{r})$ vanishes anyway. From (20.10) and (20.7) we obtain

$$\Theta_{ik} = \int d^3r\; \varrho(\boldsymbol{r}) \left(r^2\, \delta_{ik} - x_i\, x_k\right) \tag{20.11}$$

For an ensemble of mass points, we may insert

$$\varrho(\boldsymbol{r}) = \sum_{\nu=1}^{N} m_\nu\, \delta(\boldsymbol{r} - \boldsymbol{r}_\nu) \tag{20.12}$$

in order to regain (20.7).

In this section, we have considered a time-independent mass density of a rigid body in the body-fixed system. With respect to the inertial frame IS, however, the mass density depends in general on time, $\varrho = \varrho(\boldsymbol{r}, t)$.

Angular Momentum

The angular momentum (4.11) for a system of mass points depends on the reference point. For this reference point, we choose the centers of the respective coordinate systems:

$$\boldsymbol{L} = \begin{cases} \sum\limits_{\nu} m_\nu \left(\boldsymbol{r}_{\nu,\,\text{IS}} \times \dot{\boldsymbol{r}}_{\nu,\,\text{IS}}\right) & \text{(origin of IS)} \\ \sum\limits_{\nu} m_\nu \left(\boldsymbol{r}_\nu \times \dot{\boldsymbol{r}}_\nu\right) & \text{(origin of BS)} \end{cases} \tag{20.13}$$

With

$$\dot{\boldsymbol{r}}_\nu = \left(\frac{d\boldsymbol{r}_\nu}{dt}\right)_{\text{IS}} \overset{(19.7)}{=} \boldsymbol{\omega} \times \boldsymbol{r}_\nu \tag{20.14}$$

we get:

$$\boldsymbol{L} = \sum_{\nu=1}^{N} m_\nu \, \boldsymbol{r}_\nu \times (\boldsymbol{\omega} \times \boldsymbol{r}_\nu) \tag{20.15}$$

We evaluate this in the Cartesian coordinates x_i of BS. Using

$$\boldsymbol{r} \times (\boldsymbol{\omega} \times \boldsymbol{r}) = \boldsymbol{\omega}\, r^2 - \boldsymbol{r}\,(\boldsymbol{\omega} \cdot \boldsymbol{r}) = \sum_{i=1}^{3} \sum_{k=1}^{3} \left(r^2\, \delta_{ik} - x_i\, x_k\right) \omega_k\, \boldsymbol{e}_i \tag{20.16}$$

and (20.7) we get

$$\boldsymbol{L} = \sum_{i=1}^{3} \sum_{k=1}^{3} \Theta_{ik}\, \omega_k\, \boldsymbol{e}_i = \sum_{i=1}^{3} L_i\, \boldsymbol{e}_i \tag{20.17}$$

For an alternative notation, we may use the matrix Θ and the column vectors for L and ω,

$$L = \begin{pmatrix} L_1 \\ L_2 \\ L_3 \end{pmatrix}, \quad \Theta = \begin{pmatrix} \Theta_{11} & \Theta_{12} & \Theta_{13} \\ \Theta_{21} & \Theta_{22} & \Theta_{23} \\ \Theta_{31} & \Theta_{32} & \Theta_{33} \end{pmatrix}, \quad \omega = \begin{pmatrix} \omega_1 \\ \omega_2 \\ \omega_3 \end{pmatrix} \tag{20.18}$$

With this, we can write (20.17) in matrix form:

$$L = \Theta\,\omega \tag{20.19}$$

Another possible notation arises if we use the *dyad* $\widehat{\Theta}$:

$$\widehat{\Theta} = \sum_{i,k=1}^{3} \Theta_{ik}\, \boldsymbol{e}_i \circ \boldsymbol{e}_k \tag{20.20}$$

The dyadic product "$\circ$" is defined in such a way that the scalar multiplication of $\boldsymbol{e}_i \circ \boldsymbol{e}_k$ with any vector $\boldsymbol{a}$ produces the vector $\boldsymbol{e}_i\,(\boldsymbol{e}_k \cdot \boldsymbol{a})$; more details will be given in the following chapter. We can now write down the relation between angular momentum and angular velocity alternatively in component, vector and matrix notation:

$$L_i = \sum_{k=1}^{3} \Theta_{ik}\, \omega_k, \quad \boldsymbol{L} = \widehat{\Theta} \cdot \boldsymbol{\omega}, \quad L = \Theta\,\omega \tag{20.21}$$

These relations imply that the directions of $\boldsymbol{L}$ and $\boldsymbol{\omega}$ are in general different.

For the kinetic energy $T_{\rm rot}$ (20.6) the component, vector and matrix notation reads

$$T_{\rm rot} = \frac{1}{2} \sum_{i,k=1}^{3} \Theta_{ik}\, \omega_i\, \omega_k = \frac{1}{2}\, \boldsymbol{\omega} \cdot \big(\widehat{\Theta} \cdot \boldsymbol{\omega}\big) = \frac{1}{2}\, \omega^{\rm T}\, \Theta\, \omega \tag{20.22}$$

The upper index T denotes the transposed matrix (Roman font T in contrast to the italic font T for the kinetic energy). Then $\omega^{\rm T}$ is the row vector $(\omega_1,\, \omega_2,\, \omega_3)$. By

$$\Theta_{nn} = \boldsymbol{n} \cdot \big(\widehat{\Theta} \cdot \boldsymbol{n}\big) = n^{\rm T}\, \Theta\, n = \sum_{i,k=1}^{3} \Theta_{ik}\, n_i\, n_k \tag{20.23}$$

we define the *moment of inertia* for the rotation around an axis defined by the unit vector $\boldsymbol{n} := n^{\rm T} = (n_1, n_2, n_3)$. The diagonal elements of Θ_{ik} are such moments of inertia with respect to the coordinate axes. These moments of inertia occur, for example, when one considers the rotation around a fixed axis, i.e. for

$$\boldsymbol{\omega} = \omega_n\, \boldsymbol{n} \tag{20.24}$$

The $\boldsymbol{n}$-component of the angular momentum is then

$$L_n = \boldsymbol{n} \cdot \boldsymbol{L} = \boldsymbol{n} \cdot \big(\widehat{\Theta} \cdot \boldsymbol{\omega}\big) = \boldsymbol{n} \cdot \big(\widehat{\Theta} \cdot \boldsymbol{n}\big)\, \omega_n = \Theta_{nn}\, \omega_n \qquad (20.25)$$

Principal Axis Transformation

To introduce the inertia tensor, we referred to a body-fixed system, so that $\Theta_{ik} = \text{const}$. The concrete values Θ_{ik} depend on the orientation of the body-fixed BS. The eigenvalues (principal moments of inertia) of the matrix $\Theta = (\Theta_{ik})$ are, however, independent from this orientation. Usually, we will take the center of mass as the origin of BS. Under this precondition, the moments of inertia are properties of the rigid body, like its mass. (For a shift of BS's origin, the changes in Θ_{ik} are discussed in Exercise 20.1.) In this section, we discuss the procedure for determining the principal moments of inertia.

We consider two different body-fixed systems, BS and BS$'$, which are rotated against each other (by constant angles), but which are supposed to have the same origin. The components of angular velocity $\boldsymbol{\omega}$ are denoted by ω_i in BS and by ω_i' in BS$'$. The transformation between these components is

$$\omega_i' = \sum_{k=1}^{3} \alpha_{ik}\, \omega_k, \qquad \omega' = \alpha\, \omega \qquad (20.26)$$

On the right, we used the matrix notation with the column vector ω and the 3×3 matrix $\alpha = (\alpha_{ik})$. The transformation is orthogonal, $\alpha\, \alpha^{\mathrm{T}} = 1$. Therefore, the reverse transformation is.

$$\omega = \alpha^{\mathrm{T}}\, \omega' \qquad (20.27)$$

The orthogonal transformations will be discussed in more detail in the following chapter.

We use the transformation (20.27) in the expression (20.22) for the rotational kinetic energy:

$$2\, T_{\text{rot}} = \omega^{\mathrm{T}}\, \Theta\, \omega = \big(\alpha^{\mathrm{T}} \omega'\big)^{\mathrm{T}}\, \Theta\, \big(\alpha^{\mathrm{T}} \omega'\big) = \omega'^{\mathrm{T}} \big(\alpha\, \Theta\, \alpha^{\mathrm{T}}\big)\, \omega' = \omega'^{\mathrm{T}}\, \Theta'\, \omega' \qquad (20.28)$$

One can now choose α such that the inertia matrix Θ' in BS$'$ is diagonal:

$$\Theta' = \left(\Theta'_{ik}\right) = \alpha\,\Theta\,\alpha^{\mathrm{T}} = \begin{pmatrix} \Theta_1 & 0 & 0 \\ 0 & \Theta_2 & 0 \\ 0 & 0 & \Theta_3 \end{pmatrix} \tag{20.29}$$

The system BS$'$ in which Θ' is diagonal is called the *principal axis system*. The diagonal elements Θ'_{ii} have been denoted by Θ_i. According to (20.23), they are the moments of inertia with respect to the rotation of the axes of BS$'$. They are called *principal moments of inertia*.

The moments of inertia Θ_i are positive. To show this formally, we evaluate the defining equation (20.7) for Θ_1:

$$\Theta_1 = \Theta'_{11} = \sum_{\nu=1}^{N} m_\nu \left(r_\nu'^{2} - \left(x_1'^{\nu}\right)^2\right) = \sum_{\nu=1}^{N} m_\nu \left(\left(x_2'^{\nu}\right)^2 + \left(x_3'^{\nu}\right)^2\right) > 0 \tag{20.30}$$

All masses are weighted by the square of their perpendicular distance to the x_1 axis. This applies accordingly to the other Θ_i.

A symmetric matrix Θ can be *diagonalized* by an orthogonal transformation. That is, $\alpha = (\alpha_{ik})$ in (20.29) can be chosen such that $\alpha\,\Theta\alpha^{\mathrm{T}}$ is diagonal. We prove constructively that this is possible; i.e. we determine the α_{ik} that diagonalizes Θ. We multiply (20.29), $\alpha\,\Theta\alpha^{\mathrm{T}} = \Theta'$ from left with α^{T}:

$$\Theta\,\alpha^{\mathrm{T}} = \alpha^{\mathrm{T}}\,\Theta' \tag{20.31}$$

We used $\alpha\,\alpha^{\mathrm{T}} = \alpha^{\mathrm{T}}\,\alpha = 1$. We denote the kth column of α^{T} by the column vector $\omega^{(k)}$:

$$\alpha^{\mathrm{T}} = \left(\omega^{(1)}, \omega^{(2)}, \omega^{(3)}\right) \quad \text{so } \omega^{(k)} = \begin{pmatrix} \alpha_{k1} \\ \alpha_{k2} \\ \alpha_{k3} \end{pmatrix} \tag{20.32}$$

This allows the kth column of (20.31) to be expressed as

$$\Theta\,\omega^{(k)} = \Theta_k\,\omega^{(k)} \tag{20.33}$$

This is the *eigenvalue equation* of the matrix Θ. From it the eigenvalues Θ_k and the corresponding eigenvectors $\omega^{(k)}$ can be determined. The eigenvalues Θ_k are the principal moments of inertia Θ_k, whereas the eigenvectors $\omega^{(k)}$ define the orthogonal matrix α. We now present the procedure for how to obtain these eigenvalues and eigenvectors.

Using the unit matrix I, we can express (20.33) in the form

$$\big(\Theta - \Theta_k I\big)\,\omega^{(k)} = 0 \tag{20.34}$$

This represents a linear, homogeneous system of equations for the three components of $\omega^{(k)}$. It has the trivial solution $\omega^{(k)} = 0$. If the determinant of the coefficient matrix is non-zero then the solution is unique. For the existence of a non-trivial solution we must, therefore, require that the determinant vanishes:

$$\det\big(\Theta - \Theta_k I\big) = 0 \tag{20.35}$$

This is a third-degree polynomial in Θ_k with the three solutions Θ_1, Θ_2, and Θ_3. For physical reasons, all eigenvalues Θ_k are positive, as shown in (20.30). The eigenvectors $\omega^{(k)}$ and $\omega^{(k')}$ are orthogonal for $\Theta_k \neq \Theta_{k'}$; for coincident eigenvalues (like $\Theta_k = \Theta_{k'}$) they can be made orthogonal. With $\omega^{(k)}$ also const. $\cdot\, \omega^{(k)}$ solves the eigenvalue equation (20.33). We determine the constant so that the eigenvectors are normalized. The now orthonormalized vectors $\omega^{(k)}$ form the columns of α^{T} (i.e., the rows of α). Therefore, they represent the basis vectors of the coordinate system BS′ in which Θ is diagonal.

Equation (20.35) for the principal moments of inertia is independent of the orientation of the body-fixed coordinate system, because of

$$\det\big(\Theta' - \Theta_k I\big) = \det\big(\alpha\,(\Theta - \Theta_k I)\,\alpha^{\mathrm{T}}\big) = \det\big(\Theta - \Theta_k I\big) \tag{20.36}$$

Here, $\det(AB) = \det(A)\det(B)$ and $\det(\alpha\,\alpha^{\mathrm{T}}) = 1$ have been used. If in addition one chooses a certain origin of BS (for example, the center of mass of the rigid body), then the principal moments of inertia are fixed. The so-defined principal moments of inertia are *properties* of the rigid body (like its mass).

Physically, the principal axes are distinguished by the fact that for a rotation around a major axis (i.e. for $\omega = \omega^{(k)}$) the angular momentum $\boldsymbol{L}$ is parallel to $\boldsymbol{\omega}$. We show this in the matrix notation:

$$L = \Theta\,\omega^{(k)} = \Theta_k\,\omega^{(k)}, \quad \text{or} \quad \boldsymbol{L} \parallel \boldsymbol{\omega}^{(k)} \tag{20.37}$$

For symmetric rigid bodies (a cuboid, for example), the principal axes coincide with the symmetry axes.

If all three principal moments of inertia are different, we also speak of an asymmetric gyroscope. If two of them are equal, the gyroscope is called symmetric. For a sphere or a cube, all principal moments of inertia are equal (provided that the origin of BS is chosen as the center of mass).

Exercises

20.1 *Steiner's theorem*

A rigid body has the inertia tensor Θ_{ik}. This tensor refers to a body-fixed coordinate system BS whose origin is equal to the center of mass. Calculate the inertia tensor Θ'_{ik} in a coordinate system BS′ whose origin is shifted by the vector $\boldsymbol{a}$ relative to BS, and whose axes are parallel to BS. The resulting relation is called *Steiner's theorem*.

20.2 *Inertia tensor of the homogeneous cube*

A homogeneous cube (with the density ϱ_0 and the edge length a) rotates around an axis, which passes through the center point and through a corner. How large is the moment of inertia with respect to this axis?

20.3 *Inertia tensor of the homogeneous cuboid*

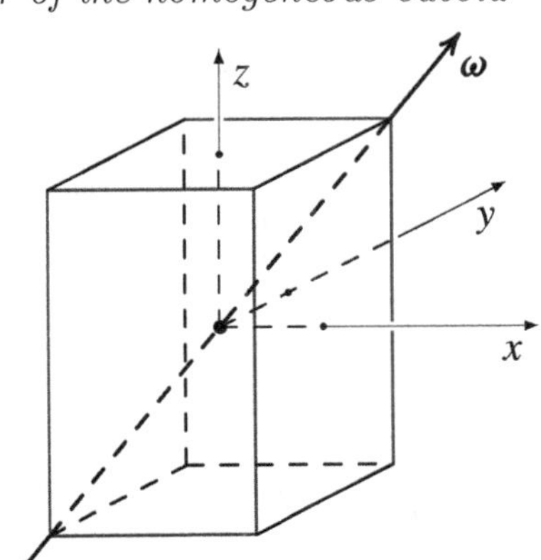

Determine the principal moments of inertia Θ_i of a cuboid with constant mass density ϱ_0 and the side lengths a, b and c. The cuboid rotates with the angular velocity $\omega = |\boldsymbol{\omega}|$ around an axis which coincides with a space diagonal of the cuboid. How large is the moment of inertia Θ_0 with respect to this axis? Express the kinetic energy of rotation alternatively by Θ_i or by Θ_0.

20.4 *Inertia tensor of a homogeneous ellipsoid*

Determine the principal moments of inertia Θ_i of a homogeneous ellipsoid (mass M_{E}) with semi-axes a, b, and c.

The mass distribution of the earth can be approximated by a homogeneous rotational ellipsoid ($a = b$). The oblateness of the earth is given by $(a - c)/a \approx 1/300$. How large is then $\Delta\Theta/\Theta = (\Theta_3 - \Theta_1)/\Theta_3$?

Determine the inertia tensor of a homogeneous sphere (radius R, mass M_{S}), on whose equator there are four additional point masses m are attached at $\phi = 0$, $\pi/2$, π, and $3\pi/2$. Choose the parameters R and m so that the inertia tensor is equal to that of a homogeneous rotational ellipsoid with mass $M_{\mathrm{E}} = M_{\mathrm{S}} + 4m$.

20.5 *Oblateness of the Earth*

The earth is in good approximation an oblate rotational ellipsoid with the semi-major axes $a = b > c$. The gravitational energy W_{grav} and the mass M of a rotational ellipsoid with homogeneous mass density ϱ_0 are

$$W_{\text{grav}} = -\frac{3\,GM^2}{5\,a}\,\frac{\arcsin\varepsilon}{\varepsilon} \quad \text{and} \quad M = \frac{4\pi}{3}\,\varrho_0\,a^2\,c = \frac{4\pi}{3}\,\varrho_0\,R^3$$

Here, the eccentricity $\varepsilon = (1 - c^2/a^2)^{1/2}$ and the mean radius $R = (a^2\,c)^{1/3}$ were used. The rotational ellipsoid rotates in the body-fixed system with the angular momentum $L_3 = \Theta_3\,\omega_3$ around the figure axis. Then its total energy is

$$W_{\text{total}} = T_{\text{rot}} + W_{\text{grav}} = \frac{L_3^2}{2\,\Theta_3} + W_{\text{grav}}$$

The deformable earth now adjusts itself so, that W_{total} as a function of the eccentricity ε becomes minimal; the angular momentum L_3 and the mass M are fixed. Calculate the earth oblateness $(a-c)/a$. Expand for this W_{grav} and the moment of inertia Θ_3 into powers of ε. Consider this expansion up to the first non-vanishing order in ε.

Chapter 21

Tensors

The inertia tensor introduced in the previous chapter is the motivation for discussing tensors in some detail, in particular their formal definition and the corresponding calculation rules. This more formal discussion is not a necessary prerequisite for the following chapters; for these chapters, a basic knowledge of vector calculus is sufficient.

The tensors of the three-dimensional space are defined by their behavior under orthogonal transformations. We, therefore, call them 3-tensors (in particular to distinguish them from Lorentz tensors or 4-tensors in Part IX). From the definition arise the rules for the construction of new tensors by addition, multiplication, or contraction. We discuss a number of applications like the scalar and vector products and finally define tensor fields. Later in Chapter 37, we will see analogous concepts for Lorentz tensors (or 4-tensors) in the four-dimensional Minkowski space.

Orthogonal Transformations

We consider Cartesian coordinate systems (S, S′) in ordinary three-dimensional space. The coordinates are denoted by x_i and the basis vectors by $\boldsymbol{e}_i$. The S of this chapter could be a body-fixed as well as an inertial system.

Let two coordinate systems S and S′ be rotated relative to each other (by constant angles), Figure 21.1. A position vector $\boldsymbol{r}$ can

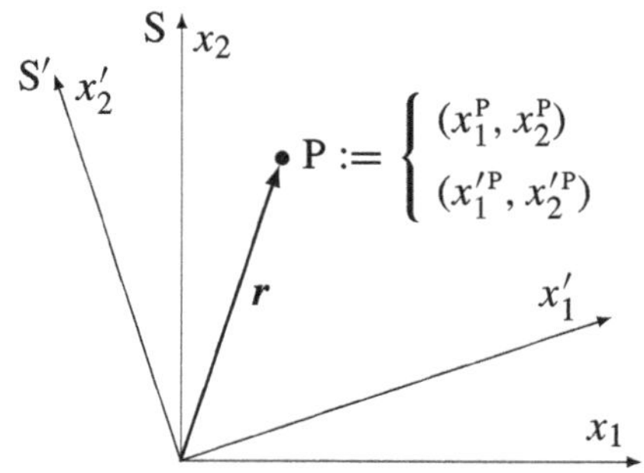

Figure 21.1 A mass point P has in system S (here two-dimensionally represented) the coordinate values $(x_1^{\mathrm{P}}, x_2^{\mathrm{P}})$, and in the rotated system S′ other values $(x_1'^{\mathrm{P}}, x_2'^{\mathrm{P}})$. The position vector $\boldsymbol{r}$ of the mass point P is determined by these coordinate values, (21.1).

alternatively be given by the basis vectors $\boldsymbol{e}_n$ of S or by $\boldsymbol{e}'_j$ of S′:

$$\boldsymbol{r} = \sum_{n=1}^{3} x_n\, \boldsymbol{e}_n = \sum_{j=1}^{3} x'_j\, \boldsymbol{e}'_j \tag{21.1}$$

The coefficients are called components of the vector. The representation of the vector $\boldsymbol{r}$ by its components, $\boldsymbol{r} := (x_1, x_2, x_3)$, obviously depends on the chosen basis (i.e. on the chosen coordinate system).

For Cartesian coordinate systems, the basis vectors are orthogonal:

$$\boldsymbol{e}_i \cdot \boldsymbol{e}_k = \delta_{ik} = \begin{cases} 1 & \text{for } i = k \\ 0 & \text{for } i \neq k \end{cases} \tag{21.2}$$

Accordingly, $\boldsymbol{e}'_i \cdot \boldsymbol{e}'_k = \delta_{ik}$. We multiply (21.1) by $\boldsymbol{e}'_i$ and obtain

$$x'_i = \sum_{n=1}^{3} (\boldsymbol{e}'_i \cdot \boldsymbol{e}_n)\, x_n = \sum_{n=1}^{3} \alpha_{in}\, x_n \tag{21.3}$$

The $\alpha_{in} = \boldsymbol{e}'_i \cdot \boldsymbol{e}_n$ are coordinate-independent coefficients that depend on the relative orientation of S and S′.

We calculate the scalar product $\boldsymbol{r}^2 = \boldsymbol{r} \cdot \boldsymbol{r}$ for the representations given in (21.1):

$$\boldsymbol{r}^2 = \begin{cases} \displaystyle\sum_n x_n^2 \\ \displaystyle\sum_i x_i'^2 = \sum_m \sum_n \left(\sum_i \alpha_{in}\, \alpha_{in} \right) x_m\, x_n \end{cases} \tag{21.4}$$

In the second line, $x_i' = \sum_n \alpha_{in}\, x_n$ has been inserted. Both lines must agree for any x_n. From this follows:

$$\sum_{i=1}^{3} \alpha_{im}\, \alpha_{in} = \delta_{mn} \quad \text{(orthogonal transformation)} \tag{21.5}$$

From the nine coefficients α_{in} we can form the three vectors $a_n = (\alpha_{1n}, \alpha_{2n}, \alpha_{3n})$ with $(n = 1, 2, 3)$. Then (21.5) means that these vectors a_n are orthogonal to each other (and also normalized). Therefore, the transformation (21.3) with (21.5) is called *orthogonal transformation*.

The inverse transformation corresponding to (21.3) is obtained by multiplying (21.3) by α_{im}, summing over i and using (21.5):

$$x_m = \sum_{i=1}^{3} \alpha_{im}\, x_i' \tag{21.6}$$

The considered relations can be given also in the short and clear-cut matrix notation. For this we introduce the matrix α, the column vector x, and the transposed vector x^{T}:

$$\alpha = \begin{pmatrix} \alpha_{11} & \alpha_{12} & \alpha_{13} \\ \alpha_{21} & \alpha_{22} & \alpha_{23} \\ \alpha_{31} & \alpha_{32} & \alpha_{33} \end{pmatrix}, \quad x = \begin{pmatrix} x_1 \\ x_2 \\ x_3 \end{pmatrix}, \quad x^{\mathrm{T}} = (x_1, x_2, x_3) \tag{21.7}$$

The forward and back transformations are

$$x' = \alpha\, x \quad \text{and} \quad x = \alpha^{\mathrm{T}} x' \tag{21.8}$$

From the invariance of the scalar product, $x'^{\mathrm{T}} x' = x^{\mathrm{T}} \alpha^{\mathrm{T}} \alpha\, x = x^{\mathrm{T}} x$, follows (21.5) in the form

$$\alpha^{\mathrm{T}} \alpha = 1, \quad \text{hence } \alpha^{-1} = \alpha^{\mathrm{T}} \tag{21.9}$$

As an example, we consider a rotation around the z-axis by the angle φ. From

$$\begin{aligned} \boldsymbol{e}_1' &= \boldsymbol{e}_1 \cos\varphi + \boldsymbol{e}_2 \sin\varphi \\ \boldsymbol{e}_2' &= -\boldsymbol{e}_1 \sin\varphi + \boldsymbol{e}_2 \cos\varphi \\ \boldsymbol{e}_3' &= \phantom{-\boldsymbol{e}_1 \sin\varphi +}\ \boldsymbol{e}_3 \end{aligned} \tag{21.10}$$

we get the α_{in}, which we combine into the matrix α:

$$\alpha = (\alpha_{in}) = (\boldsymbol{e}_i' \cdot \boldsymbol{e}_n) = \begin{pmatrix} \cos\varphi & \sin\varphi & 0 \\ -\sin\varphi & \cos\varphi & 0 \\ 0 & 0 & 1 \end{pmatrix} \tag{21.11}$$

According to Chapter 19, the relative position of two coordinate systems is in general defined by three angles. Therefore, the general orthogonal matrix α depends on three continuous parameters.[1]

Since two successive rotations result again in a rotation, they form a group (one can easily convince oneself that all group axioms are fulfilled). This statement is also valid for the representation of the rotations by the matrices α.

We have considered two coordinate systems rotated relative to each other and a single vector $\boldsymbol{r}$, Figure 21.1. Rotations can also be considered from the following alternative point of view: In a given S, the vector $\boldsymbol{r}$ is rotated. The rotated vector $\boldsymbol{r}' = \sum x_i' \boldsymbol{e}_i$ with $x_i' = \sum_n \alpha_{in} x_n$ is then in general different from $\boldsymbol{r}$. Both points of view are possible and common. The present point of view (one vector and two coordinate systems) corresponds to the case that the vector represents a physical quantity, for example, the momentum

[1]This can also be seen like this: Equation (21.5) holds for $m = 1, 2, 3$ and $n = 1, 2, 3$, thus representing nine conditions. But the conditions for (m, n) and (n, m) are the same. This leaves six independent conditions for the determination of the nine quantities α_{ik}. So three quantities remain undetermined.

of a particle. An observer describes this quantity with respect to an S chosen by him, i.e. by the three numbers x_i. The same physical quantity is described by another observer with a different coordinate system by other numbers x'_i. This point of view is particularly useful when events are considered by observers in different IS (as in the passive Galilean transformation, Chapter 5). The alternative point of view (one coordinate system and two vectors) corresponds to an active Galilean transformation.

Definition

We consider quantities that have no, one, or more indices. Let these indices take the values 1, 2, and 3; they are denoted by Latin letters $(i, k, \ldots)$.

A quantity without an index, which does not change under orthogonal transformations is called a *scalar* or *tensor of rank zero* or zeroth rank tensor. An example is the scalar product of the position vector, $\boldsymbol{r}^2 = x^{\mathrm{T}} x = \sum x_i^2$.

A quantity V_i with one index which transforms like the components x_i of the position vector (21.3), i.e.

$$V'_i = \sum_{n=1}^{3} \alpha_{in} V_n \tag{21.12}$$

is called *first rank tensor* (or tensor of first rank). Analogous to (21.6) we obtain the back transformation

$$V_m = \sum_{i=1}^{3} \alpha_{im} V'_i \tag{21.13}$$

An N-fold indexed quantity is called a *N th rank tensor* (or tensor of rank N) if it transforms componentwise like the x_i:

$$T'_{i_1 i_2 \ldots i_N} = \sum_{n_1=1}^{3} \sum_{n_2=1}^{3} \cdots \sum_{n_N=1}^{3} \alpha_{i_1 n_1} \alpha_{i_2 n_2} \cdot \cdots \cdot \alpha_{i_N n_N} T_{n_1 n_2 \ldots n_N} \tag{21.14}$$

The inverse transformation is formed analogously to (21.13).

For a first and second rank tensor, the transformations can be written in matrix form:

$$V' = \alpha V, \quad V = \alpha^{\mathrm{T}} V', \quad T' = \alpha T \alpha^{\mathrm{T}}, \quad T = \alpha^{\mathrm{T}} T' \alpha \qquad (21.15)$$

Where $V = (V_i)$ is a column vector and $T = (T_{ik})$ is a 3×3 matrix. For tensors of higher rank, this matrix notation is not possible.

With the term *tensor*, we always mean the entirety of the indexed quantity. In a more detailed terminology, the indexed quantities themselves are called "components of the tensor"; each component by itself is a number whose value depends on the index. We simplify the terminology and use the terms "tensor of first rank" and "vector" synonymously. Thus, we accept an ambiguity: in (21.1) $\boldsymbol{r}$ was the vector, and x_i was its components. In an abbreviated and common way of speaking, we now also denote the once-indexed quantity x_i itself as a vector. If we work with the indexed quantities (and this often is advantageous), it would be awkward to always call them "components of the vector". In equations, one has of course to distinguish between $\boldsymbol{r}$ and x_i.

Calculation Rules

From the definition of the tensors, the following rules for the construction of further tensors emerge. If S and T are tensors, then the following rules hold:

1. Addition of tensors of the same rank:

$$\beta S_{i_1 \ldots i_N} + \gamma T_{i_1 \ldots i_N} \quad \text{is a tensor of rank } N \qquad (21.16)$$

2. Multiplication of tensors:

$$S_{i_1 \ldots i_N} T_{j_1 \ldots j_M} \quad \text{is a tensor of the rank } N + M \qquad (21.17)$$

3. Contraction of tensor indices:

$$\sum_{k=1}^{3} S_{i_1 \ldots k \ldots k \ldots i_N} \quad \text{is a tensor of rank } N - 2 \qquad (21.18)$$

4. Tensor equations:

$$S_i = \sum_{k=1}^{3} U_{ik}\, T_k \quad \Longrightarrow \quad U_{ik} \text{ is a rank 2 tensor} \tag{21.19}$$

The first and second points follow directly from the tensor definition, the third may be proved by the reader himself. We derive the fourth assertion in matrix notation. The precondition here is the validity of $S_i = \sum U_{ik} T_k$ in all S and the vector property of S_i and T_i. From (21.15) follows:

$$S' = \alpha\, S = \alpha\, U T = \alpha\, U\, \alpha^{\mathrm{T}} \alpha\, T = \left(\alpha\, U\, \alpha^{\mathrm{T}}\right) T' \tag{21.20}$$

The comparison with $S'_i = \sum U'_{ik} T'_k$ gives

$$U' = \alpha\, U\, \alpha^{\mathrm{T}} \quad \text{or} \quad U'_{nm} = \sum_{i,k=1}^{3} \alpha_{ni}\, \alpha_{mk}\, U_{ik} \tag{21.21}$$

This shows that U_{ik} is a second rank tensor.

Tensor Equations

A vector or tensor equation is a *form invariant* or *covariant* under orthogonal transformations. For example, the following equations emerge from each other by an orthogonal transformation:

$$L_k = \sum_{n=1}^{3} \Theta_{kn}\, \omega_n \longleftrightarrow L'_k = \sum_{n=1}^{3} \Theta'_{kn}\, \omega'_n \tag{21.22}$$

The *form* of these and other tensor equations is invariant under orthogonal transformations. The values of the individual components depend, however, on the specific S.

The isotropy of space (Chapter 11) means that fundamental laws of physics are independent of the orientation of the reference frame (we consider temporarily inertial frames). The form of these laws in Cartesian coordinates must then not depend on the orientation (direction of the axes) of the chosen S. Therefore the following holds:

- From the isotropy of the space it follows that fundamental physical laws are tensor equations.

In component notation, such 3-tensor equations are form invariant (or covariant) under orthogonal transformations, such as (21.22). The vector notation (e.g. $\boldsymbol{L} = \widehat{\Theta} \cdot \boldsymbol{\omega}$) is completely coordinate independent.

The condition of covariance (synonym for form invariance) is a strong restriction to the possible form of the laws. It is therefore particularly useful for establishing new laws. In addition to the covariance under orthogonal transformations, we will also consider covariance under more general transformations, like the covariance under Galilei transformation. In relativistic mechanics, this becomes the covariance with respect to Lorentz transformations (Chapter 37).

Examples of Tensors

The standard example of a scalar is the scalar product $\sum V_i^{\,2}$ of a vector V_i with itself. Other scalars are the mass m of a particle, the time t, and the temperature T.

From (21.3) it follows that the coordinate differentials

$$dx'_i = \sum_{n=1}^{3} \alpha_{in}\, dx_n \tag{21.23}$$

are vectors. Thus, also the velocity $v_k = dx_k/dt$ is a vector, because

$$v'_i = \frac{dx'_i}{dt} = \sum_{n=1}^{3} \alpha_{in}\, \frac{dx_n}{dt} = \sum_{n=1}^{3} \alpha_{in}\, v_n \tag{21.24}$$

Similarly, we show that the acceleration $a_i = dv_i/dt$ is a vector. From the rules of calculation, it follows that the kinetic energy $m\boldsymbol{v}^2/2$ is a scalar.

The following assignments define three different onefold indexed quantities:

$$(U_i) = \begin{pmatrix} 1 \\ 2 \\ 0 \end{pmatrix}, \quad (V_i) = \begin{pmatrix} m \\ x_2 \\ t \end{pmatrix}, \quad (W_i) = \begin{pmatrix} v_1\, x_1 \\ v_2\, x_2 \\ v_3\, x_3 \end{pmatrix} \tag{21.25}$$

None of these quantities is a vector.

In the previous chapter, we have introduced the inertia tensor

$$\Theta_{ik} = \sum_{\nu=1}^{N} m_\nu \left(r_\nu^2 \, \delta_{ik} - x_i^\nu \, x_k^\nu \right) \tag{21.26}$$

We now show formally that this is a second rank tensor: The masses m_ν do not change during a rotation of the S, so they are scalars. The position vectors x_i^ν of the individual mass points are vectors. Then $m_\nu \, x_i^\nu \, x_k^\nu$ is a second rank tensor. The transformation behavior of $m_\nu \, r_\nu^2 \, \delta_{ik}$ is equal to that of δ_{ik} since r_ν^2 and m_ν are scalars. The *Kronecker symbol* δ_{ik} has been defined by a simple number assignment (21.2). Thus, this symbol is independent of the S; in any other S′ holds $\delta'_{ik} = \delta_{ik}$. The δ_{ik} defined in this way, however, satisfies the tensor definition, too:

$$\delta'_{ik} \overset{(21.14)}{=} \sum_{n=1}^{3} \sum_{l=1}^{3} \alpha_{in} \, \alpha_{kl} \, \delta_{nl} = \sum_{n=1}^{3} \alpha_{in} \, \alpha_{kn} \overset{(21.5)}{=} \delta_{ik} \tag{21.27}$$

Thus, the Kronecker symbol δ_{ik} is a tensor which is also known as *unit tensor*. Thus, both terms on the right-hand side of (21.26) are tensors of second rank; this is then also true for the sum of these terms.

By the definition

$$\epsilon_{ikl} = \left\{ \begin{array}{c} 1 \\ -1 \\ 0 \end{array} \right. \quad \text{if } (i,k,l) \text{ is } \left\{ \begin{array}{c} \text{even} \\ \text{odd} \\ \text{no} \end{array} \right\} \text{ permutation of } (1,2,3) \tag{21.28}$$

we introduce the so-called *Levi-Civita symbol* ϵ_{ikl}. Like δ_{ik}, this quantity is defined by fixed numbers, and it is thus independent of the S. We investigate whether we can interpret this quantity as a tensor, too, i.e. whether the transformation according to (21.14) leads again to the values (21.28),

$$\epsilon'_{ikl} \overset{(21.14)}{=} \sum_{i'=1}^{3} \sum_{k'=1}^{3} \sum_{l'=1}^{3} \alpha_{ii'} \, \alpha_{kk'} \, \alpha_{ll'} \, \epsilon_{i'k'l'} \tag{21.29}$$

For $i = k$, the summand on the right-hand side is antisymmetric with respect to $i' \leftrightarrow k'$. Thus, the sum vanishes, implying $\epsilon'_{iil}(21.14) = 0$. This is also true if two other indices are equal. The indices (i, k, l) must therefore be a permutation of $(1, 2, 3)$ in order that ϵ_{ikl} does not vanish. Specifically, we get

$$\epsilon'_{123} \overset{(21.14)}{=} \sum_{i'=1}^{3} \sum_{k'=1}^{3} \sum_{l'=1}^{3} \alpha_{1i'}\, \alpha_{2k'}\, \alpha_{3l'}\, \epsilon_{i'k'l'} = \det \alpha \tag{21.30}$$

The right-hand side is equal to the definition of the determinant of α. The permutation $1 \leftrightarrow 2$ on the left-hand side corresponds to $i' \leftrightarrow k'$ on the right side, so that $\epsilon'_{213} = -\det \alpha$. We can summarize these results as follows:

$$\epsilon'_{ikl} \overset{(21.14)}{=} \det \alpha\ \epsilon_{ikl} \tag{21.31}$$

Thus, the ϵ_{ikl} defined by the number assignment (21.28) is a tensor if and only if $\det \alpha = 1$.

We now define: An N-fold indexed quantity that transforms like

$$P'_{i_1 \ldots i_N} = \det\, \alpha \sum_{m_1=1}^{3} \cdots \sum_{m_N=1}^{3} \alpha_{i_1 m_1} \cdot \cdots \cdot \alpha_{i_N m_N}\, P_{m_1 \ldots m_N} \tag{21.32}$$

is called *pseudotensor* of rank N. The additional factor det α is restricted to the values $+1$ and -1, because from $\alpha^{\mathrm{T}} \alpha = 1$ follows:

$$\det\,(\alpha^{\mathrm{T}} \alpha) = \det \alpha^{\mathrm{T}}\, \det \alpha = (\det \alpha)^2 = 1, \quad \text{thus } \det \alpha = \pm 1 \tag{21.33}$$

For a rotation, $\det \alpha = 1$ holds. An additional reflection, however, leads to $\det \alpha = -1$. With the exclusion of reflections, pseudotensors behave like tensors. If we now transform ϵ_{ikl} according to (21.32), then

$$\epsilon'_{ikl} \overset{(21.32)}{=} (\det \alpha)^2\ \epsilon_{ikl} = \epsilon_{ikl} \tag{21.34}$$

So ϵ_{ikl} is a pseudotensor. Since this Levi-Civita tensor is antisymmetric under the interchange of two arbitrary indices, it is also called a *totally antisymmetric tensor*.

From two vectors V_i and W_i the following quantity can be formed:

$$U_i = \sum_{k=1}^{3} \sum_{l=1}^{3} \epsilon_{ikl}\, V_k\, W_l = \big(\boldsymbol{V} \times \boldsymbol{W}\big)_i \tag{21.35}$$

From the rules compiled above, it follows that U_i is a pseudovector. The vector $\boldsymbol{U}$ is equal to the well-known cross product $\boldsymbol{V} \times \boldsymbol{W}$.

An example of a pseudovector is the angular velocity $\boldsymbol{\omega}$. To determine the transformation property of $\boldsymbol{\omega}$ we start from

$$d\boldsymbol{r} = (\boldsymbol{\omega} \times \boldsymbol{r})\, dt \quad \text{or} \quad dx_i = \sum_{k=1}^{3} \sum_{l=1}^{3} \epsilon_{ikl}\, \omega_k\, x_l\, dt \tag{21.36}$$

The first expression may be read off from (19.7). In Equation (21.36), the dx_i and x_l are vectors, ϵ_{ikl} is a pseudotensor and dt is a scalar. With dx_i also the right-hand side must also be a vector. From this and from the known transformation properties of x_l, ϵ_{ikl} and dt, it follows that ω_k is a pseudovector.

From the pseudovector property of ω_k follows that $\omega_i\omega_k$ is a second rank tensor. Therefore, $T_{\text{rot}} = \sum \Theta_{ik}\omega_i\omega_k/2$ is a scalar.

The twofold indexed quantity α_{ij} itself is not a tensor. In order to determine whether a quantity satisfies the tensor definition, this quantity must first of all be defined in S and S$'$. But the orthogonal matrix is the quantity that connects S and S$'$; it is not a quantity that changes in some way at the transition S $\leftrightarrow$ S$'$. One can, however, consider another orthogonal matrix (β_{ij}) which describes a certain rotation within S, and the orthogonal matrix (β'_{ij}) which describes the corresponding rotation in S$'$. These quantities β_{ij} and β'_{ij} transform then like second-rank tensors.

Dyads

For any two vectors $\boldsymbol{a}$ and $\boldsymbol{b}$, we define the *dyadic product* $\boldsymbol{a} \circ \boldsymbol{b}$ by

$$(\boldsymbol{a} \circ \boldsymbol{b}) \cdot \boldsymbol{c} = \boldsymbol{a}\, (\boldsymbol{b} \cdot \boldsymbol{c}) \tag{21.37}$$

Applied to any vector $\boldsymbol{c}$, the scalar product with $\boldsymbol{b}$ has to be taken, and the resulting number $(\boldsymbol{b} \cdot \boldsymbol{c})$ is multiplied with the vector $\boldsymbol{a}$.

2. Differentiation and contraction:

$$\sum_{i=1}^{3} \frac{\partial T_{i_1 \ldots i \ldots i_N}}{\partial x_i} \quad \text{is a tensor field of rank } N-1 \tag{21.46}$$

From the vector field $\boldsymbol{V}$, we get the scalar field $\operatorname{div} \boldsymbol{V}(\boldsymbol{r})$.

3. Differentiation, multiplication by ϵ_{ikl} and contraction:

$$\sum_{k,l=1}^{3} \epsilon_{ikl} \frac{\partial V_l}{\partial x_k} \quad \text{is a pseudovector field} \tag{21.47}$$

This expression is equal to the ith component of $\operatorname{curl} \boldsymbol{V}$.

Exercises

21.1 *Contraction of a tensor*

For a tensor of Nth rank $T_{i_1, i_2, \ldots, i_N}$ prove that

$$\sum_{k=1}^{3} T_{i_1, \ldots, k, \ldots, k, \ldots, i_N}$$

is a tensor of rank $N-2$.

Chapter 22

Euler's Equations

The Euler equations[1] are the equations of motion for the rotation of a rigid body. They are differential equations for the angular velocity in the body-fixed system. The coefficients in these equations are the principal moments of inertia.

Derivation of Euler's Equations

A rigid body can be considered as a system of mass points. In (4.14) with (4.11), we derived the equation of motion

$$\frac{d}{dt}\boldsymbol{L} = \boldsymbol{M} \tag{22.1}$$

for the angular momentum

$$\boldsymbol{L} = \sum_{\nu=1}^{N} m_\nu \left(\boldsymbol{r}_\nu \times \dot{\boldsymbol{r}}_\nu\right) = \widehat{\Theta} \cdot \boldsymbol{\omega} \tag{22.2}$$

According to (20.21), the angular momentum $\boldsymbol{L}$ can also be expressed by the inertial dyad $\widehat{\Theta}$ and the angular velocity $\boldsymbol{\omega}$. The torque $\boldsymbol{M}$ is given by (4.14) by the external forces:

$$\boldsymbol{M} = \sum_{\nu=1}^{N} \boldsymbol{r}_\nu \times \boldsymbol{F}_\nu^{\text{ext}} \tag{22.3}$$

[1]There are also the Euler equations of hydrodynamics, see Chapter 32.

The internal forces $\boldsymbol{F}_{ij}$ between the mass points do not contribute to the torque. The rigid body can be considered as a system of mass points, where the internal forces $\boldsymbol{F}_{ij}$ imply fixed distances $|\boldsymbol{r}_{ij}|$.

As a reference point for the angular momentum, we choose the origin of a body-fixed system. The vectors $\boldsymbol{r}_\nu$ point from this origin to the individual mass points, Figure 19.1. The reference point shall be either the supported point of a gyro or the center of mass of the rigid body. In the first case, the system has only the three degrees of freedom of rotation, whose dynamics are described by (22.1). In the second case, there are three additional degrees of freedom of translation. Their dynamics is given by (4.7), i.e.

$$M\ddot{\boldsymbol{R}} = \sum_{\nu=1}^{N} \boldsymbol{F}_\nu^{\text{ext}} = \boldsymbol{F} \tag{22.4}$$

Here, $\boldsymbol{R}$ is the center-of-mass vector of the rigid body in the space-fixed inertial frame. The equations of motion (22.1) and (22.4) are decoupled from each other. Therefore, we can concentrate on the following on the treatment of (22.1).

A gyroscope in the narrower sense is a rotating rigid body with a support point (like a children's top). We will speak, however, of a gyroscopic motion, too, if the rotating rigid body is not supported.

We write (22.1) in the form

$$\frac{d}{dt}\left(\widehat{\Theta} \cdot \boldsymbol{\omega}\right) = \boldsymbol{M} \tag{22.5}$$

This equation can be compared with Newton's second law $d(m\boldsymbol{v})/dt = \boldsymbol{F}$. The basic difference is that the inertia for the rotational motion is a tensor (and not a scalar like the mass m).

The dyad $\widehat{\Theta}$ is independent of the orientation of the axes of S; this is also a true for the vectors $\boldsymbol{L}$ or $\boldsymbol{\omega}$. The components Θ_{ik} of $\widehat{\Theta}$ usually refer to the body-fixed system. We will continue to do so; just for the next equation, we admit temporarily a space-fixed IS, too. We may express the dyad $\widehat{\Theta}$ on the one hand by the constant basis vectors of the space-fixed IS, which we denote here exceptionally by $\boldsymbol{e}_i'$ or, on the other hand, by the time-dependent basis vectors $\boldsymbol{e}_i(t)$ of the body-fixed system S:

$$\widehat{\Theta} = \begin{cases} \sum_{i,k=1}^{3} \Theta'_{ik}(t)\, \boldsymbol{e}'_i \circ \boldsymbol{e}'_k & \text{(space-fixed IS)} \\ \sum_{i,k=1}^{3} \Theta_{ik}\, \boldsymbol{e}_i(t) \circ \boldsymbol{e}_k(t) & \text{(body-fixed BS)} \end{cases} \tag{22.6}$$

Unlike the Θ_{ik}, the $\Theta'_{ik}(t) = \sum_\nu m_\nu \,(r'^2_\nu\, \delta_{ik} - x'^\nu_i\, x'^\nu_k)$ depend on time, since the $x'^\nu_i(t)$ refer to the space-fixed IS. This time dependence is determined by the motion of the rigid body in IS, i.e. by the yet unknown solution.

The time independence of the Θ_{ik} in the body-fixed BS is a crucial advantage. Therefore, in the following, we use again this BS. Using (6.11), (22.5) becomes

$$\left(\frac{d\,(\widehat{\Theta}\cdot\boldsymbol{\omega})}{dt}\right)_{\rm BS} + \boldsymbol{\omega}\times\big(\widehat{\Theta}\cdot\boldsymbol{\omega}\big) = \boldsymbol{M} \tag{22.7}$$

Let the body-fixed system BS now be specifically the principal axis system. The components of $\widehat{\Theta}$, $\boldsymbol{\omega}$, and $\boldsymbol{M}$ in BS can be represented by matrices and column vectors:

$$\Theta = \begin{pmatrix} \Theta_1 & 0 & 0 \\ 0 & \Theta_2 & 0 \\ 0 & 0 & \Theta_3 \end{pmatrix}, \quad \omega = \begin{pmatrix} \omega_1 \\ \omega_2 \\ \omega_3 \end{pmatrix}, \quad M = \begin{pmatrix} M_1 \\ M_2 \\ M_3 \end{pmatrix} \tag{22.8}$$

Since $(d\boldsymbol{e}_i/dt)_{\rm BS} = 0$, we can immediately write (22.7) in components:

$$\begin{aligned} \Theta_1\,\dot\omega_1 + \big(\Theta_3-\Theta_2\big)\,\omega_2\,\omega_3 &= M_1 \\ \Theta_2\,\dot\omega_2 + \big(\Theta_1-\Theta_3\big)\,\omega_1\,\omega_3 &= M_2 \\ \Theta_3\,\dot\omega_3 + \big(\Theta_2-\Theta_1\big)\,\omega_1\,\omega_2 &= M_3 \end{aligned} \quad \text{Euler's equations} \tag{22.9}$$

Into these *Euler equations* we insert (19.27),

$$\omega_1 = \dot\phi\, \sin\theta\, \sin\psi + \dot\theta\, \cos\psi \tag{22.10}$$

$$\omega_2 = \dot\phi\, \sin\theta\, \cos\psi - \dot\theta\, \sin\psi \tag{22.11}$$

$$\omega_3 = \dot\phi\, \cos\theta + \dot\psi \tag{22.12}$$

This gives us three second-order differential equations for the three Euler angles $\phi(t)$, $\psi(t)$, and $\theta(t)$. These are the equations of motion for the rotation of the rigid body.

The Euler equations have the disadvantage that the components M_i of the torque are related to the body-fixed BS. The components $M_i(t)$ are therefore in general time-dependent, and this time dependence depends on the motion of the rigid body. This disadvantage does not apply to force-free motion. We will consider Euler's equations for this case in the following. In the following chapter, we will then investigate gyroscopic motions for $\boldsymbol{M} \neq 0$.

Free Rotation Around Principal Axes

We consider a rigid body without torques acting on it, $\boldsymbol{M} = 0$.

A force-free translational motion is uniform, i.e. it proceeds with constant velocity. We may ask whether Euler equations analogously lead to a uniform rotational motion in the force-free case. We write the Euler equations for constant ω_1, ω_2 and ω_3 and $\boldsymbol{M} = 0$:

$$\big(\Theta_3 - \Theta_2\big)\,\omega_2\,\omega_3 = 0, \quad \big(\Theta_1 - \Theta_3\big)\,\omega_1\,\omega_3 = 0, \quad \big(\Theta_2 - \Theta_1\big)\,\omega_1\,\omega_2 = 0 \tag{22.13}$$

We assume that all principal moments of inertia are different, so

$$\Theta_i \neq \Theta_j \quad \text{for } i \neq j \tag{22.14}$$

Then Equation (22.13) have the following non-trivial solution.

$$\omega_1 = \omega_1^0 = \text{const.}, \quad \omega_2 = \omega_3 = 0 \tag{22.15}$$

Interchanging the components yields two more analogous solutions. Two components must always be zero. An arbitrary constant angular velocity $(\omega_1, \omega_2, \omega_3) = \text{const.}$ is in general not a solution of the force-free Euler equations.

Since $\boldsymbol{M} = 0$, the angular momentum is conserved, $\boldsymbol{L} = \text{const.}$ For (22.15), this means

$$\boldsymbol{L} = \widehat{\Theta} \cdot \boldsymbol{\omega} = \Theta_1\, \omega_1^0\, \boldsymbol{e}_1 = \text{constant vector in IS} \tag{22.16}$$

Thus, the position of the body-fixed axis $\boldsymbol{e}_1$ is constant in space:

$$\boldsymbol{e}_1 = \text{constant in IS} \tag{22.17}$$

The body rotates uniformly about a principal axis, which has a constant direction.

Stability of solution

The equations of motion admit three solutions of the type (22.15). We show that only two of these solutions are stable. For this purpose, we consider a motion that deviates slightly from (22.15):

$$\omega_1 \approx \omega_1^0, \quad \omega_2 \ll \omega_1^0, \quad \omega_3 \ll \omega_1^0 \tag{22.18}$$

In the force-free Euler equations

$$\begin{aligned} \Theta_1\,\dot\omega_1 + (\Theta_3 - \Theta_2)\,\omega_2\,\omega_3 &= 0 \\ \Theta_2\,\dot\omega_2 + (\Theta_1 - \Theta_3)\,\omega_1\,\omega_3 &= 0 \\ \Theta_3\,\dot\omega_3 + (\Theta_2 - \Theta_1)\,\omega_1\,\omega_2 &= 0 \end{aligned} \tag{22.19}$$

we now neglect the terms which are quadratic in the small deviations ω_3 and ω_2. Then the first equation gives

$$\dot\omega_1 = 0, \quad \omega_1 = \omega_1^0 = \text{const.} \tag{22.20}$$

We insert this into the other two equations. We differentiate the second equation with respect to the time and eliminate $\dot\omega_3$ using the third equation. This and the corresponding procedure with the third equation result in

$$\ddot\omega_2 + H\,\omega_2 = 0, \quad \ddot\omega_3 + H\,\omega_3 = 0 \tag{22.21}$$

where

$$H = \frac{(\Theta_1 - \Theta_3)(\Theta_1 - \Theta_2)}{\Theta_2\,\Theta_3}\,\left(\omega_1^0\right)^2 \tag{22.22}$$

Now, there are two possibilities:

1. The moment of inertia Θ_1 is the smallest or the largest of the three principal moments of inertia. Then $H > 0$ and (22.21) gives

$$\omega_2(t) = a\,\cos\left(\sqrt{H}\,t + b\right) \tag{22.23}$$

and an analogous solution for $\omega_3(t)$. Thus, small deviations ($a \ll \omega_1^0$) lead to small oscillations around the solution $\omega = (\omega_1^0, 0, 0)$.

2. The moment of inertia Θ_1 is the middle one of the three principal moments of inertia. Then $H < 0$ holds and (22.21) gives

$$\omega_2(t) = a \exp\big(- \sqrt{-H}\, t\big) + b \exp\big(+ \sqrt{-H}\, t\big) \qquad (22.24)$$

and an analogous solution for $\omega_3(t)$. Thus, small deviations ($b \ll \omega_1^0$ but $b \neq 0$) from the solution $\omega = (\omega_1^0, 0, 0)$ grow exponentially. The exponential growth is valid only in the beginning, i.e. as long as the condition (22.18) for our calculation is valid.

From (22.18)–(22.24) it follows: The rotation around the axis of the smallest or of the largest moment of inertia is stable in the sense that small deviations remain small. From

$$\boldsymbol{L} = \Theta_1 \omega_1^0\, \boldsymbol{e}_1 + \mathcal{O}(\omega_2, \omega_3) = \text{const.} \qquad (22.25)$$

follows $\boldsymbol{e}_1 \approx$ const., i.e. an (approximately) stable position of the body-fixed rotation axis $\boldsymbol{e}_1$ in the space-fixed system. In contrast, the free rotation around the axis of the middle moment of inertia is not stable, because in every real system, there are small perturbations, which then grow rapidly. Experimentally this effect can be easily demonstrated with a rotating (free-falling) book: Take a book, set it in rotation around one of the principal axes, and throw it into the air. Do this for all principal axes (one after another) and watch.

Force-Free Symmetric Gyroscope

In the previous section, the force-free rotation around a principal axis was treated. Now, we investigate the force-free rotation in general for a symmetric rigid body or gyroscope. We speak of a *symmetric* gyroscope if exactly two principal moments of inertia are equal:

$$\Theta_2 = \Theta_1, \quad \Theta_3 \neq \Theta_1 \qquad (22.26)$$

This is especially true for bodies that are rotationally symmetric with respect to their x_3-axis. The x_3-axis is called the *figure axis* or symmetry axis.

As the origin of the body-fixed BS, we choose again the center of mass or the fixed support point of the rigid body. In the first case, the center-of-mass motion and the rotational motion are decoupled. In the second case, there is only the rotational motion. In either case, we can focus on the rotational motion.

No torque should act on the gyro. For $\Theta_2 = \Theta_1$ the Euler equations are then

$$\begin{aligned} \Theta_1\,\dot\omega_1 + \big(\Theta_3 - \Theta_1\big)\,\omega_2\,\omega_3 &= 0 \\ \Theta_1\,\dot\omega_2 + \big(\Theta_1 - \Theta_3\big)\,\omega_1\,\omega_3 &= 0 \\ \Theta_3\,\dot\omega_3 \qquad\qquad\qquad &= 0 \end{aligned} \tag{22.27}$$

From the last equation, it follows $\omega_3 = \text{const.} = \omega_0$. We substitute this into the other two equations:

$$\dot\omega_1 - \Omega\,\omega_2 = 0, \quad \dot\omega_2 + \Omega\,\omega_1 = 0 \quad \text{with } \Omega = \frac{\Theta_1 - \Theta_3}{\Theta_1}\,\omega_0 \tag{22.28}$$

We differentiate the first equation and insert it into the second:

$$\ddot\omega_1 + \Omega^2\,\omega_1 = 0 \tag{22.29}$$

The solution $\omega_1(t) = a\,\sin(\Omega t + \psi_0)$ fixes $\omega_2(t) = \dot\omega_1(t)/\Omega$, too:

$$\omega_1(t) = a\,\sin\big(\Omega t + \psi_0\big), \quad \omega_2(t) = a\,\cos\big(\Omega t + \psi_0\big), \quad \omega_3 = \omega_0 \tag{22.30}$$

For $a \ll \omega_0$, we had already found such a solution in (22.23).

The magnitude of $\boldsymbol{\omega}$ is constant:

$$\boldsymbol{\omega}^2 = \omega_1(t)^2 + \omega_2(t)^2 + \omega_3(t)^2 = \omega_0^2 + a^2 = \text{const.} \tag{22.31}$$

According to (22.30), the projection of $\boldsymbol{\omega}$ onto the x_1–x_2-plane has constant length a and rotates with the angular velocity Ω. Thus, the vector $\boldsymbol{\omega}$ describes a circular cone, the *body cone* (Figure 22.1). The opening angle of the body cone is

$$\gamma = \arctan\,\frac{a}{\omega_0} = \text{const.} \tag{22.32}$$

To determine the Euler angles $\theta(t)$, $\psi(t)$, and $\phi(t)$, we substitute (22.30) into (22.10) – (22.12). The integration of these equations is

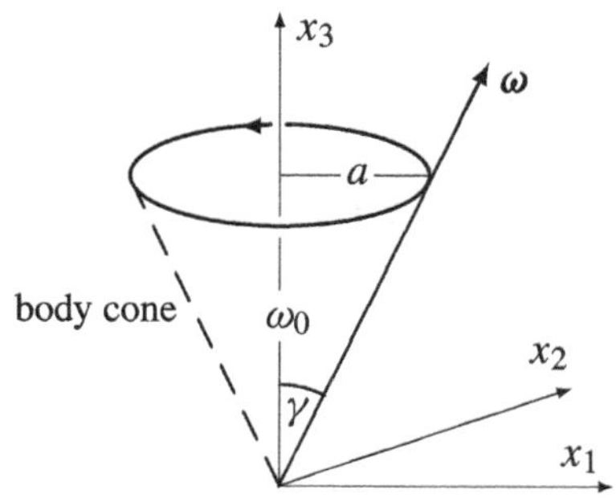

Figure 22.1 The angular velocity $\boldsymbol{\omega}$ of a force-free symmetric gyro moves in the body-fixed system on a circular cone called the *body cone.*

simplified if we orientate the inertial system such that the (conserved) angular momentum vector points in the z-direction:

$$\boldsymbol{L} = L\,\boldsymbol{e}_z = \text{const.} \tag{22.33}$$

The components L_i in the body-fixed system are:

$$\begin{aligned}(L_1, L_2, L_3) &= L\,\big(\boldsymbol{e}_z \cdot \boldsymbol{e}_1(t),\ \boldsymbol{e}_z \cdot \boldsymbol{e}_2(t),\ \boldsymbol{e}_z \cdot \boldsymbol{e}_3(t)\big) \\ &= L\,\big(\sin\theta \sin\psi,\ \sin\theta\cos\psi,\ \cos\theta\big) \\ &= \big(\Theta_1\,\omega_1,\ \Theta_1\,\omega_2,\ \Theta_3\,\omega_3\big)\end{aligned} \tag{22.34}$$

The scalar products of the basis vectors follow from (19.23). For ω_1, ω_2, and ω_3 we substitute (22.30) and write the result in components:

$$L\,\sin\theta(t)\,\sin\psi(t) = a\,\Theta_1\,\sin\left(\Omega t + \psi_0\right) \tag{22.35}$$

$$L\,\sin\theta(t)\,\cos\psi(t) = a\,\Theta_1\,\cos\left(\Omega t + \psi_0\right) \tag{22.36}$$

$$L\,\cos\theta(t) = \omega_0\,\Theta_3 \tag{22.37}$$

From (22.35) and (22.36) we find

$$\psi(t) = \Omega t + \psi_0 \tag{22.38}$$

Inserting this in (22.36) yields $L\sin\theta(t) = a\Theta_1$. We divide this by (22.37) and obtain

$$\tan\theta_0 = \frac{a}{\omega_0}\,\frac{\Theta_1}{\Theta_3} \tag{22.39}$$

This implies $\theta_0 = \text{const.}$.

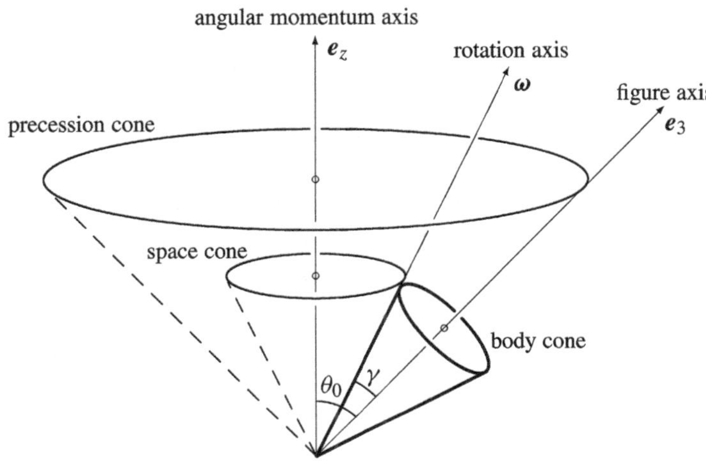

Figure 22.2 Regular precession of the force-free symmetric gyroscope. The space-fixed angular momentum direction, the axis of rotation, and the symmetry (figure) axis all lie in one plane. The motion runs as if the body cone (bold contour) rolls on the space cone; thereby the figure axis (symmetry axis of the gyro) rotates on the precession cone. The actual gyroscope could consist of a solid matter body cone.

With $\omega_1 = a \sin \psi$ and $\theta = \theta_0$, (22.10) becomes $a = \dot{\phi} \sin \theta_0$, thus

$$\phi = \frac{a}{\sin \theta_0}\, t + \phi_0 \tag{22.40}$$

Equations (22.38)–(22.40) present the general solution for the Euler angles. For the discussion of the motion of the rigid body in the inertial system (Figure 22.2) we recall the meaning of some quantities:

- θ: angle between the figure axis and the space-fixed z-axis;
- $\dot{\phi}$: rotation of the figure axis around the z-axis;
- $\dot{\psi}$: rotation of the body (i.e. of BS) around its figure axis.

Since $\theta = \theta_0$ the figure axis moves on a cone, the *precession cone*, about the z-axis. The momentary axis of rotation is defined by the direction of $\boldsymbol{\omega}$. From (19.20) and (19.21) follows:

$$\boldsymbol{\omega} = \dot{\phi}\, \boldsymbol{e}_z + \dot{\psi}\, \boldsymbol{e}_3 \tag{22.41}$$

Thus, $\boldsymbol{\omega}$ lies in the $\boldsymbol{e}_3$–$\boldsymbol{e}_z$-plane, which was chosen as the image plane in Figure 22.2. Since $\theta_0 = \sphericalangle(\boldsymbol{e}_z, \boldsymbol{e}_3) = \text{const.}$ and $\gamma = \sphericalangle(\boldsymbol{\omega}, \boldsymbol{e}_3) = \text{const.}$, also the angle between $\boldsymbol{\omega}$ and $\boldsymbol{e}_z$ must be constant. So also $\boldsymbol{\omega}$ moves on a cone which is called a *space cone*. One can then visualize the motion as the rolling of the *body cone* on the space cone; hereby the body cone could be imagined as a rigid-body cone. This motion of the figure (symmetry) axis around the z-axis describes a circular cone, which is called a *precession cone*. The so-described motion is also called *regular precession*.

In contrast to the regular precession ($\theta_0 = \text{const.}$), discussed here, the figure axis of the heavy gyroscope (Figure 23.3) oscillates between two θ-values. This is called *nutation*.[2]

Exercises

22.1 *Symmetric gyroscope with constant torque*

On a symmetrical gyroscope acts the constant torque $\boldsymbol{M} = M_0\,\boldsymbol{e}_z$. Solve the equations of motion for the initial conditions $\boldsymbol{\omega}(0) = 0$ and $\boldsymbol{e}_3(0) = \boldsymbol{e}_z$. Give the time dependence of the Euler angles.

22.2 *Torque perpendicular to angular momentum*

On a rigid body acts the torque $\boldsymbol{M} = M_0\,\boldsymbol{e}_z \times (\boldsymbol{L}/L)$. Show $|\boldsymbol{L}| = \text{const.}$ and $L_z = \text{const.}$ Derive the equation

$$\frac{d\boldsymbol{\ell}}{dt} = \boldsymbol{\omega}_{\text{prec}} \times \boldsymbol{\ell}$$

for the direction $\boldsymbol{\ell} = \boldsymbol{L}/L$ of the angular momentum. Determine the precession frequency $\boldsymbol{\omega}_{\text{prec}}$. Describe the temporal change of $\boldsymbol{\ell}$ in words.

---

[2]In some German textbooks, the regular precession is called nutation. Our designation coincides with that of Goldstein [6] and most other textbooks.

Chapter 23

Heavy Gyroscope

A gyroscope in a gravitational field is called a heavy gyroscope (or heavy gyro or top for short). Examples are a child's top in a homogeneous gravitational field, or the rotating earth (as a non-supported gyroscope) in the gravitational field of the sun and the moon. For these applications, we will use the Lagrange equations. Therefore, we start by establishing the Lagrangian function for a rigid body.

Lagrange Function

So far, we have proceeded directly from the equation of motion

$$\frac{d}{dt}\left(\widehat{\Theta}\cdot\boldsymbol{\omega}\right)=\boldsymbol{M} \tag{23.1}$$

This equation was evaluated in the body-fixed system because there the Θ_{ik} are constant. For $\boldsymbol{M}\neq 0$, this has the disadvantage that we need the components of $\boldsymbol{M}$ in the time-dependent basis $\boldsymbol{e}_i(t)$. For non-vanishing torque, it may be simpler to start from the Lagrange function.

As the origin of the body-fixed BS, we choose again the center of mass or the fixed supporting point of the rigid body. In the first case, the center of mass and the rotational motion are decoupled; in the second case, there is only the rotational motion. In any case, we can focus on the rotational motion. As generalized coordinates, we choose the three Euler angles ϕ, ψ and θ. The potential U depends only on these coordinates and on the time. The kinetic energy $T_{\rm rot}$

is given by (20.22). Thus, the Lagrangian function reads

$$\mathcal{L}(\phi,\psi,\theta,\dot{\phi},\dot{\psi},\dot{\theta},t) = T_{\text{rot}} - U = \frac{\boldsymbol{\omega}\cdot(\widehat{\Theta}\cdot\boldsymbol{\omega})}{2} - U(\phi,\psi,\theta,t) \quad (23.2)$$

We write the energy T_{rot} favorably in the body-fixed principal axis system:

$$T_{\text{rot}} = \frac{\boldsymbol{\omega}\cdot(\widehat{\Theta}\cdot\boldsymbol{\omega})}{2} = \frac{\Theta_1\,\omega_1^2 + \Theta_2\,\omega_2^2 + \Theta_3\,\omega_3^2}{2} \quad (23.3)$$

This system has the advantage that the Θ_i are properties of the rigid body. The relation between ω_1, ω_2, and ω_3 and the generalized velocities $\dot{\phi}$, $\dot{\psi}$, and $\dot{\theta}$ is known from (19.27). Thus, we obtain

$$\begin{aligned}\mathcal{L}(\phi,\psi,\theta,\dot{\phi},\dot{\psi},\dot{\theta},t) &= \frac{\Theta_1}{2}\left(\dot{\phi}\,\sin\theta\,\sin\psi + \dot{\theta}\,\cos\psi\right)^2\\ &\quad + \frac{\Theta_2}{2}\left(\dot{\phi}\,\sin\theta\,\cos\psi - \dot{\theta}\,\sin\psi\right)^2\\ &\quad + \frac{\Theta_3}{2}\left(\dot{\phi}\,\cos\theta + \dot{\psi}\right)^2 - U(\phi,\psi,\theta,t) \quad (23.4)\end{aligned}$$

The Lagrange equations follow from this. They represent a system of three coupled differential equations of second order for $\phi(t)$, $\psi(t)$ and $\theta(t)$. All the time derivatives in Equation (23.4) refer to the space-fixed IS.

Heavy Symmetric Gyro

We evaluate the equations of motion for a massive symmetric gyroscope moving under the influence of gravity, Figure 23.1. As the origin of the body-fixed system BS, we take the supporting point 0. The inertia tensor in BS is diagonal; the diagonal elements are denoted by $\Theta_1 = \Theta_2$ and Θ_3. As a gyroscope, imagine a rigid body that is rotationally symmetric with respect to the x_3-axis. The x_3 axis is called the figure or symmetry axis.

The z-coordinate of the center-of-mass CM is $z = s\cos\theta$; where θ is the Euler angle between the z-axis and the x_3 axis, and s is the distance between the support point 0 and the center-of-mass CM. The potential energy of the gyroscope is $U = mgs\cos\theta$ (mass m,

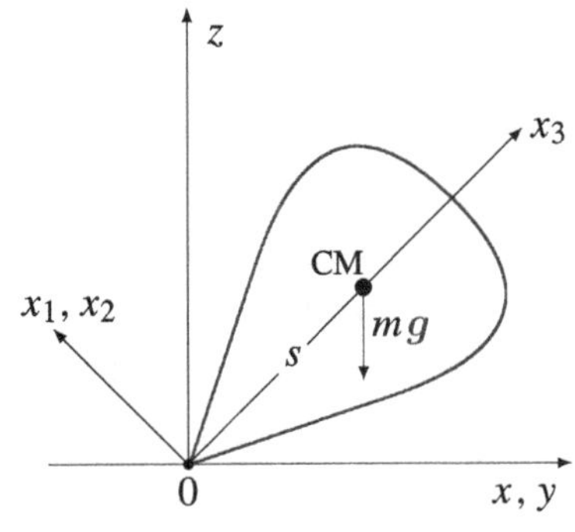

Figure 23.1 A symmetric gyroscope with the figure axis x_3 has the fixed bottom support point 0. The center-of-mass CM has the distance s from 0; the gravity force $m\boldsymbol{g}$ acts on CM. The model system differs from a child's top by the strict fixation of the supporting point 0 and by the absence of friction.

acceleration due to gravity g). Thus, the Lagrange function (23.4) becomes

$$\mathcal{L}(\theta, \dot{\phi}, \dot{\psi}, \dot{\theta}) = \frac{\Theta_1}{2}\left(\dot{\theta}^2 + \dot{\phi}^2 \sin^2\theta\right) + \frac{\Theta_3}{2}\left(\dot{\psi} + \dot{\phi}\cos\theta\right)^2 - mgs\cos\theta \tag{23.5}$$

This Lagrange function has the following symmetries:

$$\frac{\partial \mathcal{L}}{\partial t} = 0, \quad \frac{\partial \mathcal{L}}{\partial \phi} = 0, \quad \frac{\partial \mathcal{L}}{\partial \psi} = 0 \tag{23.6}$$

Therefore, instead of the three Lagrangian equations (second-order differential equations) the following three constants of motion (first-order differential equations) can be used:

$$\begin{aligned} E &= \frac{\partial \mathcal{L}}{\partial \dot{\phi}}\,\dot{\phi} + \frac{\partial \mathcal{L}}{\partial \dot{\psi}}\,\dot{\psi} + \frac{\partial \mathcal{L}}{\partial \dot{\theta}}\,\dot{\theta} - \mathcal{L} \\ &= \frac{\Theta_1}{2}\left(\dot{\theta}^2 + \dot{\phi}^2 \sin^2\theta\right) + \frac{\Theta_3}{2}\left(\dot{\psi} + \dot{\phi}\cos\theta\right)^2 + mgs\cos\theta = \text{const.} \end{aligned} \tag{23.7}$$

$$L_z = \frac{\partial \mathcal{L}}{\partial \dot{\phi}} = \Theta_1\,\dot{\phi}\,\sin^2\theta + \Theta_3\left(\dot{\psi} + \dot{\phi}\cos\theta\right)\cos\theta = \text{const.} \tag{23.8}$$

$$L_3 = \frac{\partial \mathcal{L}}{\partial \dot{\psi}} = \Theta_3\left(\dot{\psi} + \dot{\phi}\cos\theta\right) = \text{const.} \tag{23.9}$$

The constants are denoted according to their physical meaning: Since $\mathcal{L}$ does not explicitly depend on t, the energy E is conserved; since $\mathcal{L}$ does not depend on ϕ (angle of rotation about the z-axis), $M_z = 0$ holds and the angular momentum component L_z is conserved; since $\mathcal{L}$ does not depend on ψ (angle of rotation about the x_3 axis), $M_3 = 0$ and the angular momentum component L_3 is conserved.

Using (23.8) and (23.9) we eliminate $\dot{\phi}$ and $\dot{\psi}$ in (23.7):

$$E = \frac{\Theta_1}{2}\,\dot{\theta}^2 + \frac{(L_z - L_3\,\cos\theta)^2}{2\,\Theta_1\,\sin^2\theta} + \frac{L_3^2}{2\,\Theta_3} + m\,g\,s\cos\theta = \text{const.} \tag{23.10}$$

This is a first-order differential equation for $\theta(t)$. We can express it in the form

$$E = \frac{\Theta_1}{2}\,\dot{\theta}^2 + U_{\text{eff}}(\theta) \tag{23.11}$$

with the effective potential

$$U_{\text{eff}}(\theta) = \frac{(L_z - L_3\,\cos\theta)^2}{2\,\Theta_1\,\sin^2\theta} + m\,g\,s\,(1 - \cos\theta) + \frac{L_3^2}{2\,\Theta_3} - m\,g\,s \tag{23.12}$$

Equation (23.11) can be solved for $d\theta/dt = f(\theta)$ and integrated:

$$t = t_0 + \int_{\theta_0}^{\theta} d\theta' \sqrt{\frac{\Theta_1/2}{E - U_{\text{eff}}(\theta')}} = \int \frac{du}{\sqrt{P_3(u)}} \tag{23.13}$$

This determines $t = t(\theta)$ and hence $\theta = \theta(t)$. The substitution $u = \cos\theta$ leads to a polynomial P_3 of the third degree, as indicated in the last expression. Thus, we have obtained an elliptic integral, which cannot be solved analytically.

The differential equation (23.11) has the same form as the one-dimensional motion in a potential. The solution can therefore be discussed graphically. Figure 23.2 depicts the effective potential U_{eff}. From (23.12) we read off the limiting cases

$$U_{\text{eff}}(\theta) \overset{\theta\to 0}{\longrightarrow} \infty \quad \text{and} \quad U_{\text{eff}}(\theta) \overset{\theta\to\pi}{\longrightarrow} \infty \tag{23.14}$$

In between, U_{eff}, as sketched in Figure 23.2, has a minimum. The condition $E = U_{\text{eff}}$ yields two reversal points, θ_1 and θ_2. The angle θ

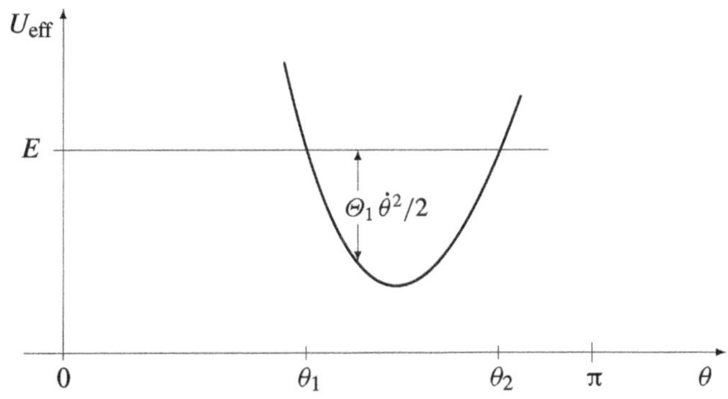

Figure 23.2 Sketch of the effective potential for the θ motion. The kinetic energy of the θ-motion equals the distance between $U_{\rm eff}(\theta)$ and the horizontal line at E. The θ-motion oscillates between the reversal points θ_1 and θ_2.

between the figure axis and the space-fixed z-axis oscillates between θ_1 and θ_2; this motion is called *nutation.*

While the figure axis (x_3) oscillates between θ_1 and θ_2, it precedes with the angular velocity

$$\dot\phi\,\boldsymbol{e}_z = \frac{L_z - L_3\cos\theta}{\Theta_1\sin^2\theta}\;\boldsymbol{e}_z \tag{23.15}$$

about the space-fixed z-axis; this follows from (23.8) and (23.9). The motion of the figure axis is determined by the oscillation between θ_1 and θ_2 and the simultaneous rotation with $\dot\phi$. For a rotationally symmetric gyroscope, it is mainly this movement of the figure axis, which catches the eye; the additional rotation of the body with $\dot\psi$ around the figure axis itself is less obvious because of its symmetry.

The velocities $\dot\theta$ and $\dot\phi$, which determine the motion of the figure axis, depend on a given gyroscope (with parameters m, s, Θ_1, and Θ_3) on the integration constants E, L_z, and L_3. In Figure 23.3, possible types of motion are shown. According to (23.15), the precession of the figure axis with $\dot\phi$ can maintain or change its sign in the range $\theta_1 < \theta < \theta_2$.

The given solution contains the special case $\theta_1 = \theta_2$ for which the nutations vanish. The figure axis then precedes a cone with the opening angle θ_1. Such a *regular* precession is also obtained in

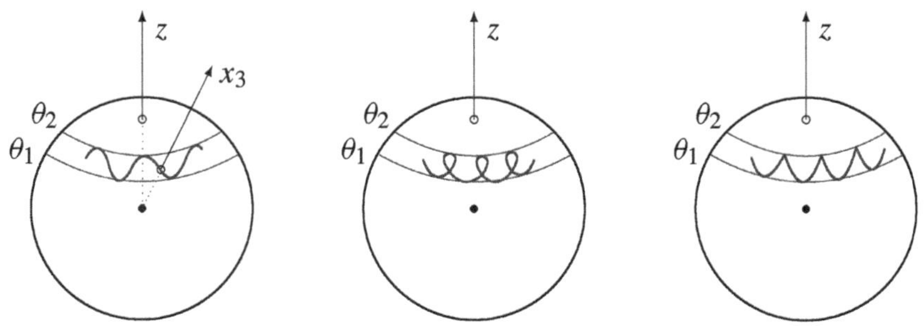

Figure 23.3 Precession and nutation of the figure axis of a heavy, symmetric gyroscope. The schematic sketch shows the path of the intersection of the figure axis with a spherical surface. On the left this path is shown for $\dot{\phi} > 0$, in the middle for the case that $\dot{\phi}$ changes its sign in the range between θ_1 and θ_2, and on the right for $\dot{\phi}(\theta_2) = 0$.

the limiting case $g \to 0$. This force-free case has been sketched in Figure 22.2 (with $\theta = \theta_0$ and $\dot{\phi} = \text{const.}$).

The limiting case $g \to 0$ can be approximately realized by

$$T_{\text{rot}} \gg |U_{\text{pot}}| \quad \text{(nearly force-free)} \tag{23.16}$$

For a child's top, this means that if it is put into vigorous rotation (T_{rot} large), (23.16) holds at first and the figure axis precedes approximately regularly, i.e. on a cone. Due to friction losses, which are not included in our model, the rotational energy gradually decreases. Then the condition (23.16) becomes less and less well fulfilled. Consequently, the nutations become stronger an stronger, and the figure axis fluctuates in an increasing range $\theta_1 \le \theta \le \theta_2$.

Earth as a Rotating Body

We discuss the intrinsic rotation of our earth. The model "rigid body" is an approximation; earthquakes or other mass shifts might violate the condition "rigid".

Free rotation

We first disregard the small torque exerted by the moon and the sun on the earth (this will be considered afterward). Then we get the motion of the free symmetric gyroscope, (22.26)–(22.41). The center-of-mass motion (free fall of the earth in the gravitational field of the sun) is decoupled from the rotational motion and will not be considered here.

The angular momentum axis of the earth almost coincides with the figure axis: At the north pole, the axis of rotation has a distance of about 10 meters from the figure axis. According to (22.28), the angular velocity Ω, with which the angular momentum axis $\boldsymbol{\omega}$ precedes around the figure axis, equals

$$\Omega = \frac{\Delta\Theta}{\Theta}\,\omega_0 \approx \frac{1}{300}\,\frac{2\pi}{\text{day}} \tag{23.17}$$

Here, we have inserted the experimental value for the oblateness of the earth (approximately $\Delta\Theta/\Theta \approx 1/300$) . The deviations from the rigid body model are in fact clearly noticeable: the actual relative motion of the figure axis and the angular momentum axis is partly irregular.

Torque on the earth

In the following, we disregard the small deviation between the figure axis and angular momentum axis (for the earth). Then we can use the simplification

$$\boldsymbol{L} = L\,\boldsymbol{e}_3 \tag{23.18}$$

We now want to investigate the influence of the gravitational field of the sun and the moon on the oblate earth. This gravitational field causes a torque, which leads to a regular precession of the figure axis x_3 around the space-fixed z-axis (precession on a circular cone with the opening angle θ_0, Figure 23.4 left). The z-axis is chosen as the direction perpendicular to the orbital plane (ecliptic). We restrict ourselves at first to the field of the sun, the influence of the moon will be taken into account later.

The mass distribution of the earth can be simulated by a spherical mass distribution and additional correction masses at the equator (Exercise 20.4). In this picture, one can easily understand the occurrence of a torque, Figure 23.4 on the right. In the following,

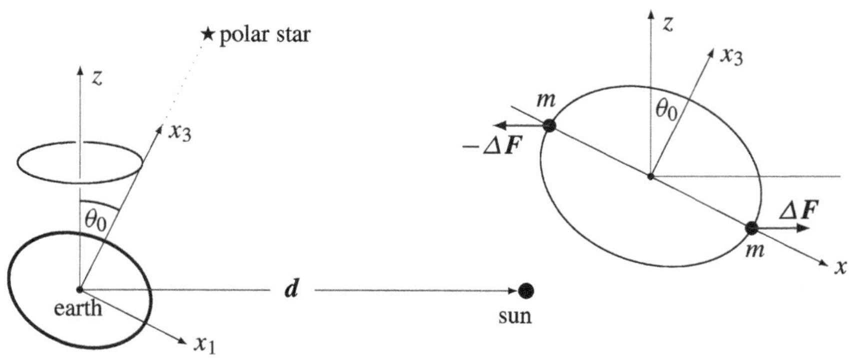

Figure 23.4 The gravitational field of the sun (and of the moon) exerts a torque on the oblate earth. This leads to a regular precession of the figure axis on a cone. This cone has an opening angle $\theta_0 = 23.5°$; and one full precession takes about 26 000 years. The present pole star will therefore in the foreseeable future no longer indicate the north direction. Right: The oblateness of the earth can be simulated by correction masses m at the equator. At the center of mass of the earth, the gravitational force of the sun and the inertia force (of the accelerated orbital motion) just cancel each other (free fall). For the two correction masses, there are extra gravitational forces ($\pm\boldsymbol{\Delta F}$) because here the gravitational force is not completely compensated by the inertial forces at these points. This results in a torque perpendicular to the image plane.

we assume that the earth is a symmetric rotator with $\Theta_1 = \Theta_2$ and $\Theta_3 > \Theta_1$.

In the gravitational field of the sun, a mass element $dm = \varrho(\boldsymbol{r})\, d^3r$ of the earth has the potential energy $dU = -G\mathcal{M}\, dm/|\boldsymbol{d} - \boldsymbol{r}|$; here G is the gravitational constant, $\mathcal{M}$ is the mass of the sun, $\boldsymbol{d}$ is the vector from the center of the earth to the sun, and $\boldsymbol{r}$ is the vector from the earth's center to the considered mass element dm. From this follows the force $d\boldsymbol{F} = -\text{grad}\, dU$ on the mass element, and the torque

$$\boldsymbol{M} = \int \boldsymbol{r} \times d\boldsymbol{F} = G\mathcal{M} \int_V d^3r\; \varrho(\boldsymbol{r})\; \boldsymbol{r} \times \frac{\boldsymbol{d} - \boldsymbol{r}}{|\boldsymbol{d} - \boldsymbol{r}|^3} \tag{23.19}$$

In the region with $\varrho \neq 0$, $|\boldsymbol{r}| \ll |\boldsymbol{d}|$ holds. Therefore, we can use the approximation $|\boldsymbol{d} - \boldsymbol{r}|^3 \approx d^3 - 3(\boldsymbol{r} \cdot \boldsymbol{d})d$ and obtain

$$\boldsymbol{M} = G\mathcal{M} \int_V d^3r\; \varrho(\boldsymbol{r})\; \frac{\boldsymbol{r} \times \boldsymbol{d}}{d^3} \left(1 + 3\,\frac{\boldsymbol{r} \cdot \boldsymbol{d}}{d^2}\right) \tag{23.20}$$

We evaluate this integral in the body-fixed system of the earth. The first term in the integrand is linear in $\boldsymbol{r}$ and averages away by the integration (for $x_i \to -x_i$ it changes its sign, but ϱ is symmetric). This leaves only the second term which is quadratic in $\boldsymbol{r}$.

Due to the orbital motion Earth–Sun, the vector $\boldsymbol{d}$ describes a circular orbit with the angular velocity $\Omega_{\mathrm{E}} = 2\pi/\text{year}$ (relative to the body-fixed system). This means:

$$\boldsymbol{d} := (d_1, d_2, d_3) = d\big(\cos\theta_0 \cos\Omega_{\mathrm{E}} t, \cos\theta_0 \sin\Omega_{\mathrm{E}} t, \sin\theta_0 \cos\Omega_{\mathrm{E}} t\big) \tag{23.21}$$

We put this into (23.20) and average over one period:

$$\begin{aligned}\langle \boldsymbol{M} \rangle = \frac{\Omega_{\mathrm{E}}}{2\pi} \int_0^{2\pi/\Omega_{\mathrm{E}}} dt\, \frac{3\,G\mathcal{M}}{d^5} \int_V d^3r\, \varrho(\boldsymbol{r})\, \big(\big(x_2^2 - x_3^2\big)\, d_2\, d_3\, \boldsymbol{e}_1 \\ + \big(x_3^2 - x_1^2\big)\, d_3\, d_1\, \boldsymbol{e}_2 + \big(x_1^2 - x_2^2\big)\, d_1\, d_2\, \boldsymbol{e}_3\big)\end{aligned} \tag{23.22}$$

Time averaging over the first and third terms yields zero; in the second term, it leads to the factor $\langle \cos^2 \Omega_{\mathrm{E}} t \rangle = 1/2$. Thus, we obtain

$$\begin{aligned}\langle \boldsymbol{M} \rangle &= \frac{3G\mathcal{M}}{2\,d^3} \cos\theta_0 \sin\theta_0 \int_V d^3r\, \varrho(\boldsymbol{r})\, \big(x_3^2 - x_1^2\big)\, \boldsymbol{e}_2 \\ &= -\frac{3G\mathcal{M}}{2\,d^3} \cos\theta_0 \sin\theta_0 \big(\Theta_3 - \Theta_1\big)\, \boldsymbol{e}_2 = \frac{3G\mathcal{M}}{2\,d^3} \cos\theta_0\, \Delta\Theta\, (\boldsymbol{e}_3 \times \boldsymbol{e}_z)\end{aligned} \tag{23.23}$$

Here, we used $\Delta\Theta = \Theta_3 - \Theta_1 > 0$, and $\sin\theta_0\, \boldsymbol{e}_2$ was replaced by $-\boldsymbol{e}_3 \times \boldsymbol{e}_z$ (see Figure 23.4).

Precession of the earth axis

We write down the equation of motion (22.1) with the torque (23.23):

$$\frac{d\boldsymbol{L}}{dt} = \frac{3G\mathcal{M}}{2\,d^3} \cos\theta_0\, \Delta\Theta\, (\boldsymbol{e}_3 \times \boldsymbol{e}_z). \tag{23.24}$$

From this, we see immediately that $d\boldsymbol{L}$ is perpendicular to $\boldsymbol{e}_3$ and to $\boldsymbol{e}_z$, i.e. perpendicular to the image plane of Figure 23.4. The change

$d\boldsymbol{L}$ is thus perpendicular to $\boldsymbol{L}$, so that the angle θ_0 between $\boldsymbol{L}$ and $\boldsymbol{e}_z$ is preserved. Therefore, $\boldsymbol{L}$ performs a regular precession on a circular cone (Figure 23.4 left).

We now show that the ansatz

$$\boldsymbol{L}(t) = L\,\boldsymbol{e}_3(t) = \Theta_3\,\omega\,\boldsymbol{e}_3(t) \tag{23.25}$$

solves the equation of motion (23.24). The solution must satisfy the initial condition (23.25); this implies that we neglect the small differences between the angular momentum axis and the figure axis (see discussion following (23.17)). The equations of motion and the initial conditions determine the solution unambiguously. If we find a solution with the ansatz (23.25), then this is also the unique solution.

We multiply (23.24) scalar by $\boldsymbol{L}$. This gives $d(\boldsymbol{L}^2)/dt = 0$, so $|\boldsymbol{L}| = L = \text{const.}$ and $\omega = \text{const.}$ We now multiply (23.24) scalar by $\boldsymbol{e}_z$. This gives $\boldsymbol{e}_z \cdot (d\boldsymbol{L}/dt) = d(\boldsymbol{e}_z \cdot \boldsymbol{L})/dt = 0$, i.e.

$$\boldsymbol{e}_z \cdot \boldsymbol{L} = L\,\boldsymbol{e}_z \cdot \boldsymbol{e}_3 = L\cos\theta_0 = \text{const.} \tag{23.26}$$

The angle θ_0 between the figure axis $\boldsymbol{e}_3$ and the space-fixed z-axis is constant. For the earth we have

$$\theta_0 = 23.5^\circ \tag{23.27}$$

This angle determines the inclination of the figure and the angular momentum axis of the earth relative to the orbital plane (the ecliptic).

We now substitute (23.25) with $\Theta_3 = \Theta$ into (23.24):

$$\frac{d\boldsymbol{e}_3}{dt} = \frac{\Delta\Theta}{\Theta}\,\frac{3\,G\mathcal{M}}{2\,d^3\omega}\cos\theta_0\,(\boldsymbol{e}_3 \times \boldsymbol{e}_z) = \boldsymbol{\omega}_{\text{prec}} \times \boldsymbol{e}_3 \tag{23.28}$$

The quantities ω and θ_0 are constant, as well as the precession frequency (index prec)

$$\boldsymbol{\omega}_{\text{prec}} = -\frac{\Delta\Theta}{\Theta}\,\frac{3\,G\mathcal{M}}{2d^3\omega^2}\cos\theta_0\,\omega\,\boldsymbol{e}_z \tag{23.29}$$

Equation (23.28) is solved by the unit vector $\boldsymbol{e}_3(t)$, which precedes on a circular cone around the $\boldsymbol{e}_z$ axis. This means that the figure axis (south–north) of the earth precedes this circular cone.

The minus sign in (23.29) implies, that the precession movement is opposite to the rotation of the earth around itself. In detail, the motion is rather complicated due to various disturbances. This means that there are in fact small deviations (nutations) from the regular precession (i.e. from $\theta_0 = \text{const.}$).

For a circular orbit, we can equate the gravitational force $G\mathcal{M}M_\text{E}/d^2$ and the centrifugal force $\mu_\text{rot}\,\Omega_\text{E}^2\,d$; where $\mu_\text{rot} = \mathcal{M}M_\text{E}/(\mathcal{M} + M_\text{E})$ and M_E is the mass of the earth. This gives $G\mathcal{M}/d^3 = \Omega_\text{E}^2\,\mathcal{M}/(\mathcal{M} + M_\text{E})$. Thus, (23.29) becomes

$$\omega_\text{prec} = \frac{3}{2}\,\frac{\Delta\Theta}{\Theta}\,\frac{\mathcal{M}\,\cos\theta_0}{\mathcal{M} + M_\text{E}}\,\frac{\Omega_\text{E}^2}{\omega} \tag{23.30}$$

The orbital plane of the moon lies approximately in the orbital plane of the earth. Therefore, the gravitational field of the moon exerts a torque on the earth in a similar way as the sun. For the following estimate, we assume that both torques add up. The corresponding ratios of the masses and orbital frequencies are known:

$$\frac{\mathcal{M}}{\mathcal{M} + M_\text{E}} \approx \begin{cases} 1 & (\mathcal{M} = M_\odot) \\ 1/81 & (\mathcal{M} = M_\text{M}) \end{cases} \qquad \frac{\omega}{\Omega_\text{E}} = \begin{cases} 365 & \text{Earth–Sun} \\ 28 & \text{Moon–Earth} \end{cases} \tag{23.31}$$

We use these values, $\Delta\Theta/\Theta \approx 1/300$ and $\cos\theta_0 \approx 0.92$ in (23.30):

$$\omega_\text{prec} = \omega_\text{prec}^{\odot} + \omega_\text{prec}^{\text{M}} \approx 10^{-7}\,\omega. \tag{23.32}$$

The contribution of the moon is about twice that of the sun. With $\omega = 2\pi/\text{day}$ we get $T_\text{prec} = 2\pi/\omega_\text{prec} \approx 10^7$ days for the period of precession. This is in good agreement with the experimental value $T_\text{prec,exp} \approx 25\,850$ years; this period of time is also called the Platonic year. The precession of the earth's axis means that in the course of time, different stars will play the role of the "polar star", i.e. the role as a pivot point of the starry sky and as a guide to the north. This precession was already discovered by Hipparchus around 150 BC.

Exercises

23.1 *Cylinder with unbalance*

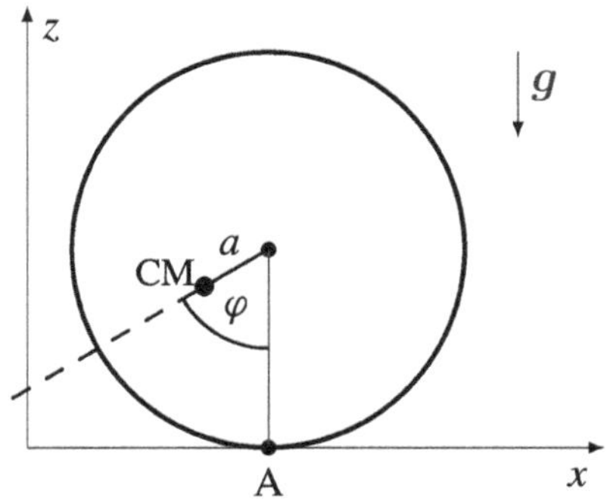

The mass M of a non-homogeneous cylindrical wheel of radius R is distributed such that the center-of-mass CM is shifted by the distance a (relative to the cylinder's symmetry axis). A straight line through CM and parallel to the cylinder axis constitutes a principal axis. The moment of inertia with respect to this axis is Θ_{CM}. The cylinder rolls under the influence of gravity on a horizontal plane. The initial conditions are $\varphi(0) = 0$ and $\dot{\varphi}(0) = \omega_0$.

Set up the Lagrange function in the coordinate φ. What is the associated equation of motion? Give the solution $\varphi_{a=0}(t)$ in the special case $a = 0$. Now, set $\varphi(t) = \varphi_{a=0}(t) + a\xi(t)$ and expand the equation of motion for a small *unbalance* a in powers of a up to the first order. Calculate $\xi(t)$. Sketch the angular velocity $\omega(t) = \dot{\varphi}(t)$ as a function of time.

23.2 *Rocking motion of a hemisphere*

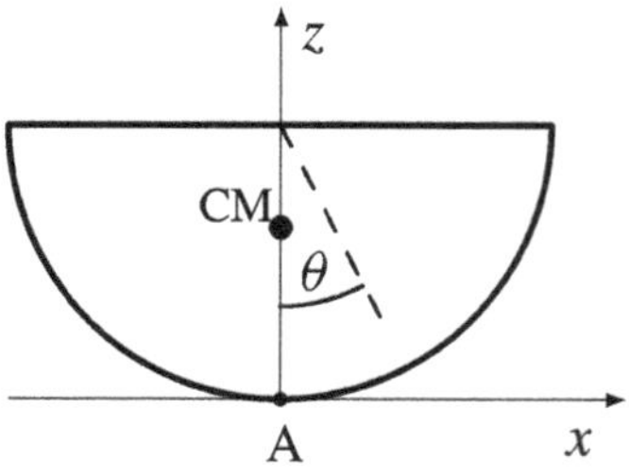

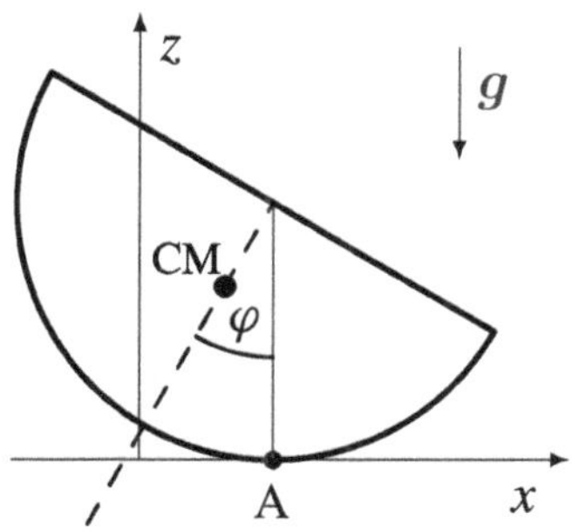

A rigid hemisphere with radius R and constant mass density ϱ_0 performs a rocking motion in the gravitational field on a horizontal plane; the sphere rolls on the plane. On the upper left, the rest position of the hemisphere is shown, on the lower left a deviation from it is depicted.

Calculate the moment of inertia of the hemisphere with respect to an axis through the center-of-mass CM which is perpendicular to the image plane. Calculate the position of the center-of-mass CM as a function of the angle φ. Determine the Lagrange function for small deviations from the equilibrium position. Give the general solution of the corresponding Lagrange equations.

PART VI
Small Oscillations

Chapter 24

Forced Oscillations

Many systems can perform oscillations around an equilibrium position; examples are a double pendulum or a polyatomic molecule. If the displacement from the rest position is sufficiently small, then these oscillations are harmonic. Part VI investigates such small oscillations. In this chapter, we treat the one-dimensional case, allowing for dissipative and external forces. In the following chapter, we study the eigenoscillations of a system with many degrees of freedom.

Equation of Motion

We consider a system with one degree of freedom described by the generalized coordinate q. Let the Lagrangian function be of the form

$$\mathcal{L}(q, \dot{q}, t) = \mathcal{L}_0 - U_\mathrm{e} = \frac{a(q)}{2}\, \dot{q}^2 - U(q) - U_\mathrm{e}(q, t) \tag{24.1}$$

Here, U_e is an external perturbation, such as an external electromagnetic field for a molecule. Let the undisturbed system with $\mathcal{L}_0$ be time-independent. The form of the kinetic energy results from (9.19).

The undisturbed system shall have a stable equilibrium position at $q = q_0$. We expand the potential $U(q)$ around this point:

$$U(q) = U(q_0) + \left(\frac{dU}{dq}\right)_0 (q - q_0) + \frac{1}{2}\left(\frac{d^2U}{dq^2}\right)_0 (q - q_0)^2 + \cdots \tag{24.2}$$

The subscript zero means that the derivative is taken at the point q_0. We assume a stable equilibrium position, i.e. $(dU/dq)_0 = 0$ and $k = (d^2U/dq^2)_0 > 0$. The displacements

$$x = q - q_0 \tag{24.3}$$

from the rest position should be so small that the expansion (24.2) can be terminated at the quadratic term:

$$U(q) \approx U(q_0) + \frac{1}{2}\left(\frac{d^2U}{dq^2}\right)_0 (q-q_0)^2 = U(q_0) + \frac{k}{2}\,x^2 \tag{24.4}$$

The kinetic energy is also calculated up to the quadratic order in $x = q - q_0$:

$$T_{\text{kin}}(q,\dot{q}) = \frac{1}{2}\left[a(q_0) + \left(\frac{da}{dq}\right)_0 x + \cdots\right]\dot{x}^2 \approx \frac{a(q_0)}{2}\,\dot{x}^2 = \frac{m}{2}\,\dot{x}^2 \tag{24.5}$$

For an oscillation $x \sim A\cos(\omega t)$, x as well as $\dot{x}$ are proportional to the amplitude A. The neglect of the term $x\,\dot{x}^2$ is therefore consistent with the restriction to x^2 in (24.4). In the last step, we abbreviated $a(q_0)$ by the "mass" m. This can be the usual mass or a mass parameter, for example, $a(q_0) = ml^2$ for a plane pendulum with $q = \varphi$.

We expand also the external potential $U_{\text{e}}(q,t)$ around the equilibrium point,

$$U_{\text{e}}(q,t) = U_{\text{e}}(q_0,t) + \left(\frac{\partial U_{\text{e}}}{\partial q}\right)_{q_0} (q-q_0) + \cdots = U_{\text{e}}(q_0,t) - F(t)\,x + \cdots \tag{24.6}$$

The term $U_{\text{e}}(q_0,t)$ depends only on time and can be omitted in the Lagrangian function (it could be written as a total time derivative of a function). The next term of the expansion results in the external force $F(t) = -\partial U_{\text{e}}/\partial x$. We assume that the quadratic and higher terms are small with respect to $x\,F(t)$ and $kx^2/2$; they are neglected in the following. For a molecule in an external electromagnetic field, this assumption is usually very well satisfied.

Thus, the Lagrange function (24.1) for *small* displacements becomes

$$\mathcal{L}(x, \dot{x}, t) = \frac{m}{2}\,\dot{x}^2 - \frac{k}{2}\,x^2 + F(t)\,x \tag{24.7}$$

This is the Lagrange function of the harmonic oscillator with an external force. This model may be applied, for example, to the oscillation of a diatomic molecule in an electromagnetic field.

From (24.7) follows the equation of motion

$$\ddot{x} + \omega_0^2\,x = f(t) \tag{24.8}$$

with $f(t) = F(t)/m$. We introduced the *natural frequency*, also called *eigenfrequency*, of the oscillator

$$\omega_0 = \sqrt{k/m} \tag{24.9}$$

As a generalization, we allow for a frictional force in (24.8):

$$\ddot{x} + 2\,\lambda\,\dot{x} + \omega_0^2\,x = f(t) \tag{24.10}$$

The frictional force can also be described by Rayleigh's dissipation function $D = \lambda m \dot{x}^2$ in the generalized Lagrangian equations (9.52).

General Solution

We determine the general solution of the differential equation (24.10), i.e. the solution of the damped harmonic oscillator. To do this we first consider a periodic force $f(t) = f_\omega \cos\omega t$:

$$\ddot{x} + 2\,\lambda\,\dot{x} + \omega_0^2\,x = f_\omega\,\cos(\omega t) \tag{24.11}$$

All quantities in this equation are real. Therefore, the real part (denoted by Re) of

$$\ddot{X} + 2\lambda\dot{X} + \omega_0^2\,X = f_\omega\,\exp(\mathrm{i}\omega t) \tag{24.12}$$

is identical to (24.11) if we set

$$x(t) = \mathrm{Re}\,\big(X(t)\big) \tag{24.13}$$

So we can solve (24.12) instead of (24.11), and take subsequently the real part of this solution. This procedure is common and somewhat simpler to handle. It is based on the fact that (24.11) is linear in x and that the coefficients are real.

The general solution of the differential equation (24.12) is of the form

$$X(t) = X_{\rm hom}(t) + X_{\rm part}(t) \tag{24.14}$$

where $X_{\rm part}(t)$ is a particular solution (i.e., any special solution), and $X_{\rm hom}(t)$ is the general solution of the homogeneous differential equation:

$$\ddot{X}_{\rm hom} + 2\lambda\dot{X}_{\rm hom} + \omega_0^2\, X_{\rm hom} = 0 \tag{24.15}$$

We insert

$$X_{\rm hom}(t) = C\,\exp\left(-\mathrm{i}\nu t\right) \tag{24.16}$$

with a complex amplitude C into (24.15) and obtain

$$-\nu^2 - 2\,\mathrm{i}\lambda\nu + \omega_0^2 = 0 \tag{24.17}$$

This equation has the solutions

$$\nu_{1,2} = \pm\sqrt{\omega_0^2 - \lambda^2} - \mathrm{i}\lambda = \pm w_0 - \mathrm{i}\lambda \tag{24.18}$$

Here, we introduced $w_0 = (\omega_0^2 - \lambda^2)^{1/2}$.

We look separately at the cases $\omega_0 > \lambda$, $\omega_0 < \lambda$, and $\omega_0 = \lambda$. We start with $\omega_0 > \lambda$. In this case, w_0 is real, and the solution $x(t)$ reads

$$x_{\rm hom}(t) = \mathrm{Re}\left(C\,\exp(-\lambda t \mp \mathrm{i}w_0 t)\right). \tag{24.19}$$

With real constants A_1 and A_2 in $C = A_2 + \mathrm{i}A_1$ this becomes

$$x_{\rm hom}(t) = A_1\,\sin(w_0 t)\,\exp(-\lambda t) + A_2\,\cos(w_0 t)\,\exp(-\lambda t) \tag{24.20}$$

The sign in $\sin(\pm w_0 t) = \pm\sin(w_0 t)$ is absorbed into the constant A_1. Instead of the constants A_1 and A_2, we can also use an amplitude A_0 and a phase δ_0:

$$x_{\rm hom}(t) = A_0\,\exp\left(-\lambda t\right)\,\cos\left(w_0 t + \delta_0\right)\quad(\lambda < \omega_0) \tag{24.21}$$

This solution describes a *damped oscillation.*

For $\lambda > \omega_0$, (24.18) has two imaginary solutions, $\nu_1 = -\mathrm{i}\,\lambda_1$ and $\nu_2 = -\mathrm{i}\,\lambda_2$, where λ_1 and λ_2 are both positive. This gives us an exponentially decreasing solution of the form

$$\begin{aligned} x_{\text{hom}}(t) &= A_0 \exp(-\lambda_1 t) + B_0 \exp(-\lambda_2 t) \\ \lambda_{1,2} &= \lambda \pm \sqrt{\lambda^2 - \omega_0^2} \qquad (\lambda > \omega_0) \end{aligned} \tag{24.22}$$

This case is denoted as an *overdamped* oscillation (in fact, there is no longer any oscillation). Between the damped oscillation for $\lambda < \omega_0$ and the overdamped one for $\lambda > \omega_0$ there is the limiting case of $\lambda = \omega_0$. For this *critically damped* oscillator, the general solution of the homogeneous equation reads

$$x_{\text{hom}}(t) = (A + Bt)\, \exp(-\lambda t) \qquad (\lambda = \omega_0) \tag{24.23}$$

In this case, the rest position $x = 0$ is approached particularly fast. Therefore, the damping of the pointer of a measuring instrument is preferably adjusted to this limit.

The solutions (24.20)–(24.23) contain two integration constants. Therefore, they constitute the *general homogeneous solution.*

It is easy to check that

$$X_{\text{part}}(t) = f_\omega\, \chi(\omega)\, \exp(\mathrm{i}\omega t) \tag{24.24}$$

solves (24.12) if

$$\chi(\omega) = \frac{1}{\omega_0^2 - \omega^2 + 2\mathrm{i}\lambda\omega} \tag{24.25}$$

Thus, we have found a *particular solution.* The function $\chi(\omega)$ is called *dynamic susceptibility*, It determines the dynamic (i.e. time-dependent) response of the system to the external force. The limit $\chi(0)$ is called static susceptibility.

The susceptibility is the ratio between the displacement $X_{\text{part}}(t)$ of the system and the driving force $f_\omega \exp(\mathrm{i}\omega t)$. The term susceptibility is generally used when a system is subjected to an external disturbance (like a force or a field). The relationship between the reaction or response of the system and the external disturbance is the susceptibility; the function χ is also called a response function. A common example is the polarization of matter when an electromagnetic field is applied. Thus, the electric susceptibility χ_e determines the polarization $\boldsymbol{P} = \chi_e \boldsymbol{E}$ in the electric field $\boldsymbol{E}$. For sufficiently small disturbances, the response of the system is generally proportional to the disturbance. The response function is then independent of strength of the perturbation. This means, for example, that χ_e does not depend on the applied electric field.

The general solution for an arbitrary force $f(t)$ is obtained by the decomposition of the force into periodic parts, i.e. by the Fourier transform

$$f(t) = \int_{-\infty}^{\infty} d\omega \; f_\omega \; \exp(\mathrm{i}\omega t), \quad f_\omega = \frac{1}{2\pi} \int_{-\infty}^{\infty} dt \; f(t) \; \exp(-\mathrm{i}\omega t) \tag{24.26}$$

Since the equation of motion is linear in $X(t)$, we obtain a particular solution as a superposition of the solutions (24.24):

$$x_{\text{part}}(t) = \mathrm{Re}\left(\int_{-\infty}^{\infty} d\omega \; f_\omega \; \chi(\omega) \; \exp(\mathrm{i}\omega t) \right) \tag{24.27}$$

For $\lambda < \omega_0$, the general solution of (24.10) is then

$$x(t) = \mathrm{Re}\left(C \; \exp(-\lambda t + \mathrm{i}\omega_0 t) + \int_{-\infty}^{\infty} d\omega \; f_\omega \; \chi(\omega) \; \exp(\mathrm{i}\omega t) \right) \tag{24.28}$$

This solution contains two integration constants, $\mathrm{Re}\, C$ and $\mathrm{Im}\, C$. For $\lambda \geq \omega_0$ the homogeneous part (first term on the r.h.s.) must be replaced by (24.22) or (24.23).

Forced Oscillation

For a periodic force $f(t) = f_\omega \exp(\mathrm{i}\omega t)$ and for $\lambda < \omega_0$ we discuss the solution in more detail. The complex-valued function $\chi(\omega)$ can be expressed by its magnitude $|\chi|$ and its phase δ

$$\chi(\omega) = \big|\chi(\omega)\big| \; \exp\big(\mathrm{i}\delta(\omega)\big). \tag{24.29}$$

where

$$\big|\chi(\omega)\big| = \frac{1}{\sqrt{\left(\omega_0^2 - \omega^2\right)^2 + 4\lambda^2\omega^2}}, \qquad \tan\delta(\omega) = \frac{2\lambda\omega}{\omega^2 - \omega_0^2} \tag{24.30}$$

For the periodic force, the general (real) solution is of the form

$$x(t) = A_0 \exp(-\lambda t) \cos(w_0 t + \delta_0) + A(\omega) \cos\big(\omega t + \delta(\omega)\big) \tag{24.31}$$

with

$$A(\omega) = f_\omega \, \big|\chi(\omega)\big| \tag{24.32}$$

The constants A_0 and δ_0 are determined by the initial conditions. For large times the solution becomes independent of the initial conditions:

$$x(t) = A(\omega) \cos\big(\omega t + \delta(\omega)\big) \quad (t \gg 1/\lambda) \tag{24.33}$$

This is a steady-state solution. It describes the *forced oscillation* due to the external harmonic force. The phase $\delta(\omega)$ and the amplitude $A(\omega)$ are given by (24.30). Both functions are sketched in Figure 24.1. The maximum of the amplitude $A(\omega)$ lies at the *resonance frequency*

$$\omega_{\rm res} = \sqrt{\omega_0^2 - 2\,\lambda^2} \tag{24.34}$$

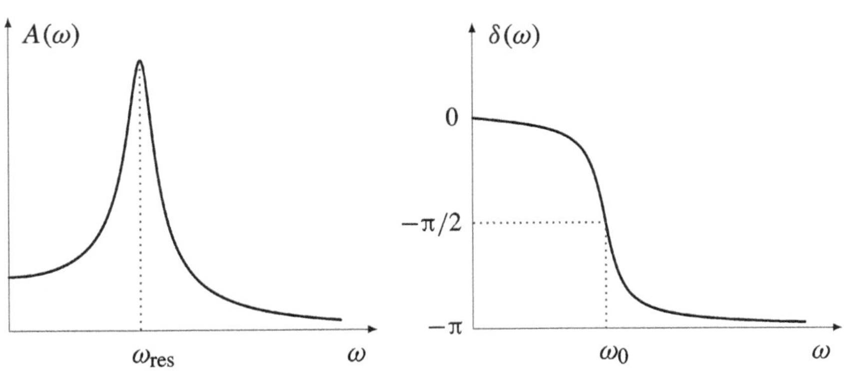

Figure 24.1 Amplitude $A(\omega)$ and phase $\delta(\omega)$ of the forced oscillation as a function of the frequency ω. The weaker the damping is, the higher and narrower becomes the resonance curve; for the illustration we used $\omega_0/\lambda = 10$. For small frequencies, the oscillator oscillates in phase with the exciting force ($\delta \approx 0$), for large frequencies it oscillates with opposite phase ($\delta \approx -\pi$); the transition occurs in the range $\omega \sim \omega_0 \pm \lambda$.

For $\lambda \ll \omega_0$, this resonance frequency lies slightly below the natural frequency (or eigenfrequency), $\omega_{\rm res} \approx \omega_0 - \lambda^2/\omega_0$. The width of the resonance curve is proportional to λ, and the height of the maximum is proportional to $1/\lambda$.

In the static case ($\omega = 0$), the displacement becomes $A(0) = f_0/\omega_0^2$. The displacement is in the direction of the force; the phase is zero.

From (24.30) follows $\delta \approx -2\lambda\omega/\omega_0^2$ for small ω. The small negative phase means that the solution $x(t) \propto \cos(\omega t + \delta)$ lags a little behind the driving force $f(t) \propto \cos(\omega t)$. The slower the force oscillates, the smaller is this phase shift. For $\omega = \omega_0$ we get $\delta = \pi/2$, the displacement $x(t) \propto \cos(\omega t + \pi/2) = -\sin(\omega t)$ is out of phase with the driving force $f(t) \propto \cos(\omega t)$. For $\omega \gg \omega_0$ the oscillation has a phase opposite to that of the driving force (counter phase). The transition between in phase, out of phase, and counter phase occurs in a range of the size λ around ω_0.

We consider the case of weak damping, i.e. $\lambda \ll \omega_0$. In the neighborhood of the maximum, $|\epsilon| \ll \omega_0$ with $\epsilon = \omega - \omega_0$, we neglect the terms $\mathcal{O}(\epsilon^2,\, \epsilon\lambda)$ and obtain from (24.25)

$$\chi(\epsilon) \approx -\frac{1}{2\,(\epsilon - \mathrm{i}\lambda)\,\omega_0} \qquad (\epsilon = \omega - \omega_0,\ |\epsilon| \ll \omega_0,\ \lambda \ll \omega_0) \tag{24.35}$$

For this, we get the magnitude and the phase

$$|\chi(\epsilon)| = \frac{1}{2\,\omega_0\,\sqrt{\epsilon^2 + \lambda^2}}, \qquad \tan\delta(\epsilon) = \frac{\lambda}{\epsilon} \tag{24.36}$$

Energy balance

The x-component of the friction force is

$$F_{\rm diss} = -2\,\lambda\, m\,\dot{x} \tag{24.37}$$

Via the friction force, the energy of the oscillator is transferred to other degrees of freedom; these other degrees of freedom are not treated explicitly. Eventually, this energy is dissipated into heat (disordered motion of the atoms). In the steady state, the absorbed power (from the driving force) is continuously fed into the system and dissipated via the friction force. We calculate this power P.

During a movement about dx, the (positive) energy $-F_{\rm diss}\, dx$ is absorbed. The absorbed power P is, therefore $-F_{\rm diss}\,\dot{x}$. Since $P \propto \dot{x}^2$ this power oscillates. We average P over an oscillation period:

$$\begin{aligned} P(\omega) &= -\,\overline{F_{\rm diss}\,\dot{x}} = 2\,\lambda\, m A(\omega)^2 \omega^2\ \overline{\sin^2(\omega t + \delta)} \\ &= \lambda\, m A(\omega)^2 \omega^2 = -\frac{m\,\omega\, f_\omega^2}{2}\ {\rm Im}\,\chi(\omega) \end{aligned} \qquad (24.38)$$

The averaging is indicated by the horizontal bar. The imaginary part of χ follows from (24.25).

The result $P = \lambda m\, A(\omega)^2 \omega^2$ is to be understood as follows: $1/\lambda$ is the time during which the amplitude of the oscillator is substantially damped, and $E = m\,\omega^2 A(\omega)^2/2$ is the energy of the oscillation with the amplitude $A(\omega)$.

The result of (24.38) means that the imaginary part of the susceptibility determines the absorbed power. From (24.35) we obtain ${\rm Im}\,\chi = -(\lambda/2\omega_0)/(\epsilon^2 + \lambda^2)$ and

$$P(\epsilon) = \frac{m\, f_\omega^2}{4}\ \frac{\lambda}{\epsilon^2 + \lambda^2} \qquad (\epsilon = \omega - \omega_0,\ |\epsilon| \ll \omega_0,\ \lambda \ll \omega_0) \qquad (24.39)$$

The energy absorption occurs predominantly in the frequency range $\omega \approx \omega_0 \pm \lambda$.

If the driving force contains a continuum of frequencies, then the steady state solution (24.33) has to be integrated over these frequencies:

$$x(t) = \int_{-\infty}^{\infty} d\omega\ A(\omega)\, \cos\big(\omega t + \delta(\omega)\big), \qquad (t \gg 1/\lambda) \qquad (24.40)$$

For this we obtain

$$\begin{aligned} \overline{\dot{x}^2} &= \int d\omega \int d\omega'\ \omega\,\omega'\ A(\omega)\, A(\omega')\ \overline{\sin(\omega t + \delta)\ \sin(\omega' t + \delta')} \\ &= \frac{1}{2}\int d\omega\ \omega^2\, A(\omega)^2 \end{aligned} \qquad (24.41)$$

Hereby, $\int_0^T dt\, \sin(\omega t)\sin(\omega' t)/T = \delta(\omega - \omega')/2$ was used. The total absorbed power is then

$$P = 2\lambda m\, \overline{\dot{x}^2} = \lambda m \int_{-\infty}^{\infty} d\omega\, \omega^2\, A(\omega)^2 \qquad (24.42)$$

We evaluate this for the case (24.39) of weak damping:

$$P \overset{(24.39)}{=} \int_{-\infty}^{\infty} d\epsilon\, \frac{m\, f_\omega^2}{4}\, \frac{\lambda}{\epsilon^2 + \lambda^2} \qquad (24.43)$$

The function f_ω shall change only weakly in the relevant range $\epsilon \sim \pm\lambda$ so that we can use $f_\omega \approx f_{\omega_0}$ in the integrand. The absorbed power is then independent of the friction coefficient λ

$$P = \frac{m\, f_{\omega_0}^2}{4} \int_{-\infty}^{\infty} d\epsilon\, \frac{\lambda}{\epsilon^2 + \lambda^2} = \frac{\pi}{4}\, m\, f_{\omega_0}^2 \qquad (24.44)$$

Exercises

24.1 *Impulse on an oscillator*

A damped oscillator has the equation of motion $\ddot{x} + 2\lambda\dot{x} + \omega_0^2\, x = f(t)$ with $\lambda < \omega_0$. The oscillator rests in its equilibrium position. Then it gets an impulse which is described by the force

$$f(t) = \begin{cases} v_0/T & 0 \le t \le T \\ 0 & \text{other} \end{cases}$$

Determine the displacement $x(t)$. Discuss the limiting cases $T \to 0$ and $T \gg 1/\lambda$, and sketch the corresponding solutions.

Chapter 25

System with Many Degrees of Freedom

We study small oscillations of a system with many degrees of freedom. Starting from a general form of the Lagrange function we determine the natural frequencies and oscillations of the system. Finally, we introduce normal coordinates; these are the amplitudes of the eigenoscillations (the oscillations belonging to specific natural frequencies).

Natural Frequencies

The system under consideration has f degrees of freedom given by the generalized coordinates $q_1, \ldots, q_f$. Let the Lagrange function be of the form

$$\begin{aligned}\mathcal{L}_0 &= \mathcal{L}_0(q_1, \ldots, q_f, \dot{q}_1, \ldots, \dot{q}_f)\\ &= \frac{1}{2} \sum_{i,j=1}^{f} m_{ij}(q_1, \ldots, q_f)\, \dot{q}_i\, \dot{q}_j - U(q_1, \ldots, q_f) \qquad (25.1)\end{aligned}$$

An external perturbation potential U_{e} will be allowed later. At $q_1^0, \ldots, q_f^0$ the system shall have a stable equilibrium position. We expand the potential energy U around this position:

$$\begin{aligned}U(q_1, \ldots, q_f) &= U(q_1^0, \ldots, q_f^0) + \sum_{i=1}^{f} \left(\frac{\partial U}{\partial q_i}\right)_0 (q_i - q_i^0)\\ &\quad + \frac{1}{2} \sum_{i,j=1}^{f} \left(\frac{\partial^2 U}{\partial q_i\, \partial q_j}\right)_0 (q_i - q_i^0)\,(q_j - q_j^0) + \cdots\end{aligned} \qquad (25.2)$$

The index zero means that the quantity is evaluated at the equilibrium position. The linear term can be omitted because at the equilibrium position the generalized forces $Q_i = -\partial U/\partial q_i$ vanish. For small displacements

$$x_i = q_i - q_i^0 \tag{25.3}$$

from the equilibrium position, we terminate the expansion with the quadratic term. Furthermore, we discard the constant term (no influence on equations of motion) and obtain

$$U \approx \frac{1}{2}\sum_{i,j=1}^{f} V_{ij}\, x_i\, x_j \quad \text{with } V_{ij} = V_{ji} = \left(\frac{\partial^2 U}{\partial q_i\, \partial q_j}\right)_0 \tag{25.4}$$

In the kinetic energy, the lowest term is quadratic in the displacement,

$$T_{\text{kin}}(q,\dot q) = \frac{1}{2}\sum_{i,j=1}^{f} m_{ij}(q_1,\ldots,q_f)\,\dot q_i\,\dot q_j \approx \frac{1}{2}\sum_{i,j=1}^{f} m_{ij}(q_1^0,\ldots,q_f^0)\,\dot q_i\,\dot q_j \tag{25.5}$$

In this, we put $\dot q_i = \dot x_i$ and denote the coefficients by T_{ij},

$$T_{\text{kin}} \approx \frac{1}{2}\sum_{i,j=1}^{f} T_{ij}\,\dot x_i\,\dot x_j \quad \text{with } T_{ij} = T_{ji} = m_{ij}(q_1^0,\ldots,q_f^0) \tag{25.6}$$

The real coefficients T_{ij} and V_{ij} are symmetric under an interchange of the indices. With the approximations (25.4) and (25.6), the Lagrange function becomes

$$\mathcal{L}_0 = \mathcal{L}_0(x_1,\ldots,x_f,\dot x_1,\ldots,\dot x_f) = \frac{1}{2}\sum_{i,j=1}^{f}\left(T_{ij}\,\dot x_i\,\dot x_j - V_{ij}\,x_i\,x_j\right) \tag{25.7}$$

The Lagrange equations are

$$\sum_{j=1}^{f} T_{ij}\,\ddot x_j = -\sum_{j=1}^{f} V_{ij}\,x_j \tag{25.8}$$

We introduce the symmetric matrices T and V,

$$T = \begin{pmatrix} T_{11} & T_{12} & \dots & T_{1f} \\ T_{21} & T_{22} & \dots & T_{2f} \\ \vdots & \vdots & \ddots & \vdots \\ T_{f1} & T_{f2} & \dots & T_{ff} \end{pmatrix}, \quad V = \begin{pmatrix} V_{11} & V_{12} & \dots & V_{1f} \\ V_{21} & V_{22} & \dots & V_{2f} \\ \vdots & \vdots & \ddots & \vdots \\ V_{f1} & V_{f2} & \dots & V_{ff} \end{pmatrix} \tag{25.9}$$

and the column vector x:

$$x = \begin{pmatrix} x_1 \\ x_2 \\ \vdots \\ x_f \end{pmatrix}, \qquad x^{\mathrm{T}} = (x_1, x_2, \dots, x_f) \tag{25.10}$$

The exponent T denotes the transposed matrix; it differs by the typeface (roman instead of italic) from the symbol T for the matrix (T_{ij}). The equations of motion (25.8) can now be written in the matrix form:

$$T\,\ddot{x} + V\,x = 0 \tag{25.11}$$

At the equilibrium point $x_i = 0$ we have $U = \sum_{i,j} V_{ij}\, x_i\, x_j = 0$. The equilibrium is stable, if for any displacements $U > 0$ holds. Furthermore, the kinetic energy (25.6) is positive for arbitrary velocities $\dot{x}_i$. This means that all eigenvalues of the matrices V and T are positive:

$$\begin{aligned} &\sum_{i,j=1}^{f} V_{ij}\, x_i\, x_j > 0 \ \text{ for } x \neq 0 \ \longleftrightarrow \ \text{eigenvalues of } V \text{ are positive} \\ &\sum_{i,j=1}^{f} T_{ij}\, \dot{x}_i\, \dot{x}_j > 0 \ \text{ for } \dot{x} \neq 0 \ \longleftrightarrow \ \text{eigenvalues of } T \text{ are positive} \end{aligned} \tag{25.12}$$

The symmetric matrices $V = (V_{ij})$ and $T = (T_{ij})$ can be diagonalized by an orthogonal transformation. This diagonalization leads to the corresponding eigenvalues.

Equations (25.8) or (25.11) are a system of f linear, homogeneous differential equations of second order with constant coefficients. The solutions are therefore of the form

$$x_j(t) = A_j \exp(-\mathrm{i}\omega t) \quad \text{or} \quad x(t) = A \exp(-\mathrm{i}\omega t) \tag{25.13}$$

Here, A is a column vector containing f constants $A_1, \ldots, A_f$. All quantities in the equation of motion (25.11) are real. It is, however, somewhat more convenient to use the complex form (25.13). Since (25.11) is linear in x and has real coefficients, with $x(t)$ also $\operatorname{Re} x(t)$ is a solution. We, therefore, start with (25.13) and turn only later to the real solution by

$$x(t) \longrightarrow \operatorname{Re} x(t) \tag{25.14}$$

We substitute (25.13) into (25.8) or (25.11):

$$\sum_{j=1}^{f} \left(V_{ij} - \omega^2 T_{ij}\right) A_j = 0 \quad \text{or} \quad \left(V - \omega^2 T\right) A = 0 \tag{25.15}$$

This is a linear, homogeneous system of equations for the amplitudes $A_1, \ldots, A_f$. If the coefficient determinant does not vanish, then the solution is unique; the trivial solution $A = 0$ would then be the only solution. For a non-trivial solution to exist, the determinant must vanish:

$$\det\left(V - \omega^2 T\right) = P^{(f)}(\omega^2) = 0. \tag{25.16}$$

This determinant is a polynomial $P^{(f)}$ of degree f in ω^2 with real coefficients. Such a polynomial has f zeros:

$$P^{(f)}(\omega^2) = 0 \longrightarrow \omega_1^2, \omega_2^2, \ldots, \omega_f^2 \tag{25.17}$$

The zeros of a polynomial are generally complex. If an ω_k^2 is complex or negative real, this would lead to an ω_k with a positive imaginary part. The corresponding solution $\exp(-\mathrm{i}\omega_k t)$ would increase exponentially. But this would be in contradiction to the stability assumption $\sum V_{ij}\, x_i\, x_j > 0$. Therefore, all ω_k^2 are real and nonnegative:

$$\left(\omega_k^2\right)^* = \omega_k^2, \quad \omega_k^2 \geq 0 \quad (k = 1, \ldots, f) \tag{25.18}$$

We have argued physically using the presupposed stability. Mathematically $\omega_k^2 > 0$ follows from (25.16) and the positivity (25.12) of the eigenvalues of V and T.

According to (25.16), the limiting case $\omega_k = 0$ implies $\det V = 0$. Then at least one eigenvalue of V is zero. We admit this limiting case in (25.18) and weaken in this respect the condition (25.12). For $\omega = 0$ the oscillation equation $\ddot{x} + \omega^2 x = 0$ becomes $\ddot{x} = 0$. Instead of $x = \mathrm{Re}\, C \exp(-\mathrm{i}\omega t)$, we then get the solution

$$x(t) = C_1 + C_2\, t \qquad \text{for } \omega = 0 \tag{25.19}$$

For $\omega_k = 0$, the potential does not provide a restoring force, and we obtain a uniform motion. A physical application to this will be discussed in the section "molecular oscillations" in the following chapter.

For every positive ω_k^2, there is a positive and a negative ω_k. Since

$$\mathrm{Re}\left[(A + \mathrm{i}B)\exp(\pm\mathrm{i}\omega t)\right] = A\,\cos(\omega t) \mp B\sin(\omega t) \tag{25.20}$$

it is sufficient to consider one of these signs. We can therefore restrict the *natural frequencies* (or, synonymously *eigenfrequencies*) ω_k by

$$\omega_k \geq 0 \quad (k = 1, \ldots, f) \tag{25.21}$$

The corresponding *eigenvectors* $A^{(k)}$ are obtained from (25.15):

$$\left(V - \omega_k^2\, T\right) A^{(k)} = 0 \quad (k = 1, \ldots, f) \tag{25.22}$$

In the next section, we will show that the matrices T and V can be diagonalized with the help of the column vectors $A^{(k)}$. These vectors are therefore also called eigenvectors of the corresponding matrices.

The condition (25.22) is a system of f linear equations for the f unknown constants $A_1^{(k)}, \ldots, A_f^{(k)}$. Due to (25.16), the column vector $A^{(k)}$ is not uniquely determined; rather we can freely choose one component. After such a choice, (25.22) is a linear inhomogeneous system of equations for the other components. For each of the f eigenvalues ω_k we then obtain an eigenvector $A^{(k)}$. If r eigenvalues ω_k coincide, the system of Equations (25.22) has rank $f - r$. Then we can freely choose r components of $A^{(k)}$ and obtain the corresponding

r independent solutions. Altogether this always gives f eigenvectors

$$A^{(k)} = \begin{pmatrix} A_1^{(k)} \\ A_2^{(k)} \\ \vdots \\ A_f^{(k)} \end{pmatrix} \qquad (k = 1, \ldots, f) \tag{25.23}$$

If we choose the free components to be real, then (25.22) is a linear, inhomogeneous system of equations with real coefficients, and all components of the eigenvectors $A^{(k)}$ are real. In the solution, we will use the amplitudes in the form $C_k A^{(k)}$ with complex numbers C_k. Due to the factors C_k, we can use an arbitrary normalization of the eigenvectors $A^{(k)}$. Then (25.13) leads to

$$x(t) = A^{(k)} \, \mathrm{Re}\left(C_k \exp\left(-\mathrm{i}\omega_k t\right)\right) \quad (k = 1, \ldots, f) \tag{25.24}$$

The complex number C_k contains two real constants which become the amplitudes of $\sin(\omega_k t)$ and $\cos(\omega_k t)$, if we take the real part.

The solution (25.24) is a special solution of the equation of motion (25.11), which is called *normal mode* or *eigenmode*, or also *natural oscillation*. It describes a periodic motion with the natural frequency ω_k. Since the equations of motion (25.8) or (25.11) are linear, an arbitrary superposition of natural oscillations is also a solution. Thus, we obtain the *general solution*

$$x(t) = \sum_{k=1}^{f} A^{(k)} \, B_k \cos(\omega_k t + \alpha_k) \quad \text{general solution} \tag{25.25}$$

Here, the f complex amplitudes C_k have been replaced by the f amplitudes B_k and by f real phases α_k. The solution contains $2f$ integration constants (B_k, α_k); it is, therefore, the general solution of the f differential equations of second order (25.8).

Normal Coordinates

By a transformation

$$x_i = x_i(Q_1, Q_2, \ldots, Q_f) \quad (i = 1, \ldots, f) \tag{25.26}$$

one can go from the f coordinates $x_i = q_i - q_i^0$ to f other coordinates Q_j. In the following, we introduce the *normal coordinates*, which are equal to the amplitudes of the normal modes (or eigen oscillations).

The procedure for determining the normal coordinates Q_i corresponds to the principal axis transformation of the inertia tensor; this connection becomes particularly clear in the first example of Chapter 26. From the eigenvectors $A^{(k)}$ we form the square matrix a:

$$a = (a_{ik}) = \left(A_i^{(k)}\right) = \begin{pmatrix} A_1^{(1)} & A_1^{(2)} & \dots & A_1^{(f)} \\ A_2^{(1)} & A_2^{(2)} & \dots & A_2^{(f)} \\ \vdots & \vdots & \ddots & \vdots \\ A_f^{(1)} & A_f^{(2)} & \dots & A_f^{(f)} \end{pmatrix} \tag{25.27}$$

With this notation, (25.22) becomes

$$\sum_{j=1}^{f} V_{ij}\, a_{jk} = \omega_k^2 \sum_{j=1}^{f} T_{ij}\, a_{jk} \tag{25.28}$$

We write down the corresponding equation for ω_l, rename the indices $(i \leftrightarrow j)$, and interchange the indices of the symmetric matrices T and V:

$$\sum_{i=1}^{f} V_{ij}\, a_{il} = \omega_l^2 \sum_{i=1}^{f} T_{ij}\, a_{il} \tag{25.29}$$

We multiply (25.28) by a_{il} and (25.29) by a_{jk}, and sum over i and j, respectively. The difference between the two equations is then

$$\left(\omega_k^2 - \omega_l^2\right) \sum_{i,j=1}^{f} T_{ij}\, a_{il}\, a_{jk} = 0 \tag{25.30}$$

We first assume that the eigenvalues are not degenerate, i.e. $\omega_l \neq \omega_k$ for $l \neq k$. For $l \neq k$ the sum in (25.30) must vanish. For $l = k$ we normalize the $A^{(k)}$ such that the sum in (25.30) equals 1. This yields

$$\sum_{i,j=1}^{f} T_{ij}\, a_{il}\, a_{jk} = A^{(l)\,\mathrm{T}}\, T\, A^{(k)} = \delta_{lk} \quad \text{or} \quad a^{\mathrm{T}} T a = 1 \tag{25.31}$$

This is a *generalized orthogonality relation* for the eigenvectors $A^{(k)}$. For $T_{ij} = \delta_{ij}$, (25.28)–(25.31) becomes the usual proof of the orthogonality of the eigenvectors of a symmetric matrix.

We now consider the case where two eigenvalues coincide, so $\omega_k = \omega_l$ for $k \neq l$. Then $\alpha A^{(k)} + \beta A^{(l)}$ is also eigenvector of (25.22). In this case, two linear combinations can be chosen such that (25.31) holds.

The eigenvectors $A^{(k)}$ (normalized according to (25.31)) diagonalize not only T, but also V. To see this, we write (25.28) in matrix form

$$V a = T a\, \lambda \tag{25.32}$$

with the diagonal matrix

$$\lambda = (\omega_k^2\, \delta_{kl}) = \begin{pmatrix} \omega_1^2 & 0 & \dots & 0 \\ 0 & \omega_2^2 & \dots & 0 \\ \vdots & \vdots & \ddots & \vdots \\ 0 & 0 & \dots & \omega_f^2 \end{pmatrix} \tag{25.33}$$

We multiply (25.32) from the left by a^{T} and use $a^{\mathrm{T}} T a = 1$:

$$a^{\mathrm{T}} V a = \lambda \quad \text{or} \quad \sum_{i,j=1}^{f} V_{ij}\, a_{il}\, a_{jk} = \omega_k^2\, \delta_{kl} \tag{25.34}$$

Thus, T and V are both diagonalized by the matrix $a = \big(A_i^{(k)}\big)$. Therefore, the following transformation leads to the wanted normal coordinates Q_j,

$$x_i(t) = x_i(Q_1, \dots, Q_f) = \sum_{j=1}^{f} a_{ij}\, Q_j(t) \quad \text{or} \quad x = a\, Q \tag{25.35}$$

We insert this transformation into the Lagrange function (25.7):

$$\begin{aligned} 2\,\mathcal{L}_0 &= \sum_{i,j=1}^{f} \big(T_{ij}\, \dot{x}_i\, \dot{x}_j - V_{ij}\, x_i\, x_j\big) = \dot{x}^{\mathrm{T}}\, T\, \dot{x} - x^{\mathrm{T}}\, V\, x \\ &= (a\,\dot{Q})^{\mathrm{T}}\, T\, (a\,\dot{Q}) - (aQ)^{\mathrm{T}}\, V\, (aQ) = \dot{Q}^{\mathrm{T}} \dot{Q} - Q^{\mathrm{T}} \lambda\, Q \\ &= \sum_{k=1}^{f} \Big(\dot{Q}_k^2 - \omega_k^2\, Q_k^2\Big) \end{aligned} \tag{25.36}$$

This Lagrange function leads to the decoupled equations of motion:

$$\ddot{Q}_k + \omega_k^2\, Q_k = 0 \quad (k = 1, \ldots, f) \tag{25.37}$$

If we insert the solutions

$$Q_k(t) = \mathrm{Re}\big(C_k\, \exp(-\mathrm{i}\omega_k t)\big) = B_k\, \cos(\omega_k t + \alpha_k) \tag{25.38}$$

into (25.35), we regain the general solution (25.25) for $x_i(t)$. In the limiting case $\omega_k = 0$, we obtain from (25.37) the solution

$$Q_k(t) = A + B\,t \quad (\omega_k = 0) \tag{25.39}$$

External forces

We add an external perturbation potential U_e in (25.1):

$$\mathcal{L} = \mathcal{L}_0 - U_\mathrm{e}(q_1, \ldots, q_f, t) \tag{25.40}$$

We expand this potential around the equilibrium position of the undisturbed system

$$U_\mathrm{e}(q_1, \ldots, q_f, t) = U_\mathrm{e}(q_1^0, \ldots, q_f^0, t) - \sum_{i=1}^{f} f_i(t)\, x_i + \cdots \tag{25.41}$$

The first term is a function of time only and therefore can be omitted in the Lagrange function. We assume that U_e is so weak that the expansion can be limited to the linear term. In this term, $f_i(t) = -(\partial U_\mathrm{e}/\partial q_i)_0$ is the force acting on the coordinate x_i. Thus, (25.40) becomes

$$\mathcal{L} = \mathcal{L}_0 + \sum_{i=1}^{f} f_i(t)\, x_i = \sum_{k=1}^{f} \left(\frac{1}{2}\,\dot{Q}_k^2 - \frac{\omega_k^2}{2}\, Q_k^2 + F_k(t)\, Q_k\right) \tag{25.42}$$

Here, we used (25.36) and (25.35), and introduced the generalized force

$$F_k(t) = \sum_{i=1}^{f} a_{ik}\, f_i(t) \tag{25.43}$$

The equations of motion for (25.42) lead to the solution

$$\ddot{Q}_k(t) + \omega_k^2\, Q_k(t) = F_k(t) \quad (k = 1, \ldots, f) \tag{25.44}$$

At this point, frictional forces could also be admitted. Each of the f equations of motion describes a forced oscillation, as treated in the previous chapter.

Exercises

25.1 *Transformation to normal coordinates*

The displacements x_i for small oscillations can be expressed by the normal coordinates Q_k:

$$x_i = x_i(Q_1, \ldots, Q_f) = \sum_{j=1}^{f} a_{ij}\, Q_j \tag{25.45}$$

What is the inverse transformation $Q_k = Q_k(x_1, \ldots, x_f)$?

Chapter 26

Applications

For three systems with several degrees of freedom, we determine the natural frequencies and oscillations. The first example (two-dimensional motion) illustrates the relation between our solution method and the principal axis transformation. The second example (coupled pendulum) is so simple, that we can immediately introduce normal coordinates. In the third example (triatomic molecule), we carry out the procedure given in the previous chapter step by step.

Two-dimensional Motion

First, we consider the two-dimensional motion of a mass point in an arbitrary potential $U(x, y)$. The Lagrangian function of this system is

$$\mathcal{L} = \frac{m}{2}\left(\dot{x}^2 + \dot{y}^2\right) - U(x, y) \tag{26.1}$$

Let the potential have a minimum at $(x_0,\, y_0)$. Then the expansion of $U(x, y)$ in powers of $x_1 = x - x_0$ and $x_2 = y - y_0$ reads

$$\begin{aligned} U &= U_0 + \frac{1}{2}\left(\frac{\partial^2 U}{\partial x^2}\right)_0 x_1^2 + \frac{1}{2}\left(\frac{\partial^2 U}{\partial y^2}\right)_0 x_2^2 + \left(\frac{\partial^2 U}{\partial x\, \partial y}\right)_0 x_1\, x_2 + \cdots \\ &= U_0 + \frac{1}{2}\left(V_{11}\, x_1^2 + V_{22}\, x_2^2 + 2\, V_{12}\, x_1\, x_2\right) + \cdots \end{aligned} \tag{26.2}$$

The index 0 means that the equilibrium positions x_0, y_0 have to be inserted for the arguments. The linear terms vanished because we presupposed an equilibrium at x_0, y_0. For small oscillations, we stop the expansion at the quadratic term. The Lagrange function becomes then

$$\mathcal{L} = \frac{1}{2} \sum_{i,j=1}^{2} \left(T_{ij}\, \dot{x}_i\, \dot{x}_j - V_{ij}\, x_i\, x_j \right) \quad \text{with } T_{ij} = m\, \delta_{ij} \tag{26.3}$$

In the approximation (26.2), the equipotential lines are given by ellipses in the x_1–x_2-plane:

$$U \approx U_0 + \frac{1}{2} \sum_{i,j=1}^{2} V_{ij}\, x_i\, x_j = \text{const.} \tag{26.4}$$

We now introduce a system rotated by the angle φ with the coordinates ξ_1 and ξ_2. The well-known orthogonal transformation to the new coordinates reads

$$\begin{pmatrix} \xi_1 \\ \xi_2 \end{pmatrix} = \begin{pmatrix} \cos\varphi & \sin\varphi \\ -\sin\varphi & \cos\varphi \end{pmatrix} \begin{pmatrix} x_1 \\ x_2 \end{pmatrix} \quad \text{or} \quad \xi = \alpha\, x \tag{26.5}$$

We denote the transformation matrix by α, the column vectors with ξ and x. Since $\alpha^{\mathrm{T}}\alpha = 1$, the inverse transformation reads

$$x = \alpha^{\mathrm{T}}\, \xi \tag{26.6}$$

We put this into (26.4):

$$\sum_{i,j=1}^{2} V_{ij}\, x_i\, x_j = x^{\mathrm{T}} V x = (\alpha^{\mathrm{T}} \xi)^{\mathrm{T}}\, V\, \alpha^{\mathrm{T}} \xi = \xi^{\mathrm{T}}\, (\alpha\, V \alpha^{\mathrm{T}})\, \xi = \xi^{\mathrm{T}}\, V'\, \xi \tag{26.7}$$

We now choose φ in (26.5) such that the nondiagonal element $V'_{12} = V'_{21}$ vanishes, i.e.

$$V' = \alpha V \alpha^{\mathrm{T}} = \begin{pmatrix} k_1 & 0 \\ 0 & k_2 \end{pmatrix} \tag{26.8}$$

We write the potential (26.4) with the new coordinates:

$$\sum_{i,j=1}^{2} V'_{ij}\, \xi_i\, \xi_j = k_1\, \xi_1^2 + k_2\, \xi_2^2 = \text{const.} \quad (k_1 > 0,\ k_2 > 0) \tag{26.9}$$

The stability of the equilibrium position implies that (26.4) and (26.9) are positive for arbitrary displacements. Thus, the eigenvalues

k_1 and k_2 of the matrix V are positive and (26.4) describes ellipses. The principal axes of these ellipses are the ξ_1- and the ξ_2-axis.

The kinetic energy matrix $T = (T_{ij})$ in (26.1) is proportional to the identity matrix. Therefore, T does not change by the orthogonal transformation (26.5), $T' = \alpha\, T\, \alpha^{\mathrm{T}} = T$. In the rotated coordinate system, (26.3) becomes

$$\mathcal{L}(\xi, \dot{\xi}) = \frac{m}{2}\left(\dot{\xi}_1^{\,2} + \dot{\xi}_2^{\,2}\right) - \frac{k_1}{2}\,\xi_1^{\,2} - \frac{k_2}{2}\,\xi_2^{\,2} \qquad (26.10)$$

The trivial transformation

$$Q_i = \sqrt{m}\;\xi_i(t) \qquad (26.11)$$

then leads to normal coordinates:

$$\mathcal{L} = \frac{1}{2}\sum_{i=1}^{2}\left(\dot{Q}_i^{\,2} - \omega_i^{\,2}\,Q_i^{\,2}\right) \quad \text{with } \omega_i = \sqrt{\frac{k_i}{m}} \qquad (26.12)$$

The eigenmodes (or normal modes) are oscillations in the direction of the principal axes of the ellipses.

We have deviated here somewhat from the general solution procedure of the previous chapter. According to that procedure, the eigenvalues $\omega_i^{\,2} = k_i/m$ are determined from the condition $\det\,(V - \omega^2\,T) = \det\,(V - m\,\omega^2\,I) = 0$. Then $V A^{(k)} = m\,\omega_k^{\,2} A^{(k)}$ yields the column vectors $A^{(k)}$. The eigenvectors $A^{(k)}$ describe the natural oscillations (along the principal axes of the ellipses). The treatment chosen here illustrates the connection to the principal axis transformation.

Double Pendulum

We consider the double pendulum sketched in Figure 26.1. For small displacements, the gravitational energy of each of the two pendulums is of the form $mgz = mgl\,(1 - \cos\varphi) \approx mgl\,\varphi^2/2$. The distance between the two masses changes for small displacements by $x = l\,(\varphi_2 - \varphi_1)$. The spring shall be relaxed for $\varphi_1 = \varphi_2 = 0$; then its potential energy is equal to $k\,x^2/2$. The Lagrange function for small

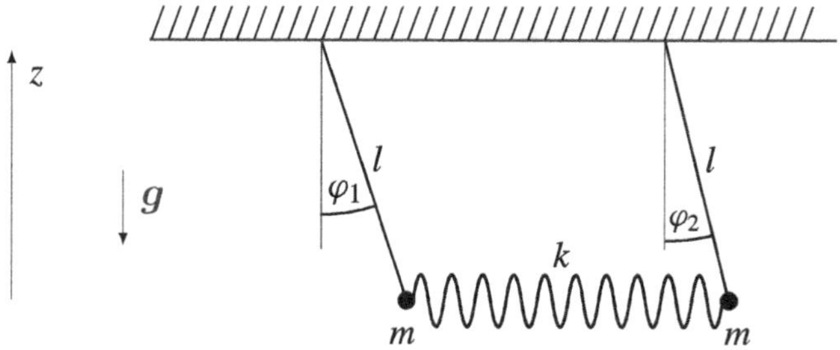

Figure 26.1 Coupled pendulums in the gravitational field. Each pendulum has the length l and the mass m. Let the two masses be coupled by a (massless) spring with the spring constant k. The spring is relaxed when both pendulums are at their lowest point. The equilibrium position is therefore given by $\varphi_1 = \varphi_2 = 0$.

oscillations is

$$\mathcal{L} = \frac{ml^2}{2}\left(\dot{\varphi}_1^2 + \dot{\varphi}_2^2\right) - \frac{mgl}{2}\left(\varphi_1^2 + \varphi_2^2\right) - \frac{kl^2}{2}\left(\varphi_1 - \varphi_2\right)^2 \quad (26.13)$$

Due to the quadratic form and the dependence of $(\varphi_1 - \varphi_2)$, it is plausible to use the new coordinates

$$\theta_1 = \frac{\varphi_1 + \varphi_2}{2}, \quad \theta_2 = \varphi_1 - \varphi_2 \quad (26.14)$$

We insert the reverse transformation

$$\varphi_1 = \theta_1 + \frac{\theta_2}{2}, \quad \varphi_2 = \theta_1 - \frac{\theta_2}{2} \quad (26.15)$$

into the Lagrange function and get

$$\mathcal{L} = ml^2\,\dot{\theta}_1^2 + \frac{ml^2}{4}\,\dot{\theta}_2^2 - mgl\,\theta_1^2 - \left(kl^2 + \frac{mgl}{2}\right)\frac{\theta_2^2}{2} \quad (26.16)$$

Then the trivial transformation

$$Q_1 = \sqrt{2ml^2}\,\theta_1 \quad \text{and} \quad Q_2 = \sqrt{\frac{ml^2}{2}}\,\theta_2 \quad (26.17)$$

leads to

$$\mathcal{L} = \frac{1}{2}\sum_{i=1}^{2}\left(\dot{Q}_i^2 - \omega_i^2\,Q_i^2\right) \quad (26.18)$$

This shows that the $Q_i(t)$ are the normal coordinates. The $Q_i(t) \propto \theta_i(t)$ describe the natural oscillations with the frequencies

$$\omega_1 = \sqrt{\frac{g}{l}}, \quad \omega_2 = \sqrt{\frac{2k}{m} + \frac{g}{l}} \tag{26.19}$$

The first natural oscillation is

$$\theta_1(t) = A\,\cos(\omega_1 t + \alpha), \quad \theta_2 = 0 \longrightarrow \varphi_1(t) = \varphi_2(t) = A\,\cos(\omega_1 t + \alpha) \tag{26.20}$$

In this case, both pendulums swing in phase. The spring remains relaxed and has no influence on the frequency.

For the other natural oscillation

$$\begin{aligned} \theta_2(t) &= B\,\cos(\omega_2 t + \beta), \\ \theta_1 = 0 \longrightarrow \varphi_1(t) &= -\varphi_2(t) = \frac{B}{2}\,\cos(\omega_2 t + \beta) \end{aligned} \tag{26.21}$$

the pendulums oscillate with opposite phases. The restoring forces of the spring and the gravity field add up in ω_2. The general solution is a superposition of the two solutions (26.20) and (26.21).

For weak coupling, $k \ll mg/l$, both frequencies are close to each other. In this case, the two pendulums initially seem to oscillate independently of each other. The exchange of oscillation energy between the pendulums requires a relatively long time $t \sim 1/(\omega_2 - \omega_1) \gg (l/g)^{1/2}$. If, for example, the second pendulum rests at the beginning, then it takes over the oscillation energy of the first pendulum after this time.

Triatomic Molecule

For the example of a triatomic molecule, we will follow step by step the solution procedure described in the previous chapter. The three atoms of the molecule shall only be able to move along a straight line (top of Figure 26.2). Their respective positions are denoted by the coordinate values y_1, y_2, and y_3. Let the equilibrium distance between neighboring atoms be b, so

$$y_2^0 - y_1^0 = b \quad y_3^0 - y_2^0 = b \tag{26.22}$$

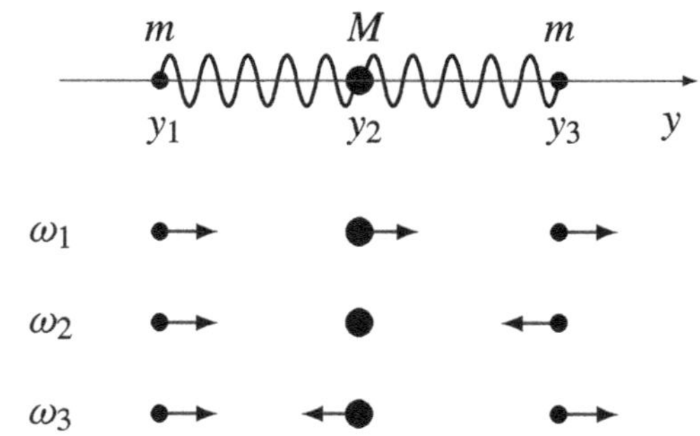

Figure 26.2 A triatomic molecule (top) with two equal edge atoms can perform one-dimensional oscillations. The displacements for the three resulting eigenmodes (or natural oscillations) are sketched below. The first solution describes the uniform translation of the entire molecule.

Thus, only two of the three quantities y_1^0, y_2^0, and y_3^0 are given which we generally need for the expansion (25.2). However, the potential is a priori quadratic in the displacements $x_i = y_i - y_i^0$:

$$V(y_1, y_2, y_3) = \frac{k}{2}[(y_2 - y_1 - b)^2 + (y_3 - y_2 - b)^2]$$
$$= \frac{k}{2}[(x_2 - x_1)^2 + (x_3 - x_2)^2] \qquad (26.23)$$

We express the kinetic energy in terms of the displacements $x_i = y_i - y_i^0$:

$$T(\dot{y}_1, \dot{y}_2, \dot{y}_3) = \frac{m}{2}\left(\dot{y}_1^2 + \dot{y}_3^2\right) + \frac{M}{2}\,\dot{y}_2^2 = \frac{m}{2}\left(\dot{x}_1^2 + \dot{x}_3^2\right) + \frac{M}{2}\,\dot{x}_2^2 \qquad (26.24)$$

The Lagrange function is thus of the form (25.7),

$$\mathcal{L}(x, \dot{x}) = \frac{1}{2}\sum_{i,j=1}^{3}\left(T_{ij}\,\dot{x}_i\,\dot{x}_j - V_{ij}\,x_i\,x_j\right) \qquad (26.25)$$

The coefficients T_{ij} and V_{ij} can be read off from (26.23) and (26.24), where the symmetry ($V_{12} = V_{21}$) must be taken into account.

This gives

$$T = \left(T_{ij}\right) = \begin{pmatrix} m & 0 & 0 \\ 0 & M & 0 \\ 0 & 0 & m \end{pmatrix}, \quad V = \left(V_{ij}\right) = \begin{pmatrix} k & -k & 0 \\ -k & 2k & -k \\ 0 & -k & k \end{pmatrix} \tag{26.26}$$

This brings the problem to the form (25.7). The condition (25.16) for the natural frequencies (or eigenfrequencies) is

$$\det\left(V - \omega^2 T\right) = \begin{vmatrix} k - \omega^2 m & -k & 0 \\ -k & 2k - \omega^2 M & -k \\ 0 & -k & k - \omega^2 m \end{vmatrix} = 0 \tag{26.27}$$

This yields

$$\begin{aligned} 0 &= \left(k - \omega^2 m\right)^2 \left(2k - \omega^2 M\right) - 2k^2 \left(k - \omega^2 m\right) \\ &= \omega^2 \left(k - \omega^2 m\right)\left(\omega^2 M m - k\,(2m + M)\right) \end{aligned} \tag{26.28}$$

with the solutions

$$\omega_1 = 0, \quad \omega_2 = \sqrt{\frac{k}{m}}, \quad \omega_3 = \sqrt{\frac{k}{m}\left(1 + \frac{2m}{M}\right)} \tag{26.29}$$

From (25.22), $(V - \omega_k^2\, T)A^{(k)} = 0$, follow the eigenvectors:

$$A^{(1)} = \begin{pmatrix} 1 \\ 1 \\ 1 \end{pmatrix}, \quad A^{(2)} = \begin{pmatrix} 1 \\ 0 \\ -1 \end{pmatrix}, \quad A^{(3)} = \begin{pmatrix} 1 \\ -2m/M \\ 1 \end{pmatrix} \tag{26.30}$$

These eigenvectors can be arbitrarily normalized, since they will be multiplied with constants C_k. Taking into account (25.19), the general solution is (25.25):

$$\begin{aligned} x(t) = {} & \begin{pmatrix} 1 \\ 1 \\ 1 \end{pmatrix} (b_1 t + a_1) + \begin{pmatrix} 1 \\ 0 \\ -1 \end{pmatrix} B_2 \cos(\omega_2 t + \alpha_2) \\ & + \begin{pmatrix} 1 \\ -2m/M \\ 1 \end{pmatrix} B_3 \cos(\omega_3 t + \alpha_3) \end{aligned} \tag{26.31}$$

The six real constants b_1, a_1, B_2, α_2, B_3, α_3 are determined by the initial conditions for $x_i(0)$ and $\dot{x}_i(0)$.

The first eigenmode (normal mode) $\omega_1 = 0$ is not an oscillation, but a uniform motion of the whole molecule with constant velocity. This motion is indeed described by the Lagrange function (26.25) because the center-of-mass coordinate

$$X = \frac{m(x_1 + x_3) + Mx_2}{2m + M} \tag{26.32}$$

is a linear combination of the considered coordinates. Since the potential (26.23) does not yield a restoring force for X, the associated frequency is zero. In a three-dimensional treatment of the molecule, an analogous situation occurs also for the rotations because in the absence of external forces, there is also no restoring force. Alternatively, to the given treatment, the trivial translational motion could be split off from the beginning; then one treat explicitly only the remaining two degrees of freedom.

In the second eigenmode, the middle atom is at rest, and the other two atoms oscillate in the opposite phase. In the third eigenmode, the two edge atoms oscillate together (in phase), but opposite with respect to the middle atom (counter phase). All modes are sketched in Figure 26.2.

Exercises

26.1 *Standard method for double pendulum*

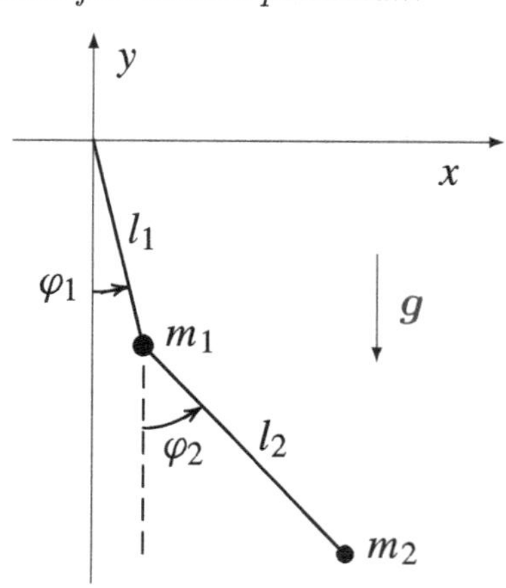

For the double pendulum already solved in Exercise 10.1: Give the matrices T and V of the general procedure for small oscillations. Establish the equation for the eigenfrequencies and the eigenvectors:

$$\left(V - \omega^2 T\right) A = 0$$

Determine the eigenvalues and eigenvectors for the case $m_1 = m_2 = m$ and $l_1 = l_2 = l$.

26.2 *Normal coordinates for molecular vibration*

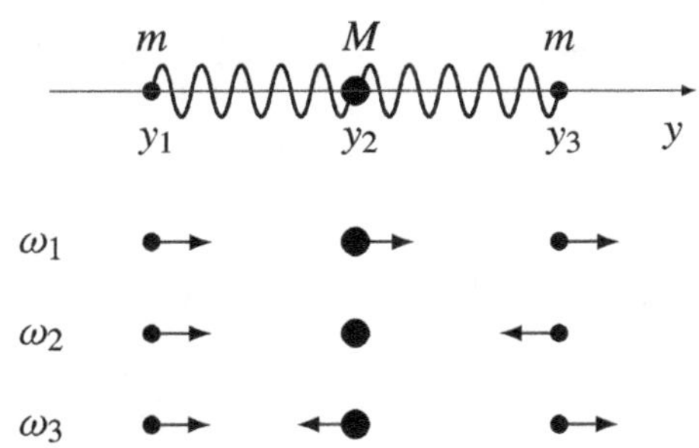

The atoms (masses m, M, and m) of a triatomic molecule can perform one-dimensional vibrations. The resulting types of vibrations are sketched in the figure. The first solution consists of the translation of the entire molecule.

The eigenvectors belonging to the oscillations are

$$A^{(1)} = c_1 \begin{pmatrix} 1 \\ 1 \\ 1 \end{pmatrix}, \quad A^{(2)} = c_2 \begin{pmatrix} 1 \\ 0 \\ -1 \end{pmatrix}, \quad A^{(3)} = c_3 \begin{pmatrix} 1 \\ -2m/M \\ 1 \end{pmatrix}$$

The eigenvectors define the square matrix

$$a = (a_{ik}) = \left(A_i^{(1)}, A_i^{(2)}, A_i^{(3)}\right)$$

Determine the coefficients c_i from the normalization condition $a^{\mathrm{T}} T a = 1$ with $T = \operatorname{diag}(m, M, m)$. Determine the normal coordinates $Q_k = Q_k(x)$.

26.3 *Linear chain with fixed boundary conditions*

Three equal masses m are connected by four equal springs (spring constant k) between two walls. They can oscillate in the y-direction.

The displacements from the equilibrium positions y_n^0 are denoted by $x_n(t) = y_n(t) - y_n^0$.

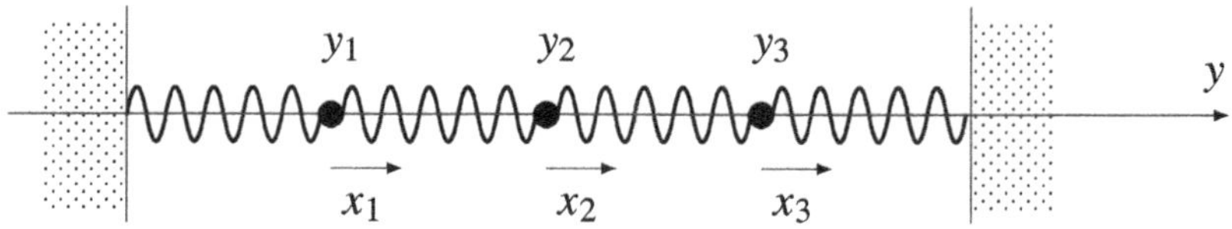

Establish the Lagrange function for this problem, and read off the matrices T and V. Determine the natural frequencies and the eigenvectors. Sketch the different types of oscillations.

26.4 *One-dimensional crystal model I*

The masses $m_n = m$ with $n = 0, \pm 1, \pm 2, \ldots$ can move along the y-axis. Harmonic springs (spring constant k) between adjacent masses lead to the equilibrium positions $y_n^0 = an$; the length a is the lattice constant of this one-dimensional crystal model. The displacements from equilibrium are given by $x_n(t) = y_n(t) - y_n^0$:

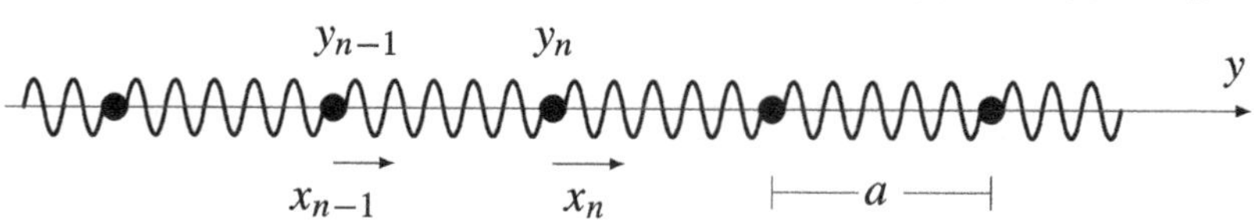

Give the Lagrange function and the equations of motion for this infinite linear chain. Solve the equations of motion using the approach $x_n(t) = Q_q(t) \exp(\mathrm{i}qna)$ and $Q_q(t) = A_q \exp(-\mathrm{i}\omega_q t)$.

This solution is characterized by the real wave number q. Justify, that q can be restricted to the range $-\pi/a \le q < \pi/a$. Sketch the natural frequencies $\omega_q = \omega(q)$ as a function of q. This relation is called the *dispersion relation*.

The physical boundary condition of a *finite* chain of N masses can be simulated by the periodic boundary condition $x_n(t) = x_{n+N}(t)$. To which discrete values of q does this boundary condition lead?

26.5 *One-dimensional crystal model II*

The masses $m_{2n} = m$ and $m_{2n+1} = M$ with $n = 0, \pm 1, \pm 2, \ldots$ can move along the y-axis. Harmonic springs (spring constant k) between adjacent masses lead to the equilibrium positions $y_\nu^0 = a \cdot \nu$; the length a is the lattice constant of this one-dimensional crystal model with two atomic species. The displacements from equilibrium are given by $x_\nu(t) = y_\nu(t) - y_\nu^0$:

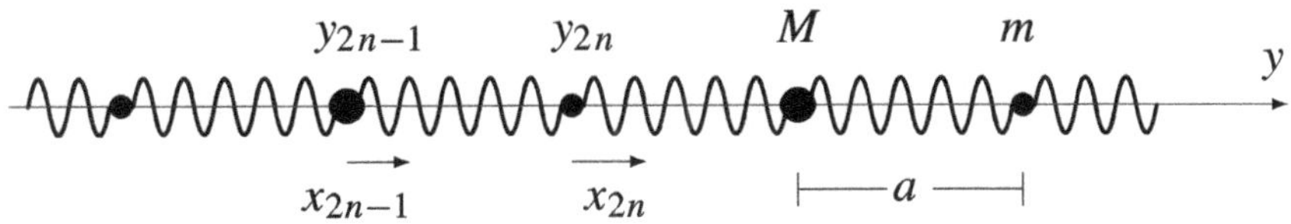

Give the Lagrange function and the equations of motion. Insert

$$x_{2n}(t) = Q_q(t)\ \exp\big(\mathrm{i}\,q(2n)a\big), \quad x_{2n+1}(t) = P_q(t)\ \exp\big(\mathrm{i}\,q(2n+1)a\big)$$

with $Q_q(t) = A_q \exp(-\mathrm{i}\,\omega_q t)$ and $P_q(t) = B_q \exp(-\mathrm{i}\,\omega_q t)$ into the equations of motion. This yields two coupled equations for A_q and B_q. Solve these equations. Calculate and sketch the dispersion relation $\omega_q = \omega(q)$ as a function of q. Choose a suitable range for the wave number q.

PART VII

Hamilton Formalism

Chapter 27

Canonical Equations

In this part, we introduce two new formulations of the fundamental laws of mechanics: The canonical equations (Chapters 27 and 28) and the Hamilton–Jacobi equation (Chapter 29). We refer to these formulations collectively as the Hamilton formalism.

The practical solution of problems is usually done in Lagrange's formalism; Hamilton formalism generally offers no advantages. Some concepts of the Hamilton formalism are, however, needed in other parts of theoretical physics: In quantum mechanics, the *Hamilton function* is the starting point for defining the Hamilton operator, in statistical physics one needs the concept of the *phase space volume*, and in chaos theory, one considers trajectories in *phase space*. All these concepts will be introduced in this chapter.

Further structures of the Hamiltonian formalism are then presented in Chapters 28 and 29; these two chapters can also be skipped. These additional concepts are, however, important for investigating the relations between mechanics, quantum mechanics and optics.

Hamilton Function

In the Lagrangian formalism, the generalized (or canonical) momenta p_i are defined by

$$p_i = \frac{\partial \mathcal{L}}{\partial \dot{q}_i} \quad (i = 1, 2, \ldots, f) \tag{27.1}$$

In the following, we will replace the generalized velocities $\dot{q}_i$ with the generalized momenta p_i. For this purpose, we extract the velocities from equation (27.1):

$$p_i = \frac{\partial \mathcal{L}(q, \dot{q}, t)}{\partial \dot{q}_i} \quad \overset{\text{solving for}}{\longrightarrow} \quad \dot{q}_k = \dot{q}_k(q, p, t) \qquad (27.2)$$

We use the abbreviations

$$q = (q_1, \ldots, q_f), \quad \dot{q} = (\dot{q}_1, \ldots, \dot{q}_f), \quad p = (p_1, \ldots, p_f) \qquad (27.3)$$

We now define the *Hamilton function* by

$$H(q, p, t) = \sum_{i=1}^{f} \dot{q}_i(q, p, t)\, p_i - \mathcal{L}\left(q,\, \dot{q}(q, p, t),\, t\right) \qquad (27.4)$$

It is essential that H is considered as a function of the arguments $q_1, \ldots, q_f, p_1, \ldots, p_f$, and t.

As a simple example, let us consider the free motion of a mass point in a plane. Using polar coordinates, the Lagrange function is

$$\mathcal{L}(\rho, \dot{\rho}, \dot{\varphi}) = \frac{m}{2}\left(\dot{\rho}^2 + \rho^2\, \dot{\varphi}^2\right) \qquad (27.5)$$

According to (27.2), we express the velocities by the momenta:

$$\left.\begin{aligned} p_\rho &= \partial\mathcal{L}/\partial\dot{\rho} = m\,\dot{\rho} \\ p_\varphi &= \partial\mathcal{L}/\partial\dot{\varphi} = m\,\rho^2\,\dot{\varphi} \end{aligned}\right\} \longrightarrow \dot{\rho} = \frac{p_\rho}{m}, \quad \dot{\varphi} = \frac{p_\varphi}{m\rho^2} \qquad (27.6)$$

For this, the Hamilton function (27.4) becomes:

$$H(\rho, p_\rho, p_\varphi) = \dot{\rho}\, p_\rho + \dot{\varphi}\, p_\varphi - \mathcal{L} = \frac{p_\rho^2}{2m} + \frac{p_\varphi^2}{2m\rho^2} \qquad (27.7)$$

In Exercises 27.1–27.3, the Hamiltonian function is to be set up for some other examples.

Using the Lagrange equations we calculate the partial derivatives of the Hamilton function (27.4):

$$\frac{\partial H}{\partial q_k} = \sum_{i=1}^{f} \frac{\partial \dot{q}_i}{\partial q_k}\, p_i - \frac{\partial \mathcal{L}}{\partial q_k} - \sum_{i=1}^{f} \frac{\partial \mathcal{L}}{\partial \dot{q}_i}\frac{\partial \dot{q}_i}{\partial q_k}$$

$$= -\frac{\partial \mathcal{L}}{\partial q_k} = -\frac{d}{dt}\left(\frac{\partial \mathcal{L}}{\partial \dot{q}_k}\right) = -\dot{p}_k \qquad (27.8)$$

$$\frac{\partial H}{\partial p_k} = \sum_{i=1}^{f} \frac{\partial \dot{q}_i}{\partial p_k}\, p_i + \dot{q}_k - \sum_{i=1}^{f} \frac{\partial \mathcal{L}}{\partial \dot{q}_i}\frac{\partial \dot{q}_i}{\partial p_k} = \dot{q}_k \qquad (27.9)$$

$$\frac{\partial H}{\partial t} = \sum_{i=1}^{f} \frac{\partial \dot{q}_i}{\partial t}\, p_i - \sum_{i=1}^{f} \frac{\partial \mathcal{L}}{\partial \dot{q}_i}\frac{\partial \dot{q}_i}{\partial t} - \frac{\partial \mathcal{L}}{\partial t} = -\frac{\partial \mathcal{L}}{\partial t} \qquad (27.10)$$

The first two equations are the *canonical* or *Hamilton equations*:

$$\dot{p}_k = -\frac{\partial H(q,p,t)}{\partial q_k}, \quad \dot{q}_k = \frac{\partial H(q,p,t)}{\partial p_k} \quad (k = 1, \ldots, f) \qquad (27.11)$$

According to (27.8) and (27.9), these equations follow from the Lagrange equations. Since the canonical equations constitute $2f$ first-order differential equations, they can replace the original f second-order differential equations (Lagrange equations). The Lagrange equations and the canonical equations are therefore alternative formulations of the equations of motion.

As an example, we consider the motion of a particle in a potential. The Lagrange function

$$\mathcal{L} = \frac{m\,\dot{\boldsymbol{r}}^2}{2} - U(\boldsymbol{r},t) = \frac{m}{2}\left(\dot{x}^2 + \dot{y}^2 + \dot{z}^2\right) - U(x,y,z,t) \qquad (27.12)$$

yields the generalized momenta $p_x = \partial\mathcal{L}/\partial\dot{x} = m\dot{x}$, $p_y = m\dot{y}$ and $p_z = m\dot{z}$. Hence, the Hamiltonian function (27.4) becomes

$$H(x,y,z,p_x,p_y,p_z) = \frac{p_x^2 + p_y^2 + p_z^2}{2m} + U(x,y,z,t) = \frac{\boldsymbol{p}^2}{2m} + U(\boldsymbol{r},t) \qquad (27.13)$$

The canonical equations for the x-motion are

$$\dot{p}_x = -\frac{\partial H}{\partial x} = -\frac{\partial U}{\partial x}, \qquad \dot{x} = \frac{\partial H}{\partial p_x} = \frac{p_x}{m} \tag{27.14}$$

The equation for the y- and z-motion are of the same form. All these equations can be combined into vector form:

$$\dot{\boldsymbol{p}} = -\operatorname{grad} U(\boldsymbol{r}), \qquad \dot{\boldsymbol{r}} = \frac{\boldsymbol{p}}{m} \tag{27.15}$$

The equivalence with the Lagrange equations $m\ddot{\boldsymbol{r}} = -\operatorname{grad} U$ is obvious.

Conservation Laws

We briefly sketch the relationship between symmetry and conservation laws for the Hamiltonian function. We calculate first

$$\frac{dH}{dt} = \frac{d}{dt}\left(\sum_{i=1}^{f} \frac{\partial \mathcal{L}}{\partial \dot{q}_i}\,\dot{q}_i - \mathcal{L}\right) \overset{(9.35)}{=} -\frac{\partial \mathcal{L}}{\partial t} \overset{(27.10)}{=} \frac{\partial H}{\partial t} \tag{27.16}$$

From this, we get the relation between the time translation invariance and the conservation of energy in the form:

$$\text{Invariance:} \quad \frac{\partial H}{\partial t} = 0 \longrightarrow \quad \begin{array}{c}\text{Conserved quantity:}\\ H = \text{const.}\end{array} \tag{27.17}$$

If the kinetic energy T depends only quadratically on the velocities $\dot{q}_i$ and the Lagrangian function is of the form $\mathcal{L} = T - U(q,t)$, then $\sum(\partial\mathcal{L}/\partial\dot{q}_i)\,\dot{q}_i = 2T$. Then the Hamiltonian function

$$H(q,p,t) = T + U \tag{27.18}$$

is equal to the energy. This is especially true if the coordinates are Cartesian and if the potential does not depend on the velocities. In the following, we will tacitly assume this case, i.e. $H = E$. For a counter example, we refer to Exercise 27.3.

For a cyclic coordinate q_k it follows from (27.8)

$$\text{Invariance:} \quad \frac{\partial H}{\partial q_k} = 0 \longrightarrow \quad \begin{array}{c}\text{Conserved quantity:}\\ p_k = \text{const.}\end{array} \tag{27.19}$$

Phase Space

Specifying $2f$ values for $q_1, \ldots, q_f$ and $\dot{q}_1, \ldots, \dot{q}_f$ defines the *state* of the considered system at a given time (Chapter 9). Equivalent to this is the specification of the $2f$ values $q_1, \ldots, q_f$ and $p_1, \ldots, p_f$. We consider now an abstract $2f$-dimensional space, which is spanned by $2f$ Cartesian coordinate axes for q_i and p_i. We assign to each state a point in this *phase space*. The time evolution of the system is then a continuous sequence of points in the phase space, i.e. a trajectory. Figure 27.1 shows the phase space for a one-dimensional system.

The phase space is used in particular in statistical mechanics for the counting of states. In the treatment of chaotic systems, phase space is used as a starting point. Also, Bohr's quantization rules (representing the first attempts in quantum mechanics) refer to the phase space. In the following, we discuss the aspect of the counting of states.

The condition $H(q_1, \ldots, q_f, p_1, \ldots, p_f) = E$ describes a $(2f-1)$-dimensional surface in phase space; we omit a possible time dependence of H in the notation. This surface encloses the $2f$-dimensional *phase space volume* $V_{\rm PR}(E)$:

$$V_{\rm PR}(E) = \underset{H(q,p)\,<\,E}{\int \cdots \int} dp_1 \cdots dp_f \, dq_1 \cdots dq_f . \tag{27.20}$$

In the example of Figure 27.1, the phase space volume is the area of the depicted ellipse. Each system state with energy less than E is represented by a point in this phase space volume. In a finite volume, there are infinitely many points, thus infinitely many classical system states. In contrast, a *quantum mechanical* state occupies a (small) finite volume in phase space. The number of quantum mechanical states with energy smaller than E is then proportional to $V_{\rm PR}(E)$. This is illustrated in the following by the example of the one-dimensional oscillator; our discussion assumes only elementary quantum mechanical knowledge.

The Hamilton function of the one-dimensional oscillator is

$$H(q,p) = \frac{p^2}{2m} + \frac{m\,\omega^2 q^2}{2} \tag{27.21}$$

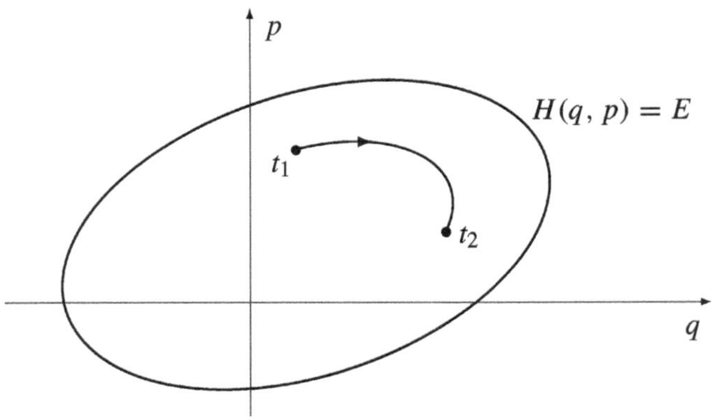

Figure 27.1 Phase space for a one-dimensional system. A point in the phase space, i.e., the specification of q and p, defines a system state. The time evolution of the system can be represented by a trajectory in the phase space, such as the sketched curve between the time points t_1 and t_2. The condition $H(q,p) = E$ is a curve on which the system states of equal energy E lie (an ellipse in the figure). The number of states with energy smaller than E is proportional to the area enclosed by the curve $H(q,p) = E$.

For this, the curve $H(q,p) = E$ is an ellipse with the semi-axes

$$a = \sqrt{\frac{2E}{m\omega^2}}, \qquad b = \sqrt{2mE} \tag{27.22}$$

The phase space volume V_{PR} equals the area of the ellipse defined by $H(q,p) = E$:

$$V_{\text{PS}}(E) = \int\!\!\int_{H(q,p)<E} dp\, dq = \pi\, ab = \frac{2\pi E}{\omega}. \tag{27.23}$$

The quantum mechanical oscillator has the discrete energy states

$$E_n = \hbar\omega\left(n + \frac{1}{2}\right) \quad (n = 0, 1, 2, \ldots) \tag{27.24}$$

The number N_E of discrete states with energy smaller than E is therefore

$$N_E = \sum_{E_n \le E} 1 \approx \frac{E}{\hbar\omega} \overset{(27.23)}{=} \frac{V_{\text{PS}}(E)}{2\pi\hbar} \quad (N_E \gg 1). \tag{27.25}$$

The result

$$N_E = \frac{V_{\rm PS}(E)}{2\pi\hbar} \quad (N_E \gg 1) \tag{27.26}$$

is valid for arbitrary one-dimensional systems. For a system with f degrees of freedom, this becomes

$$N_E = \frac{V_{\rm PS}(E)}{(2\pi\hbar)^f} \quad (N_E \gg 1) \tag{27.27}$$

The general validity of (27.26) and (27.27) is based on the quantum mechanical *uncertainty relation*: According to this relation, the coordinate q_k and the corresponding momentum p_k can only be defined within a range given by the uncertainty relation $\Delta q_k \, \Delta p_k \geq \hbar/2$. Accordingly, a quantum mechanical state occupies a small finite volume in the q_k–p_k-subspace. This explains that a quantum mechanical state (with f degrees of freedom) occupies a volume $(2\pi\hbar)^f$ in phase space.

The first quantum mechanical attempt for the calculation of discrete energy values (of an atom, for example) was Bohr's quantization rule. This rule postulated the following condition for stable closed orbits:

$$\oint_{H=E} dq\, p = 2\pi\hbar\, n, \quad n = 1, 2, 3, \ldots \tag{27.28}$$

In the integral, $p = p(q, E)$ is a function of q and E, which can be obtained by resolving $H(q, p) = E$ for p. For the harmonic oscillator, $p = p(q, E)$ describes the ellipse with the semi-axes given in (27.22). Thus, we obtain the oscillator

$$\oint dq\, p(q, E) = \pi a b = \frac{2\pi E}{\omega} = 2\pi\hbar\, n \tag{27.29}$$

This yields discrete energy values $E = n\,\hbar\omega$ for the oscillator. The correct quantum mechanical result is $E = (n + 1/2)\,\hbar\omega$, i.e. one has to add the zero point energy $\hbar\omega/2$.

Exercises

27.1 *Hamiltonian function for a mass point on a circular cone*

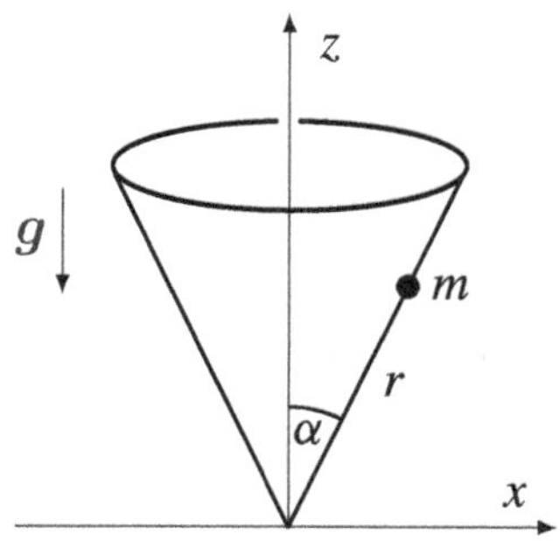

A mass point glides frictionless in a gravity field on a circular cone. The system is described by the Lagrange function

$$\mathcal{L}(r, \dot{r}, \dot{\phi}) = \frac{m}{2}\left(\dot{r}^2 + r^2\dot{\phi}^2 \sin^2\alpha\right) - mgr\cos\alpha$$

Determine the Hamilton function and set up the canonical equations.

27.2 *Hamilton function for particles in the electromagnetic field*

For a charged particle in the electromagnetic field, the Lagrangian is

$$\mathcal{L}(\dot{\boldsymbol{r}}, \boldsymbol{r}, t) = \frac{m}{2}\dot{\boldsymbol{r}}^2 - q\,\Phi(\boldsymbol{r}, t) + \frac{q}{c}\,\dot{\boldsymbol{r}} \cdot \boldsymbol{A}(\boldsymbol{r}, t)$$

Set up the corresponding Hamiltonian function.

27.3 *Mass point on a rotating rod*

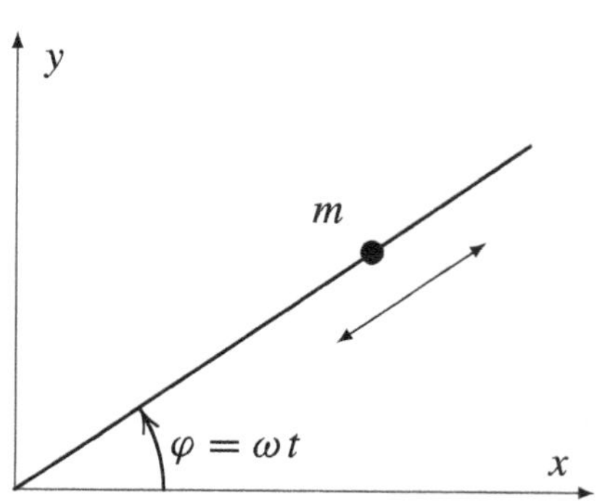

A mass point slides frictionless on a rotating rod. The Lagrange function is given by

$$\mathcal{L}(\rho, \dot{\rho}) = \frac{m}{2}\left(\dot{\rho}^2 + \omega^2 \rho^2\right)$$

Determine the Hamiltonian function. Does $\partial H/\partial t = 0$ hold? Does $H = \text{const.}$ hold? Is H equal to the energy of the mass point? Is the energy conserved?

27.4 *Plane pendulum in phase space*

A plane pendulum consists of a mass m at the end of a massless rod of length ℓ. In the gravitational field, the pendulum has the potential energy

$$U(q) = mg\ell(1 - \cos q)$$

The angle $q = \varphi$ defines the displacement from the equilibrium position. Write down the Lagrange and the Hamilton function (using q a the generalized coordinate). Sketch possible trajectories for energies $E \geq 0$ in the two-dimensional p–q phase space.

27.5 *Liouville theorem*

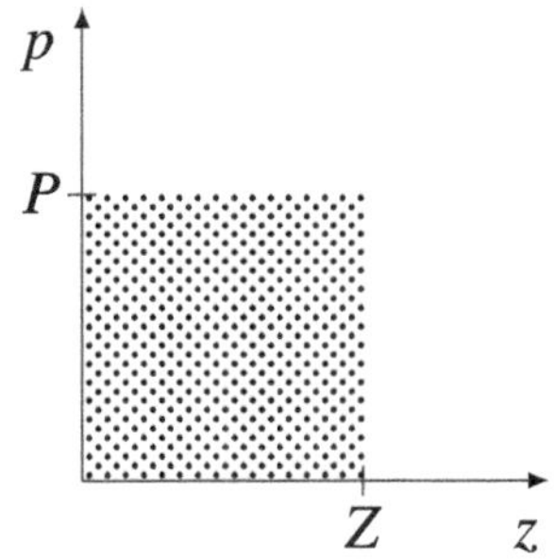

The one-dimensional motion (along the z-axis) of $N \gg 1$ particles of the same kind is considered. The state (location z and momentum $p = m\,\dot{z}$) of a singled out particle defines a point in the phase space. Considering all particles yields a certain density of points in the depicted phase space. At time $t = 0$, this density of points shall be constant in the range $0 \leq z \leq Z$ and $0 \leq p \leq P$, and zero outside this area.

Calculate and sketch how the boundaries of the occupied range shift over time, for (i) force-free motion and (ii) motion in the gravitational field $\boldsymbol{g} = g\,\boldsymbol{e}_z$. Confirm that the area of this (occupied) phase space and the density of the system points are constant.

The result can be generalized to the *Liouville theorem*: The density of the system points in the phase space is constant in time; this means that it is constant along the trajectories.

Chapter 28

Canonical Transformations

We set up the Hamiltonian principle, consider the possibility of choosing other coordinates and momenta, and finally introduce the Poisson brackets. These developments are of minor importance for the solution of practical problems of mechanics; they are therefore treated relatively briefly. The presented structures are important for the investigation of the relations between mechanics and quantum mechanics.

Hamilton's Principle

We first formulate Hamilton's principle for the canonical equations. The Hamilton principle of Chapter 14 states that the equations of motion are equivalent to the condition that the action S is stationary, thus

$$\delta S = \delta \int_{t_1}^{t_2} dt\, \mathcal{L} \stackrel{(27.4)}{=} \delta \int_{t_1}^{t_2} dt \left(\sum_{i=1}^{f} p_i\, \dot{q}_i - H(q,p,t) \right) = 0 \quad (28.1)$$

Here, the boundary values are held fixed,

$$\delta q_i(t_1) = \delta q_i(t_2) = 0 \quad (i = 1, \ldots, f) \qquad (28.2)$$

We evaluate (28.1):

$$\delta S = \int_{t_1}^{t_2} dt \sum_{i=1}^{f} \left[\dot{q}_i\, \delta p_i + p_i\, \delta \dot{q}_i - \frac{\partial H}{\partial p_i}\, \delta p_i - \frac{\partial H}{\partial q_i}\, \delta q_i \right] \qquad (28.3)$$

We transform the second term on the right-hand side:

$$\int_{t_1}^{t_2} dt\ p_i\,\delta\dot{q}_i = \int_{t_1}^{t_2} dt\ p_i\,\frac{d}{dt}\,\delta q_i \overset{\text{p.i.}}{=} p_i\,\delta q_i\Big|_{t_1}^{t_2} - \int_{t_1}^{t_2} dt\ \dot{p}_i\,\delta q_i = -\int_{t_1}^{t_2} dt\ \dot{p}_i\,\delta q_i \tag{28.4}$$

The boundary term vanishes because of (28.2). From the last two equations we get

$$\delta S[q,p] = \int_{t_1}^{t_2} dt \sum_{i=1}^{f} \left[\left(\dot{q}_i - \frac{\partial H}{\partial p_i} \right) \delta p_i - \left(\dot{p}_i + \frac{\partial H}{\partial q_i} \right) \delta q_i \right] \tag{28.5}$$

Using the canonical equations yields $\delta S = 0$. Due to the arbitrariness of δq_i and δp_i, the canonical equations follow inversely from $\delta S[q,p] = 0$. Therefore, the canonical equations of motion are equivalent to Hamilton's principle

$$\delta S[q,p] = 0 \quad \text{Hamilton's principle} \tag{28.6}$$

This principle is sometimes also called the modified Hamiltonian principle, for distinguishing it from the Hamilton principle of Chapter 14. The corresponding equations of motion (Lagrange equations or canonical equations) are, however, equivalent. And in both cases, we have $2f$ initial conditions, namely the values for $q_i(t_1)$ and $\dot{q}_i(t_1)$ in the Lagrangian formalism, or the values $q_i(t_1)$ and $p_i(t_1)$ in the Hamiltonian formalism.

Generating Function $G(q, Q, t)$

We consider transformations of the $2f$ variables p_i and q_i to $2f$ new variables P_k and Q_k:

$$\begin{aligned} Q_k &= Q_k(q_i, \ldots, q_f, p_i, \ldots, p_f, t) = Q_k(q,p,t) \\ P_k &= P_k(q_i, \ldots, q_f, p_i, \ldots, p_f, t) = P_k(q,p,t) \end{aligned} \tag{28.7}$$

These transformations shall be chosen such that canonical equations hold for the new variables, too. Such *canonical transformations* can be generated by an arbitrary (differentiable and invertible) function $G(q, Q, t)$ for which we require:

$$\sum_{i=1}^{f} p_i\, \dot{q}_i - H(q,p,t) = \sum_{k=1}^{f} P_k\, \dot{Q}_k - H'(Q,P,t) + \frac{d}{dt}\, G(q,Q,t) \tag{28.8}$$

We show: From this follows (i) the transformations (28.7), (ii) the new Hamiltonian function H', and (iii) the canonical equations for the new variables.

The total differential of the function G is

$$dG = \sum_{i=1}^{f} \frac{\partial G}{\partial q_i}\, dq_i + \sum_{k=1}^{f} \frac{\partial G}{\partial Q_k}\, dQ_k + \frac{\partial G}{\partial t}\, dt \tag{28.9}$$

We multiply (28.8) by dt (using $\dot{q}_i\, dt = dq_i$ and $\dot{Q}_k\, dt = dQ_k$) and insert dG:

$$\sum_{i=1}^{f} \left(p_i - \frac{\partial G}{\partial q_i} \right) dq_i = \sum_{k=1}^{f} \left(P_k + \frac{\partial G}{\partial Q_k} \right) dQ_k + \left(H - H' + \frac{\partial G}{\partial t} \right) dt \tag{28.10}$$

The differentials dq_i, dQ_k, and dt are independent of each other. Therefore, the coefficients of these differentials in (28.10) must vanish:

$$p_i = \frac{\partial G(q,Q,t)}{\partial q_i} = p_i(q,Q,t) \quad \text{and} \quad P_k = -\frac{\partial G(q,Q,t)}{\partial Q_k} = P_k(q,Q,t) \tag{28.11}$$

$$H'(Q,P,t) = H(q,p,t) + \frac{\partial G(q,Q,t)}{\partial t} \tag{28.12}$$

From the fist equation in (28.11) we may obtain $q = q(Q,P,t)$. We insert this $q = q(Q,P,t)$ and $p = p(Q,P,t)$ in the right-hand side of (28.12). Thereby we obtain the new Hamiltonian function $H'(Q,P,t)$. In this way, a given Hamilton function $H(q,p,t)$ and an arbitrary function $G(q,Q,t)$ lead to a new Hamilton function $H'(Q,P,t)$.

For the old Hamilton function, Hamilton's principle $\delta S = 0$ holds, (28.1). On the r.h.s of (28.1), we insert (28.8):

$$\begin{aligned} 0 = \delta S &= \delta \int_{t_1}^{t_2} dt \left(\sum_{k=1}^{f} P_k \dot{Q}_k - H'(Q,P,t) + \frac{d}{dt} G(q,Q,t) \right) \\ &= \int_{t_1}^{t_2} dt \sum_{k=1}^{f} \left[\dot{Q}_k \, \delta P_k + P_k \, \delta \dot{Q}_k - \frac{\partial H'}{\partial P_k} \, \delta P_k - \frac{\partial H'}{\partial Q_k} \, \delta Q_k \right] \\ &\quad + \delta \big(G\big(q(t_2), Q(t_2), t_2\big) - G\big(q(t_1), Q(t_1), t_1\big) \big) \end{aligned} \tag{28.13}$$

For the intended transition to the new canonical equations, we reshape the second term on the right-hand side:

$$\int_{t_1}^{t_2} dt \; P_k \, \delta \dot{Q}_k = \int_{t_1}^{t_2} dt \; P_k \, \frac{d}{dt} \, \delta Q_k \overset{\text{p.i.}}{=} P_k \, \delta Q_k \Big|_{t_1}^{t_2} - \int_{t_1}^{t_2} dt \; \dot{P}_k \, \delta Q_k \tag{28.14}$$

For the last term in (28.13), we use

$$\delta \left(G\big(q(t_2), Q(t_2), t_2\big) - G\big(q(t_1), Q(t_1), t_1\big) \right) = \sum_{k=1}^{f} \frac{\partial G}{\partial Q_k} \, \delta Q_k \Big|_{t_1}^{t_2} \tag{28.15}$$

We now substitute the last two equations into (28.13):

$$\begin{aligned} 0 = \delta S = \int_{t_1}^{t_2} dt \sum_{k=1}^{f} &\left[\left(\dot{Q}_k - \frac{\partial H'}{\partial P_k} \right) \delta P_k - \left(\dot{P}_k + \frac{\partial H'}{\partial Q_k} \right) \delta Q_k \right] \\ &+ \sum_{k=1}^{f} \left(P_k + \frac{\partial G}{\partial Q_k} \right) \delta Q_k \Big|_{t_1}^{t_2} \end{aligned} \tag{28.16}$$

The boundary term falls away because of $P_k = -\partial G / \partial Q_k$, (28.11). For arbitrary variations δP_k and δQ_k, their coefficients must vanish (in order to fulfill $\delta S = 0$). This means that we obtain the canonical equations for the new variables:

$$\dot{Q}_k = \frac{\partial H'}{\partial P_k}, \quad \dot{P}_k = -\frac{\partial H'}{\partial Q_k} \tag{28.17}$$

In this way, any function $G(q,Q,t)$ leads to a canonical transformation. Such a function is therefore called *generating function* of the transformation.

We summarize the implementation of a canonical transformation in a nutshell: The starting point is a given Hamiltonian function $H(q,p,t)$. One now chooses an arbitrary function $G(q,Q,t)$. Equations (28.11) and (28.12) imply the variable transformations $q = q(Q,P,t)$ and $p = p(Q,P,t)$, and fix the new Hamiltonian function $H'(Q,P,t)$. From $H'(Q,P,t)$ follow the new canonical equations (28.17). With a clever choice of the generators $G(q,Q,t)$ the new canonical equations might be simpler than the old ones.

Generating functions $G_2(q,P,t)$, $G_3(p,Q,t)$ or $G_4(p,P,t)$

Instead of $G(q,Q,t)$, a generating function can also be of the form $G_2(q,P,t)$, $G_3(p,Q,t)$ or $G_4(p,P,t)$. The transition to the new Hamiltonian is done by $H' = H + \partial G_i/\partial t$ as in (28.12). The transition to the new coordinates and momenta is performed similarly as in (28.11):

$$H' = H + \frac{\partial G_2(q,P,t)}{\partial t}, \qquad p_i = \frac{\partial G_2}{\partial q_i}, \qquad Q_k = \frac{\partial G_2}{\partial P_k} \tag{28.18}$$

$$H' = H + \frac{\partial G_3(p,Q,t)}{\partial t}, \qquad q_i = -\frac{\partial G_3}{\partial p_i}, \qquad P_k = -\frac{\partial G_3}{\partial Q_k} \tag{28.19}$$

$$H' = H + \frac{\partial G_4(p,P,t)}{\partial t}, \qquad q_i = -\frac{\partial G_4}{\partial p_i}, \qquad Q_k = \frac{\partial G_4}{\partial P_k} \tag{28.20}$$

The arguments are $H' = H'(P,Q,t)$ and $H = H(p,q,t)$, otherwise, they follow from the arguments of the respective generating function G_i. The method for establishing the transformation equations and the new Hamiltonian is analogous to the procedure for $G(q,Q,t)$.

In the Lagrangian formalism, one can go from f coordinates q_i to f new generalized coordinates Q_k (for example, spherical coordinates instead of Cartesian coordinates, or center of mass and relative coordinates instead of $\boldsymbol{r}_1$ and $\boldsymbol{r}_2$ for two-body systems).

This corresponds to a special canonical transformation:

$$G_2(q,P,t) = \sum_{k=1}^{f} f_k(q_1,\ldots,q_f,t)\,P_k \implies Q_k = f_k(q_1,\ldots,q_f,t) \tag{28.21}$$

Harmonic oscillator

As an example, we solve the harmonic oscillator problem by a canonical transformation. The Hamiltonian function of the oscillator is

$$H = \frac{p^2}{2m} + \frac{m\,\omega^2 q^2}{2} \tag{28.22}$$

For the generating function

$$G(q,Q) = \frac{m\,\omega}{2}\, q^2 \cot Q \tag{28.23}$$

(28.11) yields:

$$p = \frac{\partial G}{\partial q} = m\,\omega\, q \cot Q, \quad P = -\frac{\partial G}{\partial Q} = \frac{m\,\omega\, q^2}{2\,\sin^2 Q} \tag{28.24}$$

We solve these equations for the old variables:

$$q = \sqrt{\frac{2P}{m\,\omega}}\, \sin Q, \quad p = \sqrt{2m\,\omega P}\, \cos Q \tag{28.25}$$

Now, we calculate the new Hamiltonian function $H'(Q,P,t)$ according to (28.12):

$$H'(Q,P) = H\big(q(Q,P),\, p(Q,P)\big) = \omega\, P \cos^2 Q + \omega\, P \sin^2 Q = \omega\, P \tag{28.26}$$

The new variable Q is cyclic. The canonical equations are:

$$\dot{P} = -\frac{\partial H'}{\partial Q} = 0, \quad \dot{Q} = \frac{\partial H'}{\partial P} = \omega \tag{28.27}$$

These equations have the solution

$$P = \text{const.} = \frac{H'}{\omega} = \frac{E}{\omega}, \quad Q(t) = \omega t + \beta \tag{28.28}$$

Substituting this into (28.25), we regain the familiar solution in the original variables. The way described here is obviously not a standard method for problem-solving, because the simplification of the Hamiltonian function and the canonical equations depends on a suitable choice for the generating function.

Poisson bracket

We now define the *Poisson bracket*. It leads to an expression for the time derivative of an arbitrary physical quantity and to an alternative form of the equations of motion. The Poisson bracket is the counterpart of the quantum mechanical commutator.

A system state can be fixed by the coordinates q_i and the momenta p_i. Any physical quantity depends only on these variables and on time. We consider two such quantities, F and K:

$$F = F(q_1, \ldots, q_f, p_1, \ldots, p_f, t) = F(q, p, t), \quad K = K(q, p, t) \tag{28.29}$$

For these quantities, we define the so-called *Poisson bracket* by

$$\{F, K\} \equiv \sum_{i=1}^{f} \left(\frac{\partial F}{\partial q_i} \frac{\partial K}{\partial p_i} - \frac{\partial F}{\partial p_i} \frac{\partial K}{\partial q_i} \right) \tag{28.30}$$

The Poisson bracket is denoted by curly brackets $\{\ ,\ \}$ enclosing two arguments.

We first list some properties and special Poisson brackets. From the definition follows immediately

$$\{F, K\} = -\{K, F\} \quad \text{and} \quad \{F, F\} = 0 \tag{28.31}$$

In the Hamiltonian formalism, p_k, q_k, and t are independent variables. Therefore,

$$\begin{aligned} \frac{\partial q_i}{\partial q_j} &= \delta_{ij}, \quad \frac{\partial q_i}{\partial p_j} = 0, \quad \frac{\partial q_i}{\partial t} = 0, \\ \frac{\partial p_i}{\partial q_j} &= 0, \quad \frac{\partial p_i}{\partial p_j} = \delta_{ij}, \quad \frac{\partial p_i}{\partial t} = 0 \end{aligned} \tag{28.32}$$

We consider (28.30) specifically for $K = q_j$ and for $K = p_j$. This yields

$$\frac{\partial F}{\partial p_j} = -\{F, q_j\}, \quad \frac{\partial F}{\partial q_j} = \{F, p_j\} \tag{28.33}$$

The further specialization $F = q_i$ or $F = p_i$ gives

$$\{q_i, q_j\} = 0, \quad \{p_i, p_j\} = 0, \quad \{p_i, q_j\} = -\delta_{ij} \tag{28.34}$$

We now calculate the time dependence of an arbitrary physical quantity $F(q, p, t)$. In

$$\frac{dF}{dt} = \sum_{i=1}^{f} \frac{\partial F}{\partial q_i}\, \dot{q}_i + \sum_{i=1}^{f} \frac{\partial F}{\partial p_i}\, \dot{p}_i + \frac{\partial F}{\partial t} \tag{28.35}$$

we use the canonical equations. This yields

$$\frac{dF}{dt} = \{F, H\} + \frac{\partial F}{\partial t} \tag{28.36}$$

This equation determines the time dependence of an arbitrary physical quantity. The quantity F could be, for example, the angular momentum or the energy of the system.

On the basis of (28.36), one can pose the question of conservation laws in a new form. If a physical quantity F does not explicitly depend on the time, then it is a conserved quantity if its Poisson bracket with H vanishes. For $F = H$ this yields the relation already known from (27.16),

$$\frac{dH}{dt} = \frac{\partial H}{\partial t} \tag{28.37}$$

Now, $\partial H/\partial t = 0$ implies the conserved quantity $H = \text{const}$.

From (28.36) follow for $F = p_i$ or $F = q_i$ the canonical equations in the form

$$\dot{p}_i = \{p_i, H\}, \quad \dot{q}_i = \{q_i, H\} \tag{28.38}$$

Example

As a simple application of (28.36), we study the motion of the center-of-mass momentum for a system of N mass points. The Hamiltonian function is

$$H = \sum_{\nu=1}^{N} \frac{\boldsymbol{p}_\nu^2}{2m_\nu} + U(\boldsymbol{r}_1, \ldots, \boldsymbol{r}_N, t) \tag{28.39}$$

The system has $f = 3N$ degrees of freedom. The variables of the Hamiltonian function are denoted by

$$\begin{aligned} q &= (q_n) = (q_{3\nu+j-3}) = (\boldsymbol{r}_\nu \cdot \boldsymbol{e}_j) \\ p &= (p_n) = (p_{3\nu+j-3}) = (\boldsymbol{p}_\nu \cdot \boldsymbol{e}_j) \end{aligned} \tag{28.40}$$

where ν runs from 1 to N, j runs from 1 to 3, and n runs from 1 to $3N$. The center-of-mass momentum is given by

$$\boldsymbol{P} = \sum_{\nu=1}^{N} \boldsymbol{p}_\nu \tag{28.41}$$

For its x-component P_x we find

$$\frac{\partial P_x}{\partial p_{3\nu-2}} = 1, \quad \frac{\partial P_x}{\partial p_{3\nu-1}} = \frac{\partial P_x}{\partial p_{3\nu}} = 0, \quad \frac{\partial P_x}{\partial q_i} = 0, \quad \frac{\partial P_x}{\partial t} = 0 \tag{28.42}$$

Thus, we can evaluate (28.36) for P_x:

$$\begin{aligned} \dot{P}_x &= \{P_x, H\} + \frac{\partial P_x}{\partial t} = \sum_{n=1}^{3N} \left(\frac{\partial P_x}{\partial q_n} \frac{\partial H}{\partial p_n} - \frac{\partial P_x}{\partial p_n} \frac{\partial H}{\partial q_n} \right) \\ &= - \sum_{\nu=1}^{N} \frac{\partial U}{\partial q_{3\nu-2}} = \sum_{\nu=1}^{N} \boldsymbol{F}_\nu \cdot \boldsymbol{e}_1 \end{aligned} \tag{28.43}$$

Using the corresponding relation for $\dot{P}_y$ and $\dot{P}_z$. we get

$$\dot{\boldsymbol{P}} = \sum_{\nu=1}^{N} \boldsymbol{F}_\nu = \boldsymbol{F} \tag{28.44}$$

This is the well-known equation of motion for the center-of-mass momentum.

Exercises

28.1 *Example of canonical transformation*

Given is the Hamiltonian function $H(q,p) = \sum p_i^2/2m + U(q_1, \ldots, q_f)$. Proceed with the generating function

$$G(q, Q) = \sum_{i=1}^{f} q_i \, Q_i$$

to a new Hamiltonian function $H'(Q, P, t)$. Compare H' with H. What is the meaning of the new coordinates Q_i and momenta P_i?

28.2 *Generators for canonical transformation*

Given is the Hamiltonian function $H(q, p)$ for a system with one degree of freedom. Consider the transformation

$$Q = q^\alpha p^\beta, \quad P = q^\gamma p^\delta$$

Determine the parameter values $\alpha, \beta, \gamma, \delta$ for which this transformation is canonical. The necessary and sufficient condition is $\{Q, P\} = 1$, i.e. the Poisson bracket must be equal to 1. Determine the generating function $G(q, Q)$ for this case.

Chapter 29

Hamilton–Jacobi Equation

We determine the canonical transformation for which the new Hamilton function (28.12) vanishes:

$$H'(Q,P,t) = H(q,p,t) + \frac{\partial W(q,Q,t)}{\partial t} \stackrel{!}{=} 0 \qquad (29.1)$$

The generating function for this transformation is denoted by $G = W(q,Q,t)$. As we will see, W is equal to the action. The condition (29.1) leads to a partial differential equation for $W(q,Q,t)$, the Hamilton–Jacobi equation. This equation can be regarded as the basic equation of motion. This equation is used for investigating the relations between mechanics, optics, and quantum mechanics.

According to (28.11), the wanted transformation reads

$$p_k = \frac{\partial W(q,Q,t)}{\partial q_k} = p_k(q,Q,t), \quad P_k = -\frac{\partial W(q,Q,t)}{\partial Q_k} = P_k(q,Q,t) \qquad (29.2)$$

For $H' = 0$, the canonical equations give $\dot{Q}_i = 0$ and $\dot{P}_i = 0$, thus

$$Q_i = a_i = \text{const.}, \quad P_i = b_i = \text{const.} \qquad (29.3)$$

We insert $p_i = \partial W/\partial q_i$ and $Q_i = a_i$ into (29.1):

$$H\left(q_1, \ldots, q_f, \frac{\partial W(q,a,t)}{\partial q_1}, \ldots, \frac{\partial W(q,a,t)}{\partial q_f}, t\right) + \frac{\partial W(q,a,t)}{\partial t} = 0 \qquad (29.4)$$

In the following, we omit the constants $a = (a_i)$ in the argument. Then the last equation becomes

$$H\left(q_1, \dots, q_f, \frac{\partial W(q,t)}{\partial q_1}, \dots, \frac{\partial W(q,t)}{\partial q_f}, t\right) + \frac{\partial W(q,t)}{\partial t} = 0 \quad (29.5)$$

This *Hamilton–Jacobi equation* is a partial first-order differential equation for the function $W(q_1, \dots, q_f, t)$. We need a solution to this equation that satisfies (29.4), i.e., a solution $W(q,t) = W(q,a,t)$ which, in addition to the variables q and t depends on f constants $a_1, \dots, a_f$. This is a relatively special solution; because the *general* solution of (29.5) depends via the initial condition $W(q,0) = f(q)$ on an arbitrary function $f(q)$ (corresponding to a dependence on infinitely many constants).

A solution with as many integration constants as variables, i.e. $f+1$ for (29.5), is called a *complete integral*. The solution we are looking for is such a complete integral, since with $W(q,a,t)$ also $W + a_{f+1}$ is a solution. The trivial additive constant a_{f+1} does not change the canonical transformation. Often one can find a complete integral by a separation approach (see example at the end of the chapter).

We show that a solution of the special form $W(q,a,t)$ leads to the general solution of the equations of motion. From (29.2) and (29.3) follow the left equations in

$$\left.\begin{aligned} p_i &= \frac{\partial W(q,a,t)}{\partial q_i} \\ b_i &= -\frac{\partial W(q,a,t)}{\partial a_i} \end{aligned}\right\} \quad \xrightarrow{\text{resolving for}} \quad \begin{cases} q_i = q_i(a,b,t) \\ p_i = p_i(a,b,t) \end{cases} \quad (29.6)$$

The equations on the left represent $2f$ algebraic equations for the $2f$ functions $q_i(t)$ and $p_i(t)$. The solutions $q_i(t)$ and $p_i(t)$ of this system of equations depend on $2f$ constants a_i and b_i; therefore, they represent the general solution of the canonical equations.

We summarize the solution path discussed here: With the given Hamiltonian function $H(q,p,t)$ one sets up the Hamilton–Jacobi equation (29.5). One determines a solution $W(q,a,t)$ of this equation with f constants $a = (a_1, \dots, a_f)$ (without the trivial additive constant). A standard way is to try a separation expression for the solution. The solution $W(q, a, t)$ defines the algebraic system of equations on the left-hand side of (29.6). This set of equations is solved for $q_i = q_i(a, b, t)$ and $p_i = p_i(a, b, t)$. These functions $q_i = q_i(a, b, t)$ and $p_i = p_i(a, b, t)$ are then the general solution.

In the described way, the Hamilton–Jacobi equation leads to a solution of the equations of motion. Solving the Hamilton–Jacobi equation is, however, considerably more difficult than solving the original canonical equations: The Hamilton–Jacobi equation is a partial differential equation of first order for the function $W(q_1, \ldots, q_f, t)$, whereas the canonical equations are $2f$ ordinary differential equations (much simpler). The Hamilton–Jacobi equation serves less to solve concrete mechanical problems, but rather to study the relations between mechanics, optics, and quantum mechanics.

We calculate the time derivative of W, where we use (29.1)–(29.3):

$$\begin{aligned} \frac{dW(q, Q, t)}{dt} &= \sum_{i=1}^{f} \frac{\partial W}{\partial q_i}\, \dot{q}_i + \sum_{i=1}^{f} \frac{\partial W}{\partial Q_i}\, \dot{Q}_i + \frac{\partial W}{\partial t} \\ &= \sum_{i=1}^{f} p_i\, \dot{q}_i - H = \mathcal{L} \end{aligned} \tag{29.7}$$

From this follows $W = \int dt\, \mathcal{L} + \text{const}$, i.e. W is equal to the action (up to a constant).

Time-Independent Hamilton–Jacobi Equation

If H does not depend explicitly on the time t, then

$$W(q, t) = S(q) - E\, t \tag{29.8}$$

with a constant E is a solution of (29.5) if

$$H\left(q_1, \ldots, q_f, \frac{\partial S(q)}{\partial q_1}, \ldots, \frac{\partial S(q)}{\partial q_f}\right) = E \tag{29.9}$$

This *time-independent Hamilton–Jacobi equation* is a partial differential equation in the variables q, i.e. for a function $S(q)$.

As discussed above, we need a solution $W(q,a,t) = S - Et$ which depends on f integration constants. As the first constant, we can take $a_1 = E$. Thus, the wanted solution for S is of the form

$$S = S(q,a) = S(q_1,\ldots,q_f,E,a_2,\ldots,a_f), \quad a_1 = E \tag{29.10}$$

For this, we specify the equations (29.6):

$$\left.\begin{aligned} p_i &= \frac{\partial S(q,a)}{\partial q_i} \\ b_1 &= -\frac{\partial W}{\partial a_1} = -\frac{\partial S(q,a)}{\partial E} + t \\ b_{i\neq 1} &= -\frac{\partial S(q,a)}{\partial a_i} \end{aligned}\right\} \quad \underset{\longrightarrow}{\text{solving for } q_i} \quad \begin{cases} q_i = q_i(a,b,t) \\ p_i = p_i(a,b,t) \end{cases} \tag{29.11}$$

In the following, we solve the time-independent Hamilton–Jacobi equation explicitly for a special example.

Example

For the free fall in the gravity field $\boldsymbol{g} = -g\,\boldsymbol{e}_z$ the Hamiltonian function is

$$H = \frac{1}{2m}\left(p_x^2 + p_z^2\right) + mgz \tag{29.12}$$

We restrict ourselves to a motion in the x–z-plane. The time-independent Hamilton–Jacobi equation reads then

$$\frac{1}{2m}\left(\frac{\partial S(x,z)}{\partial x}\right)^2 + \frac{1}{2m}\left(\frac{\partial S(x,z)}{\partial z}\right)^2 + mgz = E \tag{29.13}$$

We need a solution of the form $S(x,z,E,a_2)$, which in addition to E depends on a second constant a_2. We try to find such a solution with a separation approach

$$S(x,z) = S_1(x) + S_2(z) \tag{29.14}$$

which we substitute into (29.13):

$$\left(\frac{dS_1}{dx}\right)^2 + \left(\frac{dS_2}{dz}\right)^2 + 2m^2 g\,z = 2mE \tag{29.15}$$

Here, $(dS_1/dx)^2 = \text{const.} = a_2$, because the other terms do not depend on x. Thus, we obtain

$$\frac{dS_1}{dx} = \sqrt{a_2} \quad \text{and} \quad \frac{dS_2}{dz} = \sqrt{2mE - a_2 - 2m^2 g z} \tag{29.16}$$

From this follows (except for insignificant additive constants)

$$S_1 = \sqrt{a_2}\, x, \quad S_2 = -\frac{1}{3m^2 g}\left(2mE - a_2 - 2m^2 g z\right)^{3/2} \tag{29.17}$$

Thus, we have found a complete integral of (29.13):

$$S = S(x, z, E, a_2) = \sqrt{a_2}\, x - \frac{1}{3m^2 g}\left(2mE - a_2 - 2m^2 g z\right)^{3/2} \tag{29.18}$$

We write down the equations of the left-hand side of (29.11):

$$p_x = \frac{\partial S}{\partial x}, \quad p_z = \frac{\partial S}{\partial z}, \quad \frac{\partial S}{\partial E} = t - b_1, \quad \frac{\partial S}{\partial a_2} = -b_2 \tag{29.19}$$

From this, $x(t)$, $z(t)$, $p_x(t)$, and $p_z(t)$ follow. We evaluate the third equation (inserting $\partial S/\partial E$ for S from 29.18):

$$-\frac{1}{mg}\sqrt{2mE - a_2 - 2m^2 g z} = t - b_1 \tag{29.20}$$

or

$$z(t) = -\frac{g}{2}(t - b_1)^2 + \frac{2mE - a_2}{2m^2 g} = -\frac{g}{2}t^2 + c_1 t + c_2 \tag{29.21}$$

In the z–t diagram, this is a parabola. From the first equation in (29.19) we get with (29.18) $p_x = \sqrt{a_2} = \text{const.}$ With $p_x = m\dot{x}$ this implies a linear time dependence of $x(t)$, and together with (29.21) we get a parabola in the x–z-plane, too.

PART VIII

Continuum Mechanics

Chapter 30

String Vibration

Continuum mechanics investigates the dynamics of elastic bodies, fluids and gases. It is an extensive field with numerous interesting applications. From a mathematical point of view, it deals with field theories; in place of the trajectories of mass points we consider fields for continuous mass distributions. In Part VIII, we will make a foray into this field.

As exemplary applications, we treat the string vibration (this chapter), the static beam bending (Chapter 31), and the basic equations of hydrodynamics (Chapter 32). Chapter 33 presents a brief outlook on other field theories.

Figure 30.1 shows a string that is stretched (by a tension force P) between two clamping points (at $x = 0$ and $x = l$); the rest position of the string coincides with the x-axis. The string might be deflected (i.e. displaced) perpendicular to the x-axis by the length u. This deflection depends on position and time:

$$u = u(x,t) = \text{deflection of the string} \tag{30.1}$$

The function $u(x,t)$ describes the motion (vibration, for example) of the string. In this chapter, we set up the equation of motion for $u(x,t)$.

The deflection of a string is a simple example of a *field*. A field is a physical quantity that depends on the position and generally also on time; examples are the temperature field $T(\boldsymbol{r},t)$ of the atmosphere, the electric field $\boldsymbol{E}(\boldsymbol{r},t)$ in an oscillating circuit, the

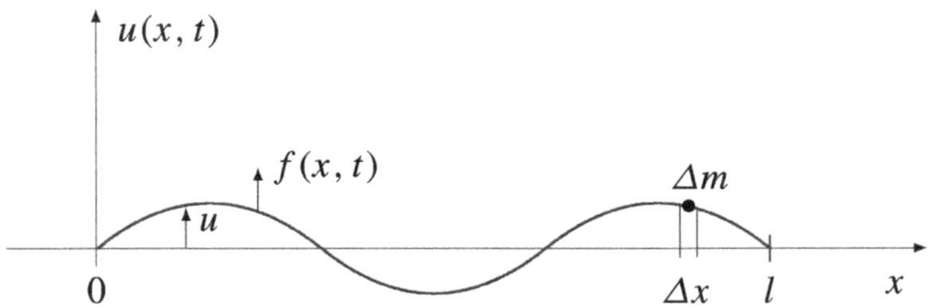

Figure 30.1 A string is clamped between the points $x = 0$ and $x = l$ on the x-axis. A tension force P on both ends of the spring implies that the rest position is equal to the x-axis (provided that there are no other forces). The length by which the string is deflected perpendicular to the x-axis, is denoted by $u(x,t)$. The figure sketches a possible deflection at a given time t.

Schrödinger wave function $\psi(\boldsymbol{r},t)$ of an electron in an atom or the velocity field $\boldsymbol{v}(\boldsymbol{r},t)$ of the ocean current. The deflection $u(x,t)$ of the string is an especially simple field because it depends only on one spatial coordinate and consists of only one quantity (in contrast to three components of $\boldsymbol{E}$ or $\boldsymbol{v}$), and because the meaning of u is immediately understandable (in contrast for instance to that of the wave function ψ). In continuum mechanics, the fields are the central quantities; instead of the equations of motion for trajectories we have to consider partial differential equations for these fields, the so-called *field equations*.

We think of the string as being divided into N mass elements, which we replace with point masses. The elastic properties of the string are simulated by spring forces between the mass points. Thus, we can start from the well-known problem of the motion of N mass points. For this, we set up the Lagrange function. Subsequently we consider the limit $N \to \infty$ and evaluate the Hamiltonian principle. This leads to the wave equation of the string.

A string can be deflected not only vertically (as considered here) but also horizontally. After the substitution by N mass points the horizontal vibrations can be treated as in Exercise 26.4.

The rest position of the string is the section $[0, l]$ of the x-axis. We divide this interval into N equal sections, $l = N\Delta x$. For $N \gg 1$

we replace the masses of the individual sections with mass points:

$$\Delta m_i \quad \text{at } x_i = \big(i - 1/2\big)\, \Delta x, \quad (i = 1, 2, \ldots, N) \tag{30.2}$$

The displacement $u_i(t)$ of the ith mass point is related to the field:

$$u_i(t) = u(x_i, t) \tag{30.3}$$

The discretization leads to N functions $u_i(t)$ instead of the field $u(x,t)$. For $N \to \infty$ the functions $u_i(t)$ (i.e. the trajectories of the mass elements) determine the field $u(x,t)$. At the same time, the errors of the discretization go to zero.

The kinetic energy of the system of N mass points is

$$T = \sum_{i=1}^{N} \frac{\Delta m_i}{2} \left(\frac{du_i}{dt}\right)^2 = \sum_{i=1}^{N} \frac{\varrho\, \Delta x}{2}\, \dot{u}_i^2 \tag{30.4}$$

In the last step, we introduced the homogeneous mass density $\varrho =$ mass/length; each section element has the mass $\varrho\, \Delta x$.

Let the string be clamped with the force P between the points $x = 0$ and $x = l$. The elastic properties of the string can be simulated by springs between neighboring mass points. The distance between the ith and $(i+1)$th mass points is

$$\Delta s = \sqrt{(\Delta x)^2 + (u_{i+1} - u_i)^2} \approx \Delta x \left[1 + \frac{(u_{i+1} - u_i)^2}{2\,(\Delta x)^2}\right] \tag{30.5}$$

Here, we have assumed that the string has a flat slope only. The tension force P of the string means that the spring is already prestressed in its rest position. An increase of the distance from Δx to Δs then results in the potential energy contribution $P\,(\Delta s - \Delta x)$. The sum of these contributions gives the potential energy

$$U = \sum_{i=1}^{N-1} P\, \Delta x\, \frac{(u_{i+1} - u_i)^2}{2\,(\Delta x)^2} \tag{30.6}$$

In addition, there is a contribution of the spring between the first and last mass point and the clamping point. For $N \to \infty$ we can neglect these extra contributions.

We perform the limit $N \to \infty$ for the kinetic energy:

$$T = \lim_{N\to\infty} \sum_{i=1}^{N} \frac{\varrho\, \Delta x}{2}\, \dot{u}_i^2 = \frac{\varrho}{2} \lim_{N\to\infty} \sum_{i=1}^{N} \Delta x \left(\frac{\partial u(x_i, t)}{\partial t}\right)^2$$

$$= \frac{\varrho}{2} \int_0^l dx \left(\frac{\partial u(x, t)}{\partial t}\right)^2 \tag{30.7}$$

Here, (30.3) was used. With

$$\lim_{\Delta x\to 0} \frac{u_{i+1}(t) - u_i(t)}{\Delta x} = \frac{\partial u(x_i, t)}{\partial x_i} \tag{30.8}$$

we evaluate the potential energy:

$$U = \lim_{N\to\infty} \sum_{i=1}^{N-1} P\, \Delta x\, \frac{(u_{i+1} - u_i)^2}{2\,(\Delta x)^2} = \frac{P}{2} \int_0^l dx \left(\frac{\partial u(x, t)}{\partial x}\right)^2 \tag{30.9}$$

The Lagrange function $\mathscr{L} = T - U$ may be expressed by the *Lagrange density* $\mathcal{L}$:

$$\mathscr{L} = T - U = \int_0^l dx\, \underbrace{\left[\frac{\varrho}{2}\left(\frac{\partial u(x, t)}{\partial t}\right)^2 - \frac{P}{2}\left(\frac{\partial u(x, t)}{\partial x}\right)^2\right]}_{\mathcal{L}(\dot{u}, u') = (\varrho\, \dot{u}^2 - P u'^2)/2} \tag{30.10}$$

Note the use of slightly different letter fonts for the Lagrange function $\mathscr{L}$ and for the Lagrangian density $\mathcal{L}(\dot{u}, u')$. In the following, we will mostly consider Lagrange densities $\mathcal{L}$; therefore, a more obvious distinction in the notation is not necessary.

Since the string can be considered as a limiting case of a system of many mass points the Hamiltonian principle is valid:

$$\delta S = \delta \int_{t_1}^{t_2} dt \int_0^l dx\ \mathcal{L}(u, \dot{u}, u', x, t) = 0 \tag{30.11}$$

For later generalizations, we allow that $\mathcal{L}$ also depends explicitly on u, x, and t. For δS we have to vary $u(x, t)$; this corresponds to the variation of N functions $u_i(t)$ in the limiting case $N \to \infty$.

The Euler–Lagrange equations for (30.11) follow from (12.32); to this end, $F = \mathcal{L}(u, \dot{u}, x, t)$, $y = u$, $x_1 = x$, $x_2 = t$ and $N = 2$ has to be inserted in (12.32). This yields

$$\frac{\partial}{\partial t}\frac{\partial \mathcal{L}}{\partial \dot{u}} + \frac{\partial}{\partial x}\frac{\partial \mathcal{L}}{\partial u'} - \frac{\partial \mathcal{L}}{\partial u} = 0 \tag{30.12}$$

The partial derivatives with respect to u, u', and $\dot{u}$ act on $\mathcal{L}(u, u', \dot{u}, x, t)$. For the following partial derivatives (like $\partial/\partial x$) these expressions (like $\partial\mathcal{L}/\partial u'$) have to be considered as functions of x and t.

For (30.10), $\mathcal{L} = (\varrho\,\dot{u}^2 - P u'^2)/2$, (30.12) becomes the wave equation

$$\frac{\partial^2 u}{\partial x^2} - \frac{1}{c^2}\frac{\partial^2 u}{\partial t^2} = 0 \qquad \text{wave equation of a free string} \tag{30.13}$$

Here, we introduced the *wave velocity* c

$$c = \sqrt{\frac{P}{\varrho}} \tag{30.14}$$

In the derivation of the Euler–Lagrange equation from $\delta S = 0$, it is assumed that the function to be varied is held fixed at the boundaries. In (30.11), the integrating region is a rectangle in the x–t-plane defined by $x = 0$, $x = l$, $t = t_1$, and $t = t_2$. On two sides of this rectangle, the clamping of the string at 0 and l implies

$$u(0,t) = u(l,t) = 0 \tag{30.15}$$

This *boundary condition* supplements the differential equation (30.13). A specification at the two other sides of the rectangle is given by two functions $F(x)$ and $H(x)$,

$$u(x,t_1) = F(x), \quad u(x,t_2) = H(x) \quad (x \in [0,l]) \tag{30.16}$$

We will use the equivalent *initial conditions*

$$u(x,t_1) = F(x), \quad \dot{u}(x,t_1) = G(x) \quad (x \in [0,l]) \tag{30.17}$$

By these conditions, the position and velocity of the string at time t_1 are given. From the wave equation then follows the deflection $u(x,t)$ at any time t.

Generalizations

The wave equation for the free string can be generalized in several respects. On the one hand, forces could act on the string. On the other hand, the wave equation can be generalized to the multidimensional case.

The force density $f(x,t)$ (force per length) shall act on the string. If f acts in the direction of u (as depicted in Figure 30.1), the deflection of a mass element Δm yields an energy contribution $\Delta U = -u\, f\, \Delta x$. Thus, the Lagrangian density becomes

$$\mathcal{L}(u, \dot{u}, u', x, t) = \frac{\varrho}{2}\, \dot{u}^2 - \frac{P}{2}\, u'^2 + u\, f(x,t) \tag{30.18}$$

In contrast to (30.10), $\mathcal{L}$ now depends explicitly on the deflection u itself, and via the external force also on x and t. From (30.12) follows the inhomogeneous wave equation

$$\frac{\partial^2 u}{\partial x^2} - \frac{1}{c^2}\frac{\partial^2 u}{\partial t^2} = -\frac{f(x,t)}{P} \tag{30.19}$$

An example of such an external force is gravity; if the acceleration due to gravity $\boldsymbol{g}$ in Figure 30.1 points downwards, then $f = -\varrho g$.

Instead of the string, we consider a clamped and prestretched membrane. The rest position of the membrane lies in the x–y-plane. For transverse deflections $u(x, y, t)$ we obtain analogously to (30.13) the two-dimensional wave equation

$$\frac{\partial^2 u}{\partial x^2} + \frac{\partial^2 u}{\partial y^2} - \frac{1}{c^2}\frac{\partial^2 u}{\partial t^2} = 0 \tag{30.20}$$

For the three-dimensional generalization, the field $u = u(x, y, z, t)$ depends on three spatial coordinates, and the wave equation becomes

$$\frac{\partial^2 u}{\partial x^2} + \frac{\partial^2 u}{\partial y^2} + \frac{\partial^2 u}{\partial z^2} - \frac{1}{c^2}\frac{\partial^2 u}{\partial t^2} = 0 \quad \text{or} \quad \Delta u - \frac{\ddot{u}}{c^2} = 0 \tag{30.21}$$

The spatial derivatives may be abbreviated by the Laplace operator Δ. Such a wave equation applies, for example, to the displacements (called deflections for the string) in a solid. The associated waves are sound waves. They can be transversal (wave vector perpendicular to the displacements, as in the case of the string) or also longitudinal (wave vector parallel to the displacements). In liquids and gases, there are only longitudinal sound waves (Chapter 32); in these cases, the considered field $u = \varrho - \varrho_0$ is the deviation of the density from the equilibrium density ϱ_0. A wave equation of the form (30.21) is also valid for each component of the electromagnetic field in the vacuum; in this case, c is then the velocity of light.

Equations (30.13) and (30.19)–(30.21) are the equations of motion for the field u; they are called *field equations*. In all cases discussed so far, the field equations are *linear* in the field u. Linear field equations are particularly easy to solve; this will become clear in the following section presenting the solution for the free string. Basic equations of physics are often linear field equations. In particular, the Maxwell, the Schrödinger, and the Dirac equations are linear. Nonlinear field equations have to be considered, for example, in hydrodynamics (Chapter 32) or in the theory of general relativity.

Vibrations of the Force-Free String

We determine the possible vibrations of the force-free string. For this purpose, we put together the equations to be solved:

$$\begin{aligned} &u''(x,t) - \ddot{u}(x,t)/c^2 = 0 && \text{wave equation} \\ &u(0,t) = u(l,t) = 0 && \text{boundary condition} \\ &u(x,0) = F(x), \quad \dot{u}(x,0) = G(x) && \text{initial condition} \end{aligned} \tag{30.22}$$

The initial time t_1 was arbitrarily set equal to zero.

For the solutions of *linear* differential equations, one can use the superposition principle. It means that with u_1 and u_2 also any linear combination of u_1 and u_2 is a solution. If we write the linear differential equation as $D(u) = 0$, then the superposition relation follows from

$$\left.\begin{aligned} D(u_1) = 0 \\ D(u_2) = 0 \end{aligned}\right\} \longrightarrow a_1\, D(u_1) + a_2\, D(u_2) = D(a_1 u_1 + a_2 u_2) = 0 \tag{30.23}$$

Here, a_1 and a_2 are arbitrary constants. We determine a set of solutions $u_n(x,t)$ that satisfy the wave equation and the boundary conditions in (30.22). This set will be complete because we can determine a linear combination $\sum a_n u_n(x,t)$ which satisfies the initial conditions. Such a linear combination, therefore, represents the *general solution* of the differential equation with boundary conditions. The presented method is also used for the solution of the time-dependent Schrödinger equation (for instance for the free motion of a particle in a box) or the free Maxwell equations (e.g. for electromagnetic oscillations in a cavity).

As we will see, the separation approach

$$u(x,t) = v(x)\, g(t) \tag{30.24}$$

leads to a complete set of solutions $u_n(x,t)$. We put (30.24) into the wave equation:

$$\frac{v''(x)}{v(x)} = \frac{1}{c^2}\frac{\ddot{g}(t)}{g(t)} \tag{30.25}$$

The partial derivatives have become ordinary derivatives (denoted by dashes or overpoints, respectively). Since the left side does not depend on t and the right side does not depend on x, both sides are independent of x and t. So they must be equal to a constant, which we denote by $-k^2$. Thus, (30.25) becomes

$$v''(x) + k^2\, v(x) = 0, \quad \ddot{g}(t) + \omega^2\, g(t) = 0 \quad (\omega = ck) \tag{30.26}$$

The general solution of the differential equation for $v(x)$ is

$$v(x) = C_1 \sin(kx) + C_2 \cos(kx) \tag{30.27}$$

The boundary condition $v(0) = v(l) = 0$ gives $C_2 = 0$ and

$$kl = n\pi, \quad k = \frac{n\pi}{l} = k_n, \quad n = 1, 2, \ldots \tag{30.28}$$

For n, the values $0,\ \pm 1,\ \pm 2, \ldots$ are possible. However, $n = 0$ leads to $u \equiv 0$, and a possible minus sign can be included in the constant C_1. Therefore, we can restrict n to positive values.

The differential equation for $g(t)$,

$$\ddot{g}(t) + \omega_n^2\, g(t), \quad \omega_n = c\, k_n, \quad n = 1, 2, \ldots \tag{30.29}$$

has the general solution $g(t) = a\cos(\omega_n t) + b\sin(\omega_n t)$. Thus, the approach (30.24) leads to the solution

$$u_n(x,t) = C_1 \sin(k_n x)\,\big[a\cos(\omega_n t) + b\sin(\omega_n t)\big] \tag{30.30}$$

The constant C_1 can be omitted in view of the other constants a and b. Due to the linearity of the wave equation, any linear combination

$$u(x,t) = \sum_{n=1}^{\infty} \sin(k_n x)\,\big[a_n\cos(\omega_n t) + b_n\sin(\omega_n t)\big] \tag{30.31}$$

is also a solution. Together with the u_n, this solution satisfies the boundary conditions. Equation (30.31) represents the *general solution* since it admits arbitrary initial conditions:

$$u(x,0) = \sum_{n=1}^{\infty} a_n \sin(k_n x) = F(x) \tag{30.32}$$

$$\dot{u}(x,0) = \sum_{n=1}^{\infty} b_n\,\omega_n \sin(k_n x) = G(x) \tag{30.33}$$

Of course, only initial conditions with $F(0) = F(l) = 0$ and $G(0) = G(l) = 0$ are allowed. From the theory of Fourier series, it is known that any such function in the interval $[0, l\,]$ can be expanded in this way. The expansion coefficients a_n and b_n are

$$a_n = \frac{2}{l}\int_0^l dx\ F(x)\,\sin(k_n x), \qquad b_n = \frac{2}{l\,\omega_n}\int_0^l dx\ G(x)\,\sin(k_n x) \tag{30.34}$$

The general solution of (30.22) is now given by (30.31) with (30.34). This solution can also be written in the form

$$u(x,t) = \sum_{n=1}^{\infty} A_n \sin(k_n x)\,\sin(\omega_n t + \varphi_n) \tag{30.35}$$

It is a superposition of the natural vibrations (or eigenvibrations)

$$u_n(x,t) = A_n \sin(k_n x)\,\sin(\omega_n t + \varphi_n) \tag{30.36}$$

In Table 30.1, some designations for the occurring quantities are compiled.

Table 30.1 Some quantities and designations, which are commonly used for wave solutions or vibrations.

Symbol	Name	Relation
A_n	Amplitude	
φ_n	Phase	
ω, ω_n	Frequency, natural frequency	
k, k_n	Wave number	$k = \omega/c$
λ	Wavelength	$\lambda = 2\pi/k$
c	Wave velocity	$c^2 = P/\varrho$

The solution (30.36) describes an *eigenvibration* or *eigenmode* of the string. Figure 30.1 can be regarded as a snapshot of the eigenmode $u_3(x,t)$, and it may depict the string at the time t for which $\sin(\omega_3 t + \varphi_3) = 1$. At a certain x-position, the string will oscillate between the shown deflection position and the corresponding negative one.

In a natural oscillation (30.36), the string has constant *nodes*, which means the points x with $u(x,t) = 0$ do not move during the oscillation. Therefore, one also speaks of a *standing wave*. A standing wave, also known as a stationary wave, is a wave that oscillates in time but whose amplitude profile does not move in space. The wavelength λ_n equals twice the nodal distance. The first oscillation $n = 1$ with $\lambda_1 = 2l$ has two nodes (the clamping points). This oscillation is also called the original or fundamental wave. The nth oscillation has $n+1$ nodes.

By coupling the string via a resonance body to the air, we can hear a tone of the frequency ω_n when the natural oscillation $u_n(x,t)$ is excited. The fundamental frequency $\omega_1 = \pi c/l = \pi\,(P/\varrho)^{1/2}/l$ of a string can be shifted by changing the tension force P (for example, when tuning a violin). All higher frequencies (overtones) change accordingly.

Wave packets can be formed by the superposition of eigen oscillations. Localized wave packets can move with velocity c between the clamping points (Exercise 30.2).

Exercises

30.1 *String vibration for given initial conditions*

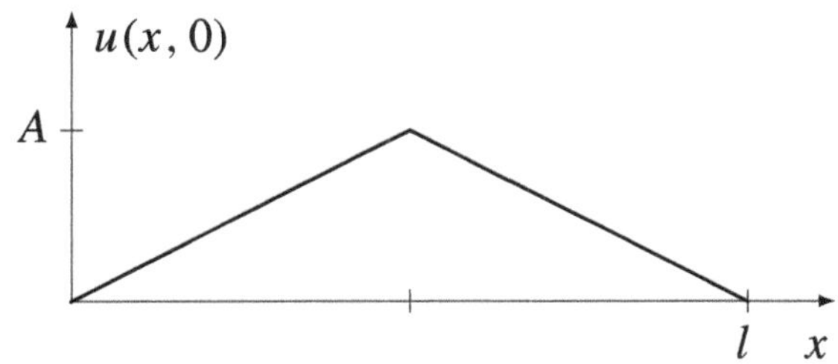

The deflection $u(x,t)$ of a string satisfies the wave equation and the boundary conditions

$$u''(x,t) - \ddot{u}(x,t)/c^2 = 0$$
$$u(0,t) = u(l,t) = 0$$

Determine the solution $u(x,t)$ for the initial condition

$$u(x,0) = A\left(1 - \left|1 - \frac{2x}{l}\right|\right), \quad \dot{u}(x,0) = 0 \tag{30.37}$$

30.2 *Solution method according to d'Alembert*

Show that the wave equation $u''(x,t) - \ddot{u}(x,t)/c^2 = 0$ is solved by

$$u(x,t) = f(x - ct) + g(x + ct) \tag{30.38}$$

At first f and g are arbitrary functions of one variable. How are f and g related to the initial conditions?

Apply this *solution method according to d'Alembert* to the string vibration from Exercise 30.1. Which relations must the functions f and g satisfy, so that the boundary conditions are fulfilled for all times? Show: Outside the definition range $[0, l]$ the boundary conditions must be continued in an antisymmetrical and periodic way (with the period $2l$).

Calculate $u(x,t)$ for $0 \le t \le T/4$ with $T = 2l/c$. The result for the full-time period follows then from symmetry considerations.

Chapter 31

Beam Bending

We investigate the bending of a beam (a massive, solid bar) in a force field. Specifically, we calculate the static deflection of the beam in the gravity field.

We consider a beam as sketched in Figure 31.1. The beam is clamped at the left wall. The rest position of the centerline coincides in the force-free case with the interval $[0, l]$ of the x-axis. The beam is now deflected downwards by the gravitational force. We denote the deflection of the beam from the force-free rest position by

$$u = u(x,t) = \text{deflection of beam} \quad \big(x \in [0,l],\ |u'| \ll 1\big) \qquad (31.1)$$

The curve $u = u(x,t)$ describes the deflected centerline of the beam. We assume that the deviations from the horizontal are small, $|u'| \ll 1$. Then the displacements of the mass elements in x-direction can be neglected.

The kinetic energy has the same form (30.7) as the string

$$T = \frac{\varrho}{2} \int_0^l dx \left(\frac{\partial u(x,t)}{\partial t}\right)^2 = \frac{\varrho}{2} \int_0^l dx\ \dot{u}^2 \qquad (31.2)$$

We abbreviate the partial derivatives with $\dot{u} = \partial u/\partial t$ and $u' = \partial u/\partial x$. The mass density ϱ is the mass per length.

In the right part of Figure 31.1, the stretch and compression due to bending are sketched. The angle δ and the stretch $\Delta x' - \Delta x$ are

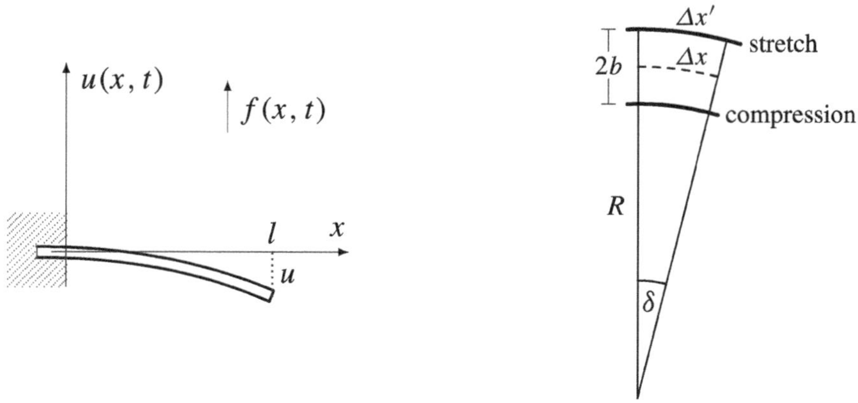

Figure 31.1 Beam bending in the gravity field: the vertical deflection from the rest position is denoted by $u(x,t)$. In contrast to the sketch, we restrict ourselves to $|u'| \ll 1$. On the right, side an enlarged section of the beam is shown. The dashed center line is not stretched by the bending. Compared to its length Δx, the upper part is stretched to $\Delta x'$, and the lower part is compressed accordingly. With Hooke's law this results in a potential energy contribution proportional to $(\Delta x' - \Delta x)^2$.

connected by

$$\delta = \frac{\Delta x' - \Delta x}{b} = \frac{\Delta x}{R} \tag{31.3}$$

Here, Δx is the unstretched length of the considered small beam section, $\Delta x'$ is the stretched length, $2b$ is the beam thickness, and R is the curvature radius. The curvature radius of a function $u(x,t)$ at the position x is given by

$$\frac{1}{R} = \frac{u''}{(1+u'^2)^{3/2}} \approx u''(x,t) \quad (|u'| \ll 1) \tag{31.4}$$

The potential energy contribution is proportional to the square of the relative stretch or compression, i.e. to $(\Delta x' - \Delta x)^2/(\Delta x)^2 = b^2/R^2 = b^2 u''^2$. The entire potential energy reads then

$$U_k = \frac{k}{2}\int_0^l dx \left(\frac{\partial^2 u(x,t)}{\partial x^2}\right)^2 = \frac{k}{2}\int_0^l dx\, u''^2 \tag{31.5}$$

Here, k is a positive parameter that depends on the material and the thickness of the beam. A contribution proportional to u'^2 as in (30.10) would also be obtained if the beam were clamped under tension (or compression). As in (30.18) we admit an external force density $f(x,t)$ acting on the beam. It gives the potential energy contribution

$$U_f = -\int_0^l dx\, u(x,t)\, f(x,t) \tag{31.6}$$

From (31.2), (31.5) and (31.6) follow the Lagrange function $\mathscr{L} = T - U_k - U_f = \int dx\, \mathcal{L}$ with the Lagrangian density

$$\mathcal{L}(u,\dot{u},u'',x,t) = \frac{\varrho}{2}\,\dot{u}^2 - \frac{k}{2}\,u''^2 + u\, f(x,t) \tag{31.7}$$

We use the Hamilton principle:

$$\delta \int_{t_1}^{t_2} dt \int_0^l dx\, \mathcal{L}(u,\dot{u},u'',x,t) = 0 \tag{31.8}$$

According to (12.32) and (12.35), the corresponding Euler–Lagrange equations are

$$\frac{\partial}{\partial t}\frac{\partial \mathcal{L}}{\partial \dot{u}} - \frac{\partial^2}{\partial x^2}\frac{\partial \mathcal{L}}{\partial u''} = \frac{\partial \mathcal{L}}{\partial u} \tag{31.9}$$

For the partial derivatives with respect to t and x, the expressions $\partial\mathcal{L}/\partial\dot{u}$ and $\partial\mathcal{L}/\partial u''$ have to be considered as functions of t and x. For (31.7) we obtain

$$\varrho\, \ddot{u}(x,t) + k\, u''''(x,t) = f(x,t) \tag{31.10}$$

This field equation allows the treatment of beam bending similarly to the string deflection. The differential equation has to be supplemented by boundary conditions, which are discussed below. The actual solution is then fixed by the initial conditions for $u(x,0)$ and $\dot{u}(x,0)$.

Static Bending in the Gravity Field

We want to calculate the *static* bending of the beam in the gravity field. Let the gravity be directed downwards in Figure 31.1,

$$f(x,t) = -\varrho\, g \tag{31.11}$$

The static equilibrium position $u = u(x)$ is determined by the condition of minimum potential energy:

$$U = U_k + U_f = \int_0^l dx \left(\frac{k}{2}\, u''^2 + \varrho\, g\, u \right) = \text{minimal} \tag{31.12}$$

This condition — or Hamilton's principle for the static case — results in

$$\delta \int_0^l dx \left(u''^2 + \frac{2\,\varrho\, g}{k}\, u \right) = 0 \longrightarrow u''''(x) = -\,\frac{\varrho\, g}{k} \tag{31.13}$$

This may also be obtained directly from (31.10) with (31.11). The general solution of this differential equation is

$$u(x) = -\,\alpha\, x^4 + C_3\, x^3 + C_2\, x^2 + C_1\, x + C_0 \tag{31.14}$$

Here, the C_i are integration constants and α is given by

$$\alpha = \frac{\varrho\, g}{24\, k} > 0 \tag{31.15}$$

The integration constants are to be determined by the boundary conditions of the beam. Initial conditions do not occur in the considered static case.

The boundary conditions depend on the type of support of the beam. We consider different bearing or clamping possibilities of the beam at its ends, i.e., at $x = 0$ and $x = l$. For the cases shown in Figure 31.2, the following holds:

$$\begin{array}{ll} \text{(i) clamped on both sides:} & u(0) = u'(0) = 0,\ u(l) = u'(l) = 0 \\ \text{(ii) clamped on one side:} & u(0) = u'(0) = 0,\ u(l),\ u'(l) \text{ arbitrary} \\ \text{(iii) underlaid on both sides:} & u(0) = u(l) = 0,\quad u'(0),\ u'(l) \ \text{ arbitrary} \end{array} \tag{31.16}$$

In case (i), four conditions are available to define the constants C_1 to C_4. In the other two cases, it is not a priori clear which boundary conditions determine the four constants C_i. To clarify this point, we reconsider the derivation of the Euler–Lagrange equation.

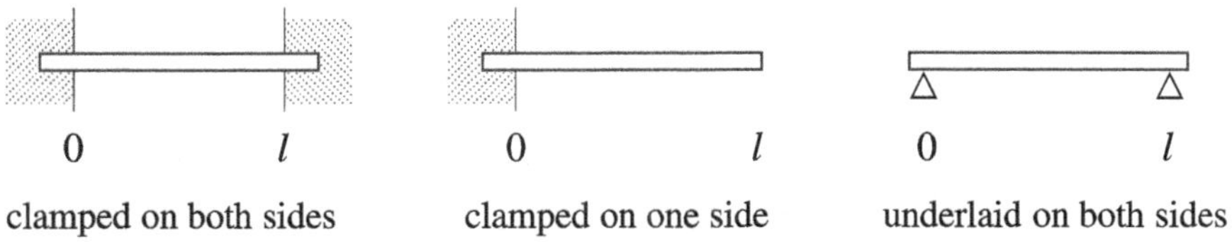

Figure 31.2 Three different bearings of a beam in the gravitational field.

In Equation (31.12), the potential energy U is a functional of $u(x)$ of the form

$$U = U[u] = \int_0^l dx\; F(u'', u) \tag{31.17}$$

We compute the variation of this functional

$$\begin{aligned} \delta U &= U[u+\delta u] - U[u] = \int_0^l dx\, \left(F_{u''}\, \delta u'' + F_u\, \delta u\right) \\ &\overset{\text{p.i.}}{=} \left(F_{u''}\, \delta u'\right)_l - \left(F_{u''}\, \delta u'\right)_0 - \left(\frac{dF_{u''}}{dx}\, \delta u\right)_l + \left(\frac{dF_{u''}}{dx}\, \delta u\right)_0 \\ &\quad + \int_0^l dx\, \left(\frac{d^2 F_{u''}}{dx^2} + F_u\right) \delta u(x) = 0 \end{aligned} \tag{31.18}$$

The term $F_{u''}\, \delta u''$ has been partially integrated twice. From the arbitrariness of $\delta u(x)$ follows the differential equation $d^2 F_{u''}/dx^2 + F_u = 0$, i.e. (31.13). In addition to this integral, the boundary terms must also vanish. For the variation $\delta u(x)$ at the boundary, we obtain according to the physical specification (31.16):

$$\begin{array}{ll} \text{(i) } \delta u(0) = \delta u'(0) = 0, & \delta u(l) = \delta u'(l) = 0 \\ \text{(ii) } \delta u(0) = \delta u'(0) = 0, & \delta u(l),\ \delta u'(l) \text{ arbitrary} \\ \text{(iii) } \delta u(0) = \delta u(l) = 0, & \delta u'(0),\ \delta u'(l) \text{ arbitrary} \end{array} \tag{31.19}$$

In the first case (i), the boundary terms in (31.18) vanish. In the second case (ii), (31.18) can only be satisfied if the coefficients of $\delta u(l)$ and $\delta u'(l)$ vanish. So, from the variation at the free end of the beam follows $F_{u''} = 2u'' = 0$ and $dF_{u''}/dx = 2u''' = 0$; thus the conditions $u''(l) = u'''(l) = 0$ are added to (31.16). In the third case (iii), it follows from (31.18) accordingly $u''(0) = u''(l) = 0$. In each case, we obtain a total of four boundary conditions:

$$\begin{aligned} &\text{(i) } u(0)=0, && u'(0)=0, && u(l)=0, && u'(l)=0 \\ &\text{(ii) } u(0)=0, && u'(0)=0, && u''(l)=0, && u'''(l)=0 \\ &\text{(iii) } u(0)=0, && u''(0)=0, && u(l)=0, && u''(l)=0 \end{aligned} \qquad (31.20)$$

The free variation of u' at the edge implies that u'' vanishes. On the other hand, a freely varying u at the boundary implies $u''' = 0$.

With (31.20) we determine the integration constants in the solution (31.14):

$$\begin{aligned} \text{(i)} \quad u(x) &= -\alpha\, x^2\, (x-l)^2 \\ \text{(ii)} \quad u(x) &= -\alpha\, x^2 \left(x^2 - 4l\,x + 6l^2\right) \\ \text{(iii)} \quad u(x) &= -\alpha\, x \left(x^3 - 2l\,x^2 + l^3\right) \end{aligned} \qquad (31.21)$$

The function $u(x)$ describes the equilibrium position of the beam in the gravity field. We specifically calculate the displacement at the center of the beam:

$$u(l/2) = -A\,l^4 \cdot \begin{cases} 1/16 & \text{(i)} \\ 17/16 & \text{(ii)} \\ 5/16 & \text{(iii)} \end{cases} \qquad (31.22)$$

Chapter 32

Hydrodynamics

We present the basic equations for the motion of gases and liquids. For this purpose, we generalize Newton's second law to the Euler equation of hydrodynamics. We discuss some simple applications (fluid in the gravitational field, Bernoulli equation, potential flow). As a solution of the linearized hydrodynamic equations, we obtain sound waves.

For liquids (gases) the mass points are replaced by the *mass density* ϱ,

$$\varrho(\boldsymbol{r}, t) = \frac{\sum_\nu m_\nu}{\Delta V} = \frac{\Delta m}{\Delta V} = \frac{\text{mass}}{\text{volume}} \tag{32.1}$$

The sum runs over all atoms or molecules (of mass m_ν, $\nu = 1, \ldots, \Delta N$), which are located in the volume element ΔV at $\boldsymbol{r}$. The volume element ΔV is chosen large enough so that $\Delta N \gg 1$, and at the same time small enough so that the macroscopic properties do not change significantly within ΔV. Under these conditions $\varrho(\boldsymbol{r}, t)$ does not depend on ΔV. The volume ΔV could contain, for example, $\Delta N = 10^7$ atoms corresponding roughly to $(\Delta V)^{1/3} \sim 10^{-5}$ cm for a liquid, or to $(\Delta V)^{1/3} \sim 10^{-4}$ cm for a gas.

The velocities $\boldsymbol{v}_\nu$ of the mass points in ΔV are replaced by the velocity field

$$\boldsymbol{v}(\boldsymbol{r}, t) = \frac{1}{\Delta N} \sum_{\nu=1}^{\Delta N} \boldsymbol{v}_\nu = \langle \boldsymbol{v}_\nu \rangle \tag{32.2}$$

This is the average velocity of the ΔN atoms in the considered volume ΔV located at $\boldsymbol{r}$. The actual velocities $\boldsymbol{v}_\nu$ of individual air molecules in the auditorium are in the range of $100\ldots1000\,\mathrm{m/s}$ (Maxwell distribution). The mean velocity $\boldsymbol{v}(\boldsymbol{r},t)$ is generally smaller by orders of magnitude; it describes, for example, the perceptible draught or wind.

The pressure[1] in the medium can be defined by

$$P = P(\boldsymbol{r},t) = \frac{\Delta F}{\Delta A} = \frac{\text{force}}{\text{area}} \tag{32.3}$$

For example, a gas is enclosed in a cylinder with a movable piston (area ΔA). Then one can measure the force ΔF on the piston, and thus the pressure $P = \Delta F/\Delta A$. Conversely, one can place a small, empty cylinder into the interior of a medium (gas, liquid), measure the force on the piston and thus determine the pressure. In the air in the lecture hall, the pressure is $P \approx 1\,\mathrm{bar}$. This matter pressure balances the gravitational pressure of the atmosphere.

Microscopically, the pressure of a gas can be explained by the impulses of the individual atoms or molecules. Enclosed in a container, the atoms are constantly reflected at the wall. Each atom thereby exerts an impulse; the time averaging over these impulses results in a certain force ΔF. The force per area yields the pressure $P = \Delta F/\Delta A$.

If the size of the volume ΔV is chosen properly, then the fields defined in (32.1)–(32.3) do not depend on this size. Fluids, which can be described by the fields $\varrho(\boldsymbol{r},t)$, $\boldsymbol{v}(\boldsymbol{r},t)$, and $P(\boldsymbol{r},t)$ are called *ideal liquids*. The model of an ideal liquid may explain some properties of real liquids. For real liquids or gases it might, however, be necessary to take into account additional effects, for example, the viscosity (friction) and the temperature field.

No matter is lost in the flow of a liquid; this is expressed by the *continuity equation*. Forces on the fluid cause changes in the velocity field; this is described by the *Euler equation*. In addition, one also

[1]The standard unit for the pressure is Pascal (denoted by Pa), where $\mathrm{Pa} \equiv \mathrm{N/m^2}$ and $1\,\mathrm{bar} = 10^5\,\mathrm{Pa}$.

considers an *equation of state*, which defines the compressibility of the matter under consideration. In the following, we present these hydrodynamic equations and discuss some simple applications.

Continuity Equation

The number of particles is conserved in the fluid. Then the following must be valid for any volume V:

$$\oint_{A(V)} d\boldsymbol{A} \cdot \big(\varrho(\boldsymbol{r},t)\, \boldsymbol{v}(\boldsymbol{r},t)\big) = -\frac{d}{dt}\int_V d^3r\, \varrho(\boldsymbol{r},t) = -\int_V d^3r\, \frac{\partial \varrho}{\partial t} \tag{32.4}$$

Here,

$$\boldsymbol{j}(\boldsymbol{r},t) = \varrho\, \boldsymbol{v} = \text{current density} = \frac{\text{mass}}{\text{area} \cdot \text{time}} \tag{32.5}$$

The left side of (32.4) equals the current (mass per time) flowing outwards through the surface $A(V)$ of V. This net current out of V must be equal to the change of mass in V per time, thus equal to the second expression in (32.4). The volume V shall be constant in time and rest in the inertial frame for which we derive the equations of motion. This allows the last step in (32.4). With Gauß's theorem, we transform the surface integral into a volume integral and obtain

$$\int_V d^3r \left(\frac{\partial \varrho}{\partial t} + \operatorname{div}(\varrho\, \boldsymbol{v})\right) = 0 \tag{32.6}$$

This is true for arbitrary volumes. Therefore, the integrand must vanish:

$$\frac{\partial \varrho(\boldsymbol{r},t)}{\partial t} + \operatorname{div}\big(\varrho(\boldsymbol{r},t)\, \boldsymbol{v}(\boldsymbol{r},t)\big) = 0 \quad \text{continuity equation} \tag{32.7}$$

The continuity equation $\dot{\varrho} + \boldsymbol{\nabla}(\varrho\, \boldsymbol{v}) = 0$ is the differential conservation law for the mass, expressed by the mass density ϱ and the mass current density $\boldsymbol{j} = \varrho\, \boldsymbol{v}$. In electrodynamics, an analogous continuity equation describes the conservation of charge.

Euler Equation

We now consider a singled-out small volume element $\varDelta V$ of the fluid and investigate the motion of the mass $\varDelta m = \varrho\, \varDelta V$. Let the volume element be so small, that the fields (32.1)–(32.3) change only slightly within $\varDelta V$. On the mass $\varDelta m$ acts the force $\varDelta \boldsymbol{F}$. Then we can use Newton's second axiom:

$$\varDelta m\, \frac{d\boldsymbol{v}(\boldsymbol{r},t)}{dt} = \varrho(\boldsymbol{r},t)\, \varDelta V\, \frac{d\boldsymbol{v}(\boldsymbol{r},t)}{dt} = \varDelta \boldsymbol{F}(\boldsymbol{r},t)$$

$$= \varDelta \boldsymbol{F}_{\text{ext}} + \varDelta \boldsymbol{F}_{\text{pressure}} \tag{32.8}$$

The force can be divided into an external and a pressure component. An external force is described by a force density $\boldsymbol{f}(\boldsymbol{r},t)$ (force per volume). An example is the gravity field with the force density $\boldsymbol{f} = \varrho\, \boldsymbol{g}$ and the force

$$\varDelta \boldsymbol{F}_{\text{ext}}(\boldsymbol{r},t) = \boldsymbol{f}(\boldsymbol{r},t)\, \varDelta V \tag{32.9}$$

To derive the pressure component $\varDelta \boldsymbol{F}_{\text{pressure}}$, we consider a cuboidal volume $\varDelta V = \varDelta x\, \varDelta y\, \varDelta z$ (Figure 32.1) and add up the forces exerted

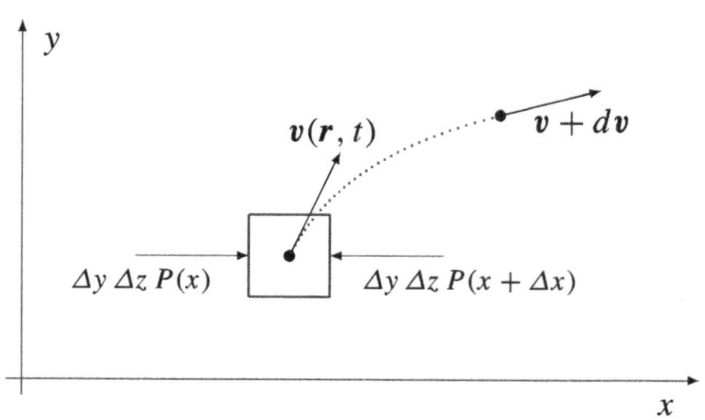

Figure 32.1 Derivation of the Euler equation: For the volume element $\varDelta V = \varDelta x\, \varDelta y\, \varDelta z$ the force is calculated, which results from the spatial dependence of the pressure. During dt, the considered fluid element moves a little bit. Therefore, both the time and the position dependence of the velocity $\boldsymbol{v}(\boldsymbol{r},t)$ contribute to the acceleration $d\boldsymbol{v}/dt$.

by the pressure on the six lateral faces of the cuboid:

$$\Delta \boldsymbol{F}_{\text{pressure}} = -\oint_{\Delta V} d\boldsymbol{A}\, P = -\boldsymbol{e}_x\, \Delta y\, \Delta z \left[P(x+\Delta x) - P(x) \right]$$
$$- \boldsymbol{e}_y\, \Delta x\, \Delta z \left[P(y+\Delta y) - P(y) \right]$$
$$- \boldsymbol{e}_z\, \Delta x\, \Delta y \left[P(z+\Delta z) - P(z) \right] \qquad (32.10)$$

Here, we have quoted only the relevant coordinates in the argument of P; for the other arguments an average value in the considered area element had to be used. We take out a factor ΔV and replace the occurring differential quotients by the derivatives:

$$\Delta \boldsymbol{F}_{\text{pressure}} = -(\operatorname{grad} P)\, \Delta V = -(\boldsymbol{\nabla} P)\, \Delta V \qquad (32.11)$$

The acceleration $d\boldsymbol{v}/dt$ in (32.8) has two contributions. First, $\boldsymbol{v}$ changes because $\boldsymbol{v}(\boldsymbol{r}, t)$ at the considered place $\boldsymbol{r}$ depends on the time t. Second, the mass element moves during dt to another point $\boldsymbol{r} + d\boldsymbol{r} = \boldsymbol{r} + \boldsymbol{v}\, dt$, where the mean velocity is different because of the position dependence of $\boldsymbol{v}(\boldsymbol{r}, t)$. Formally, these contributions show up in the total differential $d\boldsymbol{v}$:

$$d\boldsymbol{v} = \frac{\partial \boldsymbol{v}}{\partial t}\, dt + \frac{\partial \boldsymbol{v}}{\partial x}\, dx + \frac{\partial \boldsymbol{v}}{\partial y}\, dy + \frac{\partial \boldsymbol{v}}{\partial z}\, dz = \frac{\partial \boldsymbol{v}}{\partial t}\, dt + \left(\boldsymbol{v} \cdot \boldsymbol{\nabla} \right) \boldsymbol{v}\, dt \qquad (32.12)$$

Using (32.9)–(32.12), the Newtonian equation of motion (32.8) becomes

$$\varrho \left(\frac{\partial \boldsymbol{v}}{\partial t} + \left(\boldsymbol{v} \cdot \boldsymbol{\nabla} \right) \boldsymbol{v} \right) = -\boldsymbol{\nabla} P + \boldsymbol{f} \quad \text{Euler equation} \qquad (32.13)$$

This is the *Euler equation* of hydrodynamics. The term "Euler equation" by itself is not unique, since there are also the Euler equations for the gyroscope (Chapter 22).

Equation of State

The Euler equation and the continuity equation together represent four partial differential equations for the five fields $\varrho(\boldsymbol{r}, t)$, $\boldsymbol{v}(\boldsymbol{r}, t)$, and $P(\boldsymbol{r}, t)$. Therefore, in order to determine the fields, a further equation is necessary. This is the *equation of state*, which relates the pressure and the density:

$$P = P(\varrho(\boldsymbol{r}, t)) \tag{32.14}$$

This equation of state depends on the type of matter. We restrict ourselves to so-called *polytropic* equations of state:

$$P = \text{const.} \cdot \varrho^{\gamma} \tag{32.15}$$

From this equation of state, the compressibility follows:

$$\kappa = -\frac{1}{V}\frac{\partial V}{\partial P} = \frac{1}{\varrho}\frac{\partial \varrho}{\partial P} = \frac{1}{\gamma P} \tag{32.16}$$

Homogeneous systems with a fixed number of particles usually depend on two state variables, such as the pressure P and the temperature T. Therefore, when the pressure changes, it is necessary to specify which other macroscopic quantity is held fixed. Often one uses either the isothermal (temperature T constant) or the adiabatic (entropy S constant) compressibility.

As shown in Table 32.1, the behavior of various quite different systems can be described by a polytropic equation of state. For liquids, a pressure change usually results in only a small density change ($\gamma \gg 1$). For $\gamma = \infty$ the density $\varrho \propto P^{1/\gamma} = \text{const.}$ is independent of the pressure P, and the fluid is *incompressible*:

$$\varrho(\boldsymbol{r}, t) = \varrho_0 = \text{const.} \quad \text{(incompressible, homogeneous fluid)} \tag{32.17}$$

Together with the continuity equation, this implies $\operatorname{div} \boldsymbol{v} = 0$.

Table 32.1 Some examples of the exponent γ in the polytropic equation of state (32.15).

γ	System
∞	Incompressible fluid
$\gg 1$	Real fluid
1	Ideal gas, isothermal
c_P/c_V	Ideal gas, adiabatic
5/3	Fermi gas, $T \approx 0$, non-relativistic
4/3	Fermi gas, $T \approx 0$, relativistic

For an ideal gas, the equation of state is

$$PV = N k_{\mathrm{B}} T \quad \text{(ideal gas)} \tag{32.18}$$

Here, k_{B} is the Boltzmann constant and N is the number of gas particles (of mass m). For an isothermal volume change, this implies $P(\varrho) = \text{const.} \cdot \varrho$, that means $\gamma = 1$ and $\kappa_T = 1/P$. For an adiabatic volume change, thermodynamics yields $\gamma = c_P/c_V$ and

$$\kappa_S = \frac{1}{\varrho}\left(\frac{\partial \varrho}{\partial P}\right)_S = \frac{c_V}{c_P\, P} \quad \text{(ideal gas)} \tag{32.19}$$

Here, c_P and c_V are the specific heats at constant pressure and at constant volume, respectively. The index S at the partial derivative means that the entropy S is kept constant.

Ideal fluid

The following equations are the basic equations for the dynamics of ideal fluids:

$$\begin{aligned} \dot\varrho + \boldsymbol{\nabla}(\varrho\, \boldsymbol{v}) &= 0 && \text{continuity equation} \\ \varrho\, \dot{\boldsymbol{v}} + \varrho\,(\boldsymbol{v}\cdot\boldsymbol{\nabla})\,\boldsymbol{v} &= -\boldsymbol{\nabla} P + \boldsymbol{f} && \text{Euler equation} \\ P &= P(\varrho) && \text{equation of state} \end{aligned} \tag{32.20}$$

This is a system of five equations for determining the five fields $\varrho(\boldsymbol{r},t)$, $\boldsymbol{v}(\boldsymbol{r},t)$, and $P(\boldsymbol{r},t)$. As partial differential equations, the Euler equation and the continuity equation have to be complemented by boundary and initial conditions.

The special difficulty of the solution lies in the nonlinearity of (32.20). The non-linearity can lead to complicated motion courses (e.g. turbulence), whose mathematical treatment is challenging. In particular, very small differences in the initial conditions may grow exponentially fast; then the time course of the system is unpredictable. This means, for example, that long-term weather forecasts are impossible. We limit ourselves to some very simple applications in the following.

The Euler equation is valid for ideal fluids; in particular, it neglects frictional effects. In real, viscous fluids, friction forces occur, which lead to additional terms (like $\eta\Delta\boldsymbol{v}$); these terms are comparable with a friction force $-\gamma\dot{\boldsymbol{r}}$ in Newton's equation of motion. The resulting equations are called *Navier–Stokes equations*. Together with a consideration of the temperature field, one obtains the basic equations for air motions in the atmosphere. If the temperature is taken into account by a field $T(\boldsymbol{r},t)$, then one needs a further field equation like the continuity equation for the entropy (i.e. for the heat content of the gas or fluid).

Applications

Resting fluid in gravity field

In the limiting case of *hydrostatics*,

$$\varrho(\boldsymbol{r},t)=\varrho(\boldsymbol{r}),\quad \boldsymbol{v}(\boldsymbol{r},t)=0,\quad P(\boldsymbol{r},t)=P(\boldsymbol{r}) \tag{32.21}$$

the continuity equation is trivially satisfied. The Euler equation reduces to an equilibrium condition for the forces; the external and the pressure forces balance each other out.

We consider a medium at rest in a homogeneous gravitational field,

$$\boldsymbol{f}=\varrho\,\boldsymbol{g}=-\,\varrho\,g\,\boldsymbol{e}_z \tag{32.22}$$

With (32.21) and (32.22), the Euler equation yields

$$\operatorname{grad}P=\varrho\,\boldsymbol{g}\quad\text{or}\quad \frac{dP(z)}{dz}=-\,\varrho(z)\,g \tag{32.23}$$

Due to the translational invariance in the x and y directions. the pressure and density depend only on z. To determine the functions $P(z)$ and $\varrho(z)$ we still need an equation of state. We consider an incompressible fluid and an ideal gas.

For an incompressible fluid $\varrho=\varrho_0$ gives (32.23)

$$P(z)=P_0-\varrho_0\,g\,z \tag{32.24}$$

This describes, for example, how the pressure in a lake increases with depth. The integration constant P_0 is equal to the pressure at $z=0$. The pressure differences correspond to the weight Mg of the respective liquid column, $\Delta P A=\varrho_0\,g\,A\Delta z=Mg$ where M is the mass of the liquid in the volume $A\Delta z$. For water in a lake, the pressure increases with depth by about 1 bar every 10 meters.

For an ideal gas (32.18) at constant temperature (32.23) leads to

$$\frac{dP}{dz} = -\frac{mg}{k_{\mathrm{B}}T}\,P \tag{32.25}$$

The integration yields the barometric formula

$$P(z) = P_0\,\exp\left(-\frac{mgz}{k_{\mathrm{B}}T}\right) \approx P_0\,\exp\left(-\frac{z}{8\ \mathrm{km}}\right) \tag{32.26}$$

This is an expression of the height dependence of the pressure in the atmosphere. The constant P_0 is the pressure at $z = 0$, i.e. approximately at sea level. For air and $T \approx 300\,\mathrm{K}$ a numerical value for the height is given, at which the pressure and hence the density drops to the eth part. The barometric formula assumes a temperature that is independent of the altitude; more realistic is the assumption of a convective equilibrium (Exercise 20.4 in [4]).

Star equilibrium

A star is formed when an accumulation of matter (such as hydrogen gas) contracts under its own gravity. By this contraction, gravitational energy is converted into heat. For a sufficiently large mass, the heating is so strong that it comes to nuclear burning. In the course of star evolution, there are also stable phases that last for a longer or shorter period of time. Such stellar equilibria (e.g. the sun or a white dwarf) can be regarded as hydrostatic equilibria. In such an equilibrium, the gravitational and the matter pressure keep the balance.

The equilibrium of our sun is maintained by the kinetic gas pressure (32.18). In the center of the sun, where the pressure is strongest, the temperature is about $T \approx 10^7\,\mathrm{K}$. This temperature is maintained by the combustion of hydrogen into helium.

After the hydrogen has been converted into helium, the star cools down. This leaves a relatively cold accumulation of helium liquid. In the absence of kinetic gas pressure, the star continues to contract. The gravitational pressure becomes then so strong that the atomic shells of the helium atoms are crushed. This results in a dense Fermi gas of electrons, interspersed with He-nuclei. A new equilibrium arises when the Fermi pressure of the electrons balances the gravitational pressure; the last two entries in Table 32.1 refer to this Fermi pressure. This stellar equilibrium is that of a white dwarf. An elementary treatment of white dwarfs can be found in the chapter *Ideal Fermi gas* in [4].

Rotating fluid in a gravity field

We consider a cylindrical bucket filled with water; the water is treated as an incompressible liquid ($\varrho(\boldsymbol{r}, t) = \varrho_0 = \text{const.}$). The gravitational field $\boldsymbol{g} = -g\,\boldsymbol{e}_z$ acts on the liquid; the symmetry axis of the bucket coincides with the z-axis. The bucket shall rotate with constant angular velocity $\boldsymbol{\omega}$ around its symmetry axis. After some time, the velocity field becomes stationary:

$$\boldsymbol{v} = \boldsymbol{v}(\boldsymbol{r}) = \boldsymbol{\omega} \times \boldsymbol{r} \tag{32.27}$$

This is a *rigid rotation*, i.e. the velocity field is the same as for a rigid body. Individual parts of the fluid do not move against each other anymore; such an initially existing motion comes to a standstill due to the viscosity of the real fluid. After the velocity field (32.27) has been established, the friction terms no longer contribute; we can therefore apply the Euler equation.

As everyday experience shows, this experiment results in a curved water surface. With the help of the Euler equation, we determine the shape of this surface.

We use cylindrical coordinates $\rho = \sqrt{x^2 + y^2}$, φ and z and summarize the preconditions:

$$\varrho = \varrho_0, \quad \boldsymbol{\omega} = \omega\,\boldsymbol{e}_z, \quad \boldsymbol{v} = \omega\rho\,\boldsymbol{e}_\varphi, \quad \boldsymbol{f} = \varrho_0\,\boldsymbol{g} = -g\,\varrho_0\,\boldsymbol{e}_z, \quad P = P(\rho, z) \tag{32.28}$$

The continuity equation is satisfied a priori. For the velocity field holds $\partial\boldsymbol{v}/\partial t = 0$ and

$$\big(\boldsymbol{v}\cdot\boldsymbol{\nabla}\big)\,\boldsymbol{v} = \omega\rho\,\frac{1}{\rho}\,\frac{\partial}{\partial\varphi}\,(\omega\rho\,\boldsymbol{e}_\varphi) = \omega^2\rho\,\frac{\partial\boldsymbol{e}_\varphi}{\partial\varphi} = -\omega^2\rho\,\boldsymbol{e}_\rho \tag{32.29}$$

Using (32.28) and (32.29), the Euler equation becomes

$$-\varrho_0\,\omega^2 \rho\,\boldsymbol{e}_\rho = -\frac{\partial P}{\partial \rho}\,\boldsymbol{e}_\rho - \frac{\partial P}{\partial z}\,\boldsymbol{e}_z - \varrho_0\,g\,\boldsymbol{e}_z \tag{32.30}$$

From this follows $\partial P/\partial \rho = \varrho_0\,\omega^2 \rho$, $\partial P/\partial z = -\varrho_0\,g$ and thus

$$P(\rho, z) = \varrho_0\left(-g\,z + \frac{\omega^2 \rho^2}{2}\right) + \text{const.} \tag{32.31}$$

The isobars (surfaces with $P = \text{const.}$) are rotational parabola

$$z = \frac{\omega^2}{2\,g}\,\rho^2 + \text{const.} \quad \text{(isobars)} \tag{32.32}$$

On the surface of the liquid, the pressure (32.31) acts in one direction, and in the opposite direction acts the (practically constant) air pressure $P_0 \approx 1\,\text{bar}$. Therefore, the surface of the rotating fluid is of the form (32.32). The constant is determined by the condition that the liquid fits into the volume limited by the bucket and the surface.

The result can alternatively be derived in the co-rotating reference frame S′. The transformation to the rotating system leads to inertial forces. Specifically, the left-hand side of (32.30) may be transferred to the right side (with a plus sign). There it can be interpreted as the centrifugal force density.

Bernoulli's equation

We consider the force-free, stationary flow of an incompressible fluid, i.e.

$$\boldsymbol{f} = 0, \quad \varrho(\boldsymbol{r}, t) = \varrho_0, \quad \boldsymbol{v}(\boldsymbol{r}, t) = \boldsymbol{v}(\boldsymbol{r}), \quad P(\boldsymbol{r}, t) = P(\boldsymbol{r}) \tag{32.33}$$

For this the Euler equation becomes

$$\varrho_0\,(\boldsymbol{v}\cdot\boldsymbol{\nabla})\,\boldsymbol{v} = -\operatorname{grad} P \tag{32.34}$$

We consider the scalar product with $\boldsymbol{v}$:

$$\varrho_0\,\boldsymbol{v}\cdot(\boldsymbol{v}\cdot\boldsymbol{\nabla})\,\boldsymbol{v} = -(\boldsymbol{v}\cdot\boldsymbol{\nabla})\,P \tag{32.35}$$

We insert $\boldsymbol{v} \cdot (\boldsymbol{v} \cdot \boldsymbol{\nabla})\, \boldsymbol{v} = (\boldsymbol{v} \cdot \boldsymbol{\nabla})\, \boldsymbol{v}^2/2$ and obtain:

$$(\boldsymbol{v} \cdot \boldsymbol{\nabla}) \left(\frac{\varrho_0\, \boldsymbol{v}^2}{2} + P \right) = 0 \tag{32.36}$$

Now, $\boldsymbol{v} \cdot \boldsymbol{\nabla}$ means a derivative parallel to $\boldsymbol{v}$, or the *change along a streamline*. Streamlines are the curves whose tangent vector at each point is parallel to the velocity field $\boldsymbol{v}$. From (32.36) we, therefore, obtain a quantity that is constant along a streamline:

$$\frac{\varrho_0}{2}\, \boldsymbol{v}(\boldsymbol{r})^2 + P(\boldsymbol{r}) = \left\{ \begin{matrix} \text{constant along} \\ \text{a streamline} \end{matrix} \right. \qquad \text{Bernoulli-Gleichung} \tag{32.37}$$

The *Bernoulli equation* states that the sum of the common pressure $P(\boldsymbol{r})$ and the so-called dynamic pressure $\varrho_0\, \boldsymbol{v}^2/2$ is constant along a streamline.

The Bernoulli equation provides simple explanations for some well-known phenomena. In particular, the pressure $P(\boldsymbol{r})$ decreases at a narrowing or constriction of the flow cross-section, because there the flow velocity is higher. Thus, two adjacent bodies, between which water is flowing, experience an effective attraction. This applies also to moving ships in still water (one may consider the situation from the rest system of one of the ships). So it might be dangerous for ships to pass one another closely, or to drive closely parallel to each other.

Another example is the wing of an airplane: The wings are shaped such that air flows faster over the top of the wing and slower underneath. According to Bernoulli, this leads to lower pressure at the top of the wing. The result is an upward force exerted on the wing which may balance the gravitational force.

Potential flow

We obtain particularly simple equations for a stationary and *irrotational* (or vortex-free) flow:

$$\operatorname{curl} \boldsymbol{v}(\boldsymbol{r}) = 0 \quad \text{(vortex-free, stationary)} \tag{32.38}$$

Such a flow is also called *potential flow*, since because of (32.38) a scalar velocity potential $\Phi(\boldsymbol{r})$ exists, so that

$$\boldsymbol{v}(\boldsymbol{r}) = \operatorname{grad} \Phi(\boldsymbol{r}) \tag{32.39}$$

For an incompressible fluid, the continuity equation becomes $\operatorname{div}\boldsymbol{v} = 0$. Thus, we obtain the *Laplace equation*

$$\Delta\Phi = 0 \tag{32.40}$$

for the velocity potential. Herewith, the vortex-free flow around a body can be calculated. At the surface of the body, the normal component $\boldsymbol{n}\cdot\boldsymbol{v}$ must vanish, i.e.

$$\big(\boldsymbol{n}\cdot\operatorname{grad}\Phi\big)_{\text{wall}} = \left(\frac{\partial\Phi}{\partial n}\right)_{\text{wall}} = 0 \tag{32.41}$$

The index "wall" refers to the surface of the body. At a large distance from the body the current is homogeneous, i.e. $\boldsymbol{v} = v_0\,\boldsymbol{e}_z$. For Φ, this means

$$\Phi \overset{r\to\infty}{\longrightarrow} v_0\, z \tag{32.42}$$

By (32.40)–(32.42) the problem is mathematically defined. In electrostatics, the same equations would solve the problem of "conducting body in homogeneous electric field". This problem is solved analytically in Chapter 10 of [2].

Sound Waves

We explain how ordinary sound can be described by the hydrodynamic equations (32.20). A general procedure for the solution of nonlinear equations is the *linearization*. For this one starts from a known solution ϱ_0, $\boldsymbol{v}_0$, P_0 and consider small deviations, $\varrho = \varrho_0 + \delta\varrho$, $\boldsymbol{v} = \boldsymbol{v}_0 + \delta\boldsymbol{v}$ and $P = P_0 + \delta P$. We insert this approach into the hydrodynamic equations (32.20). The terms without $\delta\ldots$ satisfy the equations and are dropped. The terms which are quadratic (or higher) in the deviations are omitted. Thus, one obtains a linear system of equations in $\delta\varrho$, $\delta\boldsymbol{v}$, and δP, which can be solved by standard methods.

We start with the equilibrium solution

$$\varrho_0 = \text{const.},\qquad \boldsymbol{v}_0 = 0,\qquad P_0 = \text{const.} \tag{32.43}$$

and consider deviations of the form

$$\begin{aligned}\varrho(\boldsymbol{r},t) &= \varrho_0 + \Delta\varrho\,\exp\big(\mathrm{i}(\boldsymbol{k}\cdot\boldsymbol{r}-\omega t)\big)\\ \boldsymbol{v}(\boldsymbol{r},t) &= \boldsymbol{v}_0 + \Delta\boldsymbol{v}\,\exp\big(\mathrm{i}(\boldsymbol{k}\cdot\boldsymbol{r}-\omega t)\big)\\ P(\boldsymbol{r},t) &= P_0 + \Delta P\,\exp\big(\mathrm{i}(\boldsymbol{k}\cdot\boldsymbol{r}-\omega t)\big)\end{aligned} \tag{32.44}$$

Eventually, the real part of this is to be taken.

In the case of sound, the changes in density are so rapid that no appreciable heat exchange takes place. The relationship between density and pressure is therefore determined by the adiabatic compressibility κ_S:

$$\Delta P = \left(\frac{\partial P}{\partial \varrho}\right)_S \Delta\varrho = \frac{\Delta\varrho}{\varrho_0\,\kappa_S} \tag{32.45}$$

This equation replaces the equation of state in (32.20). We can confine ourselves to determine the four functions ϱ and $\boldsymbol{v}$ from the first two equations in (32.20).

We substitute ϱ and $\boldsymbol{v}$ from (32.44) into the continuity equation and the Euler equation, consider $\boldsymbol{v}_0 = 0$ and take into account only the terms which are linear in $\Delta\boldsymbol{v}$ and $\Delta\varrho$:

$$\big(-\mathrm{i}\omega\,\Delta\varrho + \mathrm{i}\boldsymbol{k}\cdot\Delta\boldsymbol{v}\,\varrho_0\big)\,\exp\big(\mathrm{i}(\boldsymbol{k}\cdot\boldsymbol{r}-\omega t)\big) = 0 \tag{32.46}$$

$$\left(-\mathrm{i}\omega\,\varrho_0\,\Delta\boldsymbol{v} + \frac{\mathrm{i}\boldsymbol{k}}{\varrho_0\,\kappa_S}\,\Delta\varrho\right)\exp\big(\mathrm{i}(\boldsymbol{k}\cdot\boldsymbol{r}-\omega t)\big) = 0 \tag{32.47}$$

The time derivative became $-\mathrm{i}\omega$ and the gradient became $\mathrm{i}\boldsymbol{k}$. We write the resulting coupled equations for $\Delta\varrho$ and $\Delta\boldsymbol{v}$ in matrix form:

$$\begin{pmatrix} -\omega & \varrho_0\,\boldsymbol{k} \\ \dfrac{\boldsymbol{k}}{\varrho_0\,\kappa_S} & -\omega\,\varrho_0 \end{pmatrix}\begin{pmatrix}\Delta\varrho\\ \Delta\boldsymbol{v}\end{pmatrix} = 0 \tag{32.48}$$

For $\Delta\boldsymbol{v} \perp \boldsymbol{k}$ it follows that $\omega\,\Delta\varrho = 0$, so $\omega = 0$ for $\Delta\varrho \neq 0$. From the second line then follows $k = 0$. Thus, we get only a trivial deviation from (32.43).

Of interest are therefore only the solutions with

$$\Delta\boldsymbol{v} \parallel \boldsymbol{k} \quad \text{(longitudinal wave)} \tag{32.49}$$

This solution describes *longitudinal* waves. In contrast, for example, electromagnetic waves are transversal; the electric field $\boldsymbol{E}$ is perpendicular to the wave vector $\boldsymbol{k}$. For $\boldsymbol{k} = k\,\boldsymbol{e}_z$ and $\Delta\boldsymbol{v} = \Delta v\,\boldsymbol{e}_z$ equation (32.48) becomes

$$\begin{pmatrix} -\omega & \varrho_0\,k \\ \dfrac{k}{\varrho_0\,\kappa_S} & -\omega\,\varrho_0 \end{pmatrix} \begin{pmatrix} \Delta\varrho \\ \Delta v \end{pmatrix} = 0 \tag{32.50}$$

This linear homogeneous system of equations has the trivial solution $\Delta\varrho = 0$ and $\Delta v = 0$. For a nontrivial solution, the determinant must vanish:

$$\omega^2 = \frac{k^2}{\varrho_0\,\kappa_S} = c_{\rm S}^2\,k^2 \tag{32.51}$$

In the last step, we introduced the *speed of sound* $c_{\rm S}$,

$$c_{\rm S} = \frac{\omega}{k} = \frac{1}{\sqrt{\varrho_0\,\kappa_S}} \tag{32.52}$$

For $\boldsymbol{k} = k\,\boldsymbol{e}_z$, the position and time dependence of the solution (32.44) is given by the factor $\exp[\mathrm{i}(k\,z - \omega t)]$; where $\omega = c_{\rm S}\,k$. The amplitude $\Delta\varrho$ can be chosen arbitrarily; however, it must be so small that the linear approximation is valid. The amplitudes Δv and ΔP in (32.44) then follow from (32.50) and (32.45). The so specified solution (32.44) describes *sound waves*.

The dispersion relation $\omega = c_{\rm S}\,k = c_{\rm S}\,|\boldsymbol{k}|$ is linear. Therefore, wave packets propagate without dispersion, i.e. without change of their shape.[2] For a nonlinear dispersion relation, high and low tones would have different propagation times; this would make acoustic communication difficult or impossible.

Sound waves do not only exist in gases and liquids (to which the basic hydrodynamic equations refer), but also in solids. In contrast to liquids, in solids, there are also restoring forces against shear

[2]Electromagnetic waves in the vacuum also have a linear dispersion relation, $\omega = c|\boldsymbol{k}|$, where c is the speed of light. For electromagnetic wave packets in matter or for a quantum mechanical wave packet dispersive effects occur, see Chapter 33 in [2] or Chapter 9 in [3].

displacements. Then transverse density waves are possible; they are also called sound waves. In a figurative sense, all waves which are generated from a linearization of field equations (like in (32.44)) are called sound waves. An example is the second sound in liquid helium (Chapter 38 of [4]).

Finally, we estimate the speed of sound (32.52) for air under standard conditions:

$$c_S = \frac{1}{\sqrt{\varrho_0\, \kappa_S}} \overset{(32.19)}{=} \sqrt{\frac{c_P}{c_V}\,\frac{P}{\varrho_0}} \overset{(32.18)}{=} \sqrt{\frac{c_P}{c_V}\,\frac{k_B T}{m}}$$

$$\approx \sqrt{\frac{7}{5}\,\frac{\text{eV}}{40}\,\frac{c^2}{30\,\text{GeV}}} \approx 331\ \frac{\text{m}}{\text{s}} \tag{32.53}$$

We inserted the ideal gas law $PV = Nk_B T$ and $\varrho_0 = mN/V$ (where m is the mass of a molecule). For the numerical evaluation, we used $k_B T \approx \text{eV}/40$ (room temperature), $m \approx 30\,\text{GeV}/c^2$ (for N_2- or O_2 molecules), and $c_P/c_V = 7/5$. The last value corresponds to three degrees of freedom for translation and two for rotation (vibrations are not excited at room temperature), and to $c_P = c_V + k_B$. The numerical result is quite realistic. For liquids and solids, $\varrho_0\, \kappa_S$ is about two orders of magnitude smaller; this leads, for example, for iron to $c_S \approx 5000\,\text{m/s}$.

The wave solutions considered here (32.44) are *plane waves*; $\Delta\varrho$, ΔP and $\Delta \boldsymbol{v}$ are proportional to $\exp(\mathrm{i}\,\boldsymbol{k}\cdot\boldsymbol{r})$, and this factor is constant in the planes $\boldsymbol{k}\cdot\boldsymbol{r} = k_x x + k_y y + k_z z = \text{const}$. A point source, on the other hand, would emit *spherical waves* of the form $\exp[\mathrm{i}\,(kr - \omega t)]/r$. Their squared amplitude (intensity of the sound wave) falls off like $1/r^2$. Real waves are damped, they become weaker and weaker in the course of time. In order to describe this damping, we would have to add a viscosity term in the Euler equation.

Exercises

32.1 *Generalization of Bernoulli's equation*

Repeat the derivation of Bernoulli's equation (32.37) for the case that an external force density $\boldsymbol{f} = -\operatorname{grad} u(\boldsymbol{r})$ acts on the fluid.

32.2 *Lagrangian density for incompressible fluid*

The Lagrangian density of an incompressible fluid is given by

$$\mathcal{L}(\Phi, \operatorname{grad}\Phi) = \frac{\varrho_0}{2}\,\boldsymbol{v}^2 = \frac{\varrho_0}{2}\,(\operatorname{grad}\Phi)^2 \tag{32.54}$$

Determine the field equations from the Hamiltonian principle.

Chapter 33

Field Theories

In Chapters 30–32, we have investigated some simple field equations. Field equations play an important role in physics; among others, they include the Maxwell's equations, the Schrödinger equation, and the field equations of gravitation. In this chapter, we start from a general form of the Lagrangian, establish the associated field equations, and formulate the Noether theorem. This discussion goes beyond the scope of an introduction to mechanics. It requires some knowledge of special relativity and about the respective field theory. This chapter may therefore be skipped in a first study of mechanics. The additions given here, however, point to the fundamental importance of the concepts and structures introduced in mechanics (Lagrange function or density, variational principle, Noether theorem). If these structures are encountered by the reader later in more complex theories, a recollection of simple applications, like those in mechanics, can be useful.

Lagrange Density

Fields depend on position and time. It is convenient to combine this dependence in the indexed quantity x^α:

$$(x^\alpha) = (x^0, x^1, x^2, x^3) = (ct, x, y, z) \tag{33.1}$$

The Greek indices always run from 0 to 3. In non-relativistic applications, the constant c is of no importance, in relativistic ones c is the speed of light. In Part IX, x^α is classified as a 4-vector or a Lorentz vector and supplemented by $(x_\alpha) = (ct, -x, -y, -z)$.

We sum up over equal top and bottom indices without writing down this sum explicitly.

The fields under consideration may have n components:

$$u_r = u_r(x_\alpha) = u_r(x_0, x_1, x_2, x_3) \quad (r = 1, \dots, n) \tag{33.2}$$

The argument x_α represents here the set of variables x_0, x_1, x_2, and x_3. We denote the derivatives of the fields by a vertical line in front of an index:

$$u_{r|\alpha} \equiv \frac{\partial u_r}{\partial x^\alpha} \tag{33.3}$$

We restrict ourselves to Lagrangian densities $\mathcal{L}$, which depend on the fields, their first derivatives, and on the coordinates:

$$\mathcal{L} = \mathcal{L}(u_r, u_{r|\alpha}, x_\beta) \tag{33.4}$$

Here, u_r stands for $(u_1, \dots, u_n)$ and $u_{r|\alpha}$ stands for $(u_{1|0}, \dots, u_{1|3}, u_{2|0}, \dots, u_{n|3})$. The Lagrange function reads

$$\mathscr{L} = \int d^3x \; \mathcal{L}(u_r, u_{r|\alpha}, x_\beta). \tag{33.5}$$

(Note again the use of slightly different fonts for the Lagrange function and density, $\mathscr{L}$ and $\mathcal{L}$.) We use the Hamilton principle

$$\delta \int dt \; \mathscr{L} = \frac{1}{c}\, \delta \int d^4x \; \mathcal{L} = 0 \tag{33.6}$$

Here, $d^4x = c\,dt\,dx\,dy\,dz = c\,dt\,d^3x$. From the Hamiltonian principle follow the field equations

$$\frac{\partial}{\partial x^\alpha} \frac{\partial \mathcal{L}}{\partial u_{r|\alpha}} - \frac{\partial \mathcal{L}}{\partial u_r} = 0 \quad (r = 1, \dots, n) \tag{33.7}$$

According to the summation convention, here we sum over α.

Of direct physical interest are the fields and the field equations, but not the Lagrangian density. However, the Lagrange density is often an especially simple expression (it is a scalar, and it displays the relevant symmetries). For establishing new theories or modifications of existing theories, the Lagrangian density is the suitable starting point.

Examples

A first example of such a field theory is the string oscillation with the Lagrangian density (30.18),

$$\mathcal{L}(u, \dot{u}, u', x, t) = \frac{\varrho}{2}\, \dot{u}^2 - \frac{P}{2}\, u'^2 + u\, f(x,t) \qquad (33.8)$$

Here, the number of fields is $n = 1$, and $(x^\alpha) = (t, x)$ has only two components; the constant c plays no role. By the force density $f(x,t)$ the Lagrangian density depends explicitly on x and t.

The beam bending (Chapter 31) falls out of the here considered framework because higher derivatives in $\mathcal{L}$ occur. The hydrodynamics (Chapter 32) can be treated with a Lagrangian formalism for simple cases (see Exercise 32.2). The general hydrodynamic case is more complicated because of the nonlinearity of hydrodynamics.

As a second example, we consider the Schrödinger equation of the non-relativistic quantum mechanics. The wave function $\psi(x) = \psi(\boldsymbol{r}, t)$ is complex and therefore consists of two independent fields, the real and the imaginary part. Equivalently, we may use $u_1 = \psi$ and $u_2 = \psi^*$ as independent fields. From the Lagrangian density

$$\begin{aligned}&\mathcal{L}(\psi, \psi^*, \dot{\psi}, \dot{\psi}^*, \boldsymbol{\nabla}\psi, \boldsymbol{\nabla}\psi^*, \boldsymbol{r}, t)\\ &\quad = -\frac{\hbar^2}{2m}\, \boldsymbol{\nabla}\psi^* \cdot \boldsymbol{\nabla}\psi - V(\boldsymbol{r}, t)\, \psi^*\psi + \mathrm{i}\hbar\, \psi^*\, \dot{\psi}\end{aligned} \qquad (33.9)$$

follows according to (33.7) for $r = 2$ the Schrödinger equation

$$\mathrm{i}\hbar\, \frac{\partial\psi(\boldsymbol{r}, t)}{\partial t} = -\frac{\hbar^2}{2m}\, \Delta\psi(\boldsymbol{r}, t) + V(\boldsymbol{r}, t)\, \psi(\boldsymbol{r}, t) \qquad (33.10)$$

The term with $\Delta\psi$ follows from $(\partial/\partial x^\alpha)\, \partial\mathcal{L}/\partial u_{2|\alpha} = (\partial/\partial x^i)\, \partial\mathcal{L}/\partial\psi^*_{|i}$; the other two terms result from $\partial\mathcal{L}/\partial u_2 = \partial\mathcal{L}/\partial\psi^*$. For $r = 1$, we obtain the conjugate complex equation; the time derivative in (33.10) comes from $(\partial/\partial x^0)\, \partial\mathcal{L}/\partial u_{1|0} = (\partial/\partial t)\, \partial\mathcal{L}/\partial\dot{\psi}$. By the potential $V(\boldsymbol{r}, t)$, the Lagrange density $\mathcal{L}$ may depend explicitly on $\boldsymbol{r}$ and t.

As a third example, we consider the electromagnetic field; the knowledge of the covariant formulation of Maxwell's equations is assumed. The fundamental fields u_r are the four fields of the

4-potential $A_\alpha(x)$, and the physical fields are $F_{\alpha\beta} = A_{\beta|\alpha} - A_{\alpha|\beta}$. The Lagrangian density of the electromagnetic field is

$$\mathcal{L}(A_\beta, A_{\beta|\alpha}, x_\gamma) = -\frac{1}{16\,\pi}\,F^{\alpha\beta}F_{\alpha\beta} - \frac{1}{c}\,j^\beta(x_\gamma)\,A_\beta \tag{33.11}$$

Here, $(j^\alpha) = (c\varrho, \boldsymbol{j})$ is given by the charge density ϱ and the current density $\boldsymbol{j}$, and $F^{\alpha\beta} = \eta^{\alpha\gamma}\,\eta^{\alpha\delta}\,F_{\gamma\delta}$ in the notation of Chapter 37. By $j^\alpha(x_\gamma)$, the Lagrange density $\mathcal{L}$ depends explicitly on the coordinates. According to (33.7), Maxwell's equations follow from this $\mathcal{L}$:

$$\frac{\partial F^{\alpha\beta}}{\partial x^\alpha} = \frac{4\pi}{c}\,j^\beta \tag{33.12}$$

In all three examples, $\mathcal{L}$ is maximally quadratic in the fields and in their first derivatives. The resulting field equations are then linear.

Noether Theorem

We give the generalization of Noether's theorem for the field theories. The derivation is parallel to that in Chapter 15, and it is presented here in a rather concise form.

The transformation (15.7) is replaced by

$$\begin{aligned} u_r &\longrightarrow u_r^\star = u_r + \epsilon\,\Psi_r \\ x_\alpha &\longrightarrow x_\alpha^\star = x_\alpha + \epsilon\,X_\alpha \end{aligned} \tag{33.13}$$

Since conjugate complex quantities (such as ψ^*) occur, the transformed quantities (like $u_r^\star$) are marked with a slightly different asterisk ($\star$ instead of $*$). The quantities Ψ_r and X_α can depend on the same arguments as the Lagrangian density. The symmetry of the system, i.e., the invariance under the transformation (33.13), is given by the condition

$$\frac{d}{d\epsilon}\left[\mathcal{L}(u_r^\star, u_{r|\alpha}^\star, x_\beta^\star)\,\frac{d^4x^\star}{d^4x}\right]_{\epsilon=0} = 0 \quad \text{(invariance condition)} \tag{33.14}$$

Analogous to Chapter 15, it follows that

$$\partial_\alpha\,J^\alpha = 0 \tag{33.15}$$

where

$$J^{\alpha}(x) = \sum_{r=1}^{n} \frac{\partial \mathcal{L}}{\partial u_{r|\alpha}} \left(u_{r|\beta}\, X^{\beta} - \Psi_r \right) - \mathcal{L}\, X^{\alpha} \tag{33.16}$$

Equation (33.15) is a differential conservation law for the generalized current density J^{α}. The corresponding conservation quantity Q is obtained by writing (33.15) in the form $c^{-1}\partial J^0/\partial t + \operatorname{div} \boldsymbol{J} = 0$, and by integrating over a fixed volume V:

$$\frac{1}{c}\frac{\partial}{\partial t} \int_V d^3x\; J^0 = -\int_V d^3x\; \operatorname{div} \boldsymbol{J} = -\int_{A(V)} d\boldsymbol{A} \cdot \boldsymbol{J} = 0 \tag{33.17}$$

In doing so, we used Gauß's theorem to convert the volume integral into a surface integral. We specifically consider a closed system and we choose V so large that the integrand vanishes at the surface $A(V)$. Then the surface integral is zero, and from (33.17) follows the conserved quantity

$$Q = \int d^3x\; J^0 = \text{const.} \tag{33.18}$$

We apply this to two examples.

Energy conservation for string vibration

We consider the force-free string with Lagrangian density

$$\mathcal{L}(\dot{u}, u') = \frac{\varrho}{2}\, \dot{u}^2 - \frac{P}{2}\, u'^2 \tag{33.19}$$

In this non-relativistic, one-dimensional problem, we set $x^0 = t$ and $x^1 = x$; then $d^4x \to dt\, dx$. The index r is dropped, since there is only one field.

For the time translation

$$t \longrightarrow t^{\star} = t + \epsilon \tag{33.20}$$

the quantities introduced in (33.13) are

$$\Psi = 0, \quad X^0 = 1, \quad X^1 = 0 \tag{33.21}$$

The problem is invariant under time translations because $\mathcal{L}$ does not depend explicitly on time. Formally the invariance condition (33.14) follows from

$$\mathcal{L}(\dot{u}^\star, u'^\star) = \mathcal{L}(\dot{u}, u'), \quad \frac{dx^\star}{dx} = 1, \quad \frac{dt^\star}{dt} = 1 \tag{33.22}$$

According to (33.16), we calculate

$$J^0 = \frac{\partial \mathcal{L}}{\partial \dot{u}} \left(\dot{u} X^0 + u' X^1 - \Psi\right) - \mathcal{L} X^0 = \frac{\partial \mathcal{L}}{\partial \dot{u}}\, \dot{u} - \mathcal{L} = \frac{\varrho}{2}\, \dot{u}^2 + \frac{P}{2}\, u'^2 \tag{33.23}$$

This is the sum of kinetic and potential energy densities. Thus, J^0 is the energy density of the string. Equation (33.18) expresses the conservation of energy:

$$\int_0^l dx\; J^0 = E = \text{const.} \tag{33.24}$$

Due to the boundary condition of the string, the boundary terms vanish in the partial integration in (33.17).

Gauge transformation of the Schrödinger field

In the Lagrangian density (33.9) of the Schrödinger field, the fields $u_1 = \psi$ and $u_2 = \psi^*$ are treated as independent of each other. We set $c = 1$ again.

We consider the following transformation, which is also called the gauge transformation,

$$\begin{aligned} u_1 = \psi \;\; &\longrightarrow u_1^\star = \psi \, \exp(\mathrm{i}\epsilon) = \psi + \mathrm{i}\epsilon\, \psi \\ u_2 = \psi^* &\longrightarrow u_2^\star = \psi^* \, \exp(-\mathrm{i}\epsilon) = \psi^* - \mathrm{i}\epsilon\, \psi^* \end{aligned} \tag{33.25}$$

In u_r the asterisk denotes the transformed field, in ψ the complex-conjugate field. For this transformation, the fields given in (33.13) are

$$\Psi_1 = \mathrm{i}\psi, \quad \Psi_2 = -\mathrm{i}\psi^*, \quad X^\alpha = 0 \tag{33.26}$$

For $\psi \to \psi \exp(\mathrm{i}\epsilon)$ and $\psi^* \to \psi^* \exp(-\mathrm{i}\epsilon)$ one sees immediately that (33.9) is invariant. We calculate J^α from (33.16),

$$J^0 = \frac{\partial \mathcal{L}}{\partial \dot{\psi}}\,(-\Psi_1) + \frac{\partial \mathcal{L}}{\partial \dot{\psi}^*}\,(-\Psi_2) = \hbar\,\psi^*\psi = \hbar\,\varrho \qquad (33.27)$$

$$J^k = \frac{\partial \mathcal{L}}{\partial \psi_{|k}}\,(-\Psi_1) + \frac{\partial \mathcal{L}}{\partial \psi^*_{|k}}\,(-\Psi_2) = -\frac{\hbar^2}{2m}\,\Big(\psi^*_{|k}\,(-\mathrm{i}\psi) + \psi_{|k}\,(\mathrm{i}\psi^*)\Big)$$

$$= \hbar\,\frac{\hbar}{2m\mathrm{i}}\left(\psi^*\,\frac{\partial \psi}{\partial x^k} - \psi\,\frac{\partial \psi^*}{\partial x^k}\right) = \hbar\,j^k \qquad (33.28)$$

Here, $\varrho = \psi^*\psi$ is the probability density for the particle described by ψ, and $\boldsymbol{j}$ is the corresponding probability current density. From the symmetry under the gauge transformation (33.25) now follows (33.15), i.e. the continuity equation

$$\dot{\varrho} + \operatorname{div}\boldsymbol{j} = 0 \qquad (33.29)$$

and thus, according to (33.18), the conservation of the norm

$$\int d^3x\;\psi\,\psi^* = \int d^3x\;|\psi|^2 = \text{const.} \qquad (33.30)$$

For a particle with charge q in an electromagnetic field the substitution

$$-\mathrm{i}\hbar\,\boldsymbol{\nabla}\psi \quad\longrightarrow\quad \Big(-\mathrm{i}\hbar\,\boldsymbol{\nabla} - \frac{q}{c}\,\boldsymbol{A}\Big)\,\psi \qquad (33.31)$$

has to be made in $\mathcal{L}$ (see also Exercise 27.2). In this case, we consider the transformation (33.25) with a function $\epsilon(x) = \epsilon f(x)$. Then invariance holds if $\boldsymbol{A}$ to $\boldsymbol{A} + (\hbar c/q)\,\boldsymbol{\nabla}\epsilon(x)$ is co-transformed. From this follows again the continuity equation.

PART IX
Relativistic Mechanics

Chapter 34

Relativity Principle

Newtonian mechanics treated so far is valid only for velocities that are small compared to the speed of light. In Part IX, we introduce relativistic mechanics (also called Einstein mechanics); it comprises Newtonian mechanics as a limiting case for small velocities. The basic concepts of relativistic mechanics are also those of the special relativity theory of Einstein, which concerns all areas of physics. These basic concepts are presented in Chapters 34–37, the mechanics in the narrower sense then in Chapters 38–40.

In this chapter, we replace Galileo's principle of relativity with Einstein's, and thus the Galilean transformations by the Lorentz transformations.

Newton's laws are valid only in inertial frames, i.e. in reference frames, which move with constant speed relative to the fixed-star sky. By inertial systems (abbreviated IS) we mean inertial frames with coordinates (usually denoted by x, y, z, and t). The IS are the distinguished and preferred reference systems. Among the inertial systems themselves there is no preferred system. This is expressed by the *relativity principle of Galilei*:

1. All inertial systems are equivalent.
2. Newton's laws hold in all inertial systems.

Equivalence means that fundamental physical laws have the same form in all IS. Representatively for the fundamental laws, we have taken here Newton's laws. From points 1 and 2 follows — as demonstrated in Chapter 5 — the Galilei transformation for the transition between different IS. Of course, point 2 is not a formulation of Galileo; for he lived before Newton. In the following, we will stick to point 1, but modify point 2.

Experiments with light and fast particles are in contradiction with the consequences of the Galilei transformations. Under preservation of the principle of relativity (point 1), the special relativity theory (SRT) modifies the transformation between different IS in such a way that it describes the experimental findings correctly. Thereby the Galilei transformations are replaced by the Lorentz transformations.

The *distinction* and *equivalence* of the IS remains valid in the special relativity theory. The terms space and time partly lose their absolute meaning (Chapter 35). But there is still an absolute space-time structure, which distinguishes certain reference systems, i.e. the IS. This distinction appears to be related to the fact, that the IS are *not accelerated* with respect to the fixed stars. In this context, we recall the simple experiment discussed in Chapter 5.

First, we explain at which point there is a contradiction between the Galilei transformation and experimental findings. In each IS, let us define Cartesian coordinates x, y, z and a time coordinate t. By the specification of four values for these coordinates, an *event* is defined,

$$\text{event:}\quad (x,\, y,\, z,\, t). \tag{34.1}$$

In particular, a trajectory is a sequence of events. A specific event has different coordinate values in two different inertial frames, IS and IS$'$. The transformation to be discussed (Galilean or Lorentz) links these coordinate values.

We start with the special Galilean transformation (5.10),

$$x' = x - vt, \quad y' = y, \quad z' = z, \quad t' = t \tag{34.2}$$

This is a transformation between two inertial systems with parallel axes and the relative velocity $\boldsymbol{v} = v\,\boldsymbol{e}_x$, Figure 34.1. The origin of IS$'$ coincides at $t = t' = 0$ with that of IS.

A photon is emitted at time $t = t' = 0$ from the origin of IS and IS$'$ in the x-direction. As a sequence of events, defined by (x, t) or

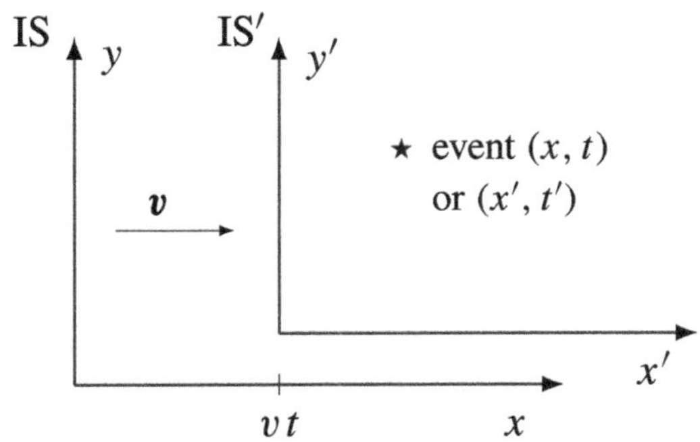

Figure 34.1 A specific event (⋆) has the coordinates x, t in the inertial system IS. Which coordinates x', t' does the same event have in IS′, which moves relative to IS with $\boldsymbol{v}$? In SRT, the Galilei transformation between (x', t') and (x, t) is replaced by the Lorentz transformation.

(x', t') respectively, we consider the position of the photon (or of the wavefront of an electromagnetic wave). The light moves in IS′ with velocity c. To this, we apply a Galilei transformation:

$$\frac{dx'}{dt'} = c \qquad \xrightarrow[\text{Galilean transformation}]{x' = x - vt,\ t' = t} \qquad \frac{dx}{dt} = c + v \tag{34.3}$$

In IS, the light then has the velocity $c+v$. If the photon moves in IS with c, then the Galilei transformation yields $c - v$ as the speed of light in IS′.

Surprisingly, however, one finds experimentally that light travels in all inertial systems with the same velocity

$$c = 2.998 \cdot 10^8 \ \frac{\text{m}}{\text{s}} \approx 3 \cdot 10^8 \ \frac{\text{m}}{\text{s}} \tag{34.4}$$

This constancy of the velocity of light is incompatible with the "natural" conception of space and time, which stands behind (34.2). In particular, the addition of velocities does not apply in general.

We therefore replace Galileo's principle of relativity with the *relativity principle of Einstein*:

1. All inertial systems are equivalent.
2′. Light propagates in every inertial frame with velocity c.

We will modify (34.2) such that the second point is satisfied. Since the transformation under consideration concerns the coordinates of arbitrary events, the second point then applies to all objects (or sequences of events) moving with the speed of light. Therefore, it is not a special property of light, but a revision of the space-time conceptions underlying (34.2).

Newtonian mechanics and (34.2) are experimentally well confirmed for velocities $v \ll c$; the to-be-found Lorentz transformation must therefore contain the Galilei transformation as a limiting case. Decisive changes, however, will arise for $v \to c$.

In the Maxwell equations, the velocity of light c occurs as a constant; therefore these equations always give the velocity c for an electromagnetic wavefront. According to the principle of relativity of Galilei, Maxwell's equations are thus non-relativistic; this opinion was also held by Maxwell himself (see last section in Chapter 5). The experimental findings of the constancy of the speed of light now suggest that Maxwell's equations are relativistic, i.e. valid in all IS; this is indeed the case. Therefore, we may describe the transition from Galileo's to Einstein's relativity principle as follows:

1. Principle of Relativity (RP): All inertial systems are equivalent.
2. Galilei's RP: Newton's laws are relativistic.

2′. Einstein's RP: Maxwell's equations are relativistic.

From 1 and 2 follow the Galilean transformations, and from 1 and 2′ the Lorentz transformations. In the following, we adopt Einstein's principle of relativity. This implies that Newton's laws must be modified (Chapter 38).

Interval of Events

We determine the Lorentz transformations that satisfy Einstein's principle of relativity. We consider two events, (t_1, x_1, y_1, z_1) and (t_2, x_2, y_2, z_2). By

$$s_{12}^2 = c^2 (t_2 - t_1)^2 - (x_2 - x_1)^2 - (y_2 - y_1)^2 - (z_2 - z_1)^2 \quad (34.5)$$

we define the square s_{12}^2 of the *interval* or distance between the events. The quantity s_{12}^2 can be negative; e.g. for two events, which take place at the same time at different positions. The square of the four-dimensional interval (34.5) may be seen in contrast to the square of the three-dimensional distance (given below in (34.9)). The interval s_{12} itself is used only if $s_{12}^2 \geq 0$.

The two events considered in (34.5) have the coordinates (t_1', x_1', y_1', z_1') and (t_2', x_2', y_2', z_2') in another IS′. Their interval in IS′ is then

$$s_{12}'^2 = c^2\,(t_2' - t_1')^2 - (x_2' - x_1')^2 - (y_2' - y_1')^2 - (z_2' - z_1')^2 \qquad (34.6)$$

Let a special event 1 consist now in the emission of a photon, the event 2 in its absorption at another place. Since the photon moves in each IS with the velocity c, it applies

$$s_{12}'^2 = s_{12}^2 = 0 \qquad (34.7)$$

Experimentally, we find that

$$s_{12}'^2 = s_{12}^2 \qquad (34.8)$$

also holds for $s_{12}^2 \neq 0$; for example, for a particle moving with constant velocity. For such a particle the squared interval between two points of its trajectory is equal to $s_{12}^2 = (c^2 - v^2)(t_2 - t_1)^2$. By measuring the velocity of the particle in two different inertial frames one can verify (34.8) As a condition for the wanted transformation we require (34.8).

We look for the transformations which leave s_{12}^2 invariant. For this purpose, we recall the orthogonal transformations, which describe the transition between Cartesian coordinate systems with differently orientated axes. The condition for these transformations is the invariance of the three-dimensional interval, i.e.

$$\begin{aligned} l_{12}^2 &= (x_2 - x_1)^2 + (y_2 - y_1)^2 + (z_2 - z_1)^2 \\ &= (x_2' - x_1')^2 + (y_2' - y_1')^2 + (z_2' - z_1')^2 \;=\; l_{12}'^2 \end{aligned} \qquad (34.9)$$

The solution is known: For a rotation, the transformation must be linear ($x' = \alpha\, x$ in matrix notation). The condition (34.9) then gives $\alpha^{\mathrm{T}} \alpha = 1$. This fixes α up to three angles of rotation. The determination of the sought Lorentz transformation is done in a quite similar way.

Lorentz Transformation

To simplify the notation, we number the space–time coordinates x^α from 0 to 3:

$$(x^\alpha) = \left(x^0, x^1, x^2, x^3\right) = (ct,\, x,\, y,\, z) \tag{34.10}$$

Whether the indices are placed at the top or at the bottom of the indexed size is in itself an arbitrary determination. In Chapter 36, however, we will introduce coordinates indexed at the bottom; this x_α will be defined somewhat differently. Greek indices shall always run from 0 to 3. We define the twofold indexed quantity $\eta_{\alpha\beta}$ by

$$\eta = (\eta_{\alpha\beta}) = \begin{pmatrix} 1 & 0 & 0 & 0 \\ 0 & -1 & 0 & 0 \\ 0 & 0 & -1 & 0 \\ 0 & 0 & 0 & -1 \end{pmatrix} \tag{34.11}$$

Then we write the interval ds between two infinitesimally adjacent events as

$$ds^2 = c^2 dt^2 - dx^2 - dy^2 - dz^2 = \sum_{\alpha=0}^{3} \sum_{\beta=0}^{3} \eta_{\alpha\beta}\, dx^\alpha\, dx^\beta \equiv \eta_{\alpha\beta}\, dx^\alpha\, dx^\beta \tag{34.12}$$

The interval ds may also be called a (four-dimensional) line element. In the last step, we used the *sum convention*: Over two equal indices, one at the top and the other at the bottom, it is summed. In the following, we will use this sum convention, i.e. the sum sign is not shown explicitly.

The condition (34.8) must be fulfilled for arbitrary intervals, so also for ds^2

$$ds^2 = ds'^2 \tag{34.13}$$

From this follows directly that the speed of light is independent of the inertial system:

$$\frac{dx'}{dt'} = c \qquad \xrightarrow[\text{Lorentz transformation}]{c^2dt'^2 - dx'^2 = c^2dt^2 - dx^2} \qquad \frac{dx}{dt} = c \tag{34.14}$$

We now set up the transformation which satisfies the condition (34.13). This transformation connects the coordinates x^α and x'^α

of the same event in IS and IS$'$. We consider a *linear* transformation:

$$x'^{\alpha} = \Lambda^{\alpha}_{\ \beta}\, x^{\beta} + b^{\alpha} \quad \text{(Lorentz transformation)} \tag{34.15}$$

The linearity implies that the $\Lambda^{\alpha}_{\ \beta}$ and b^{α} do not depend on the coordinates. The restriction to a linear transformation follows from the homogeneity of space and time: with a nonlinear transformation, the coefficients $\Lambda^{\alpha}_{\ \beta}(x)$ would depend on the coordinates. Then the transformation here would be different from there, and today different from tomorrow. This would contradict the homogeneity of space and time.

One can also write (34.15) in the matrix form

$$\begin{pmatrix} x'^0 \\ x'^1 \\ x'^2 \\ x'^3 \end{pmatrix} = \begin{pmatrix} \Lambda^0_{\ 0} & \Lambda^0_{\ 1} & \Lambda^0_{\ 2} & \Lambda^0_{\ 3} \\ \Lambda^1_{\ 0} & \Lambda^1_{\ 1} & \Lambda^1_{\ 2} & \Lambda^1_{\ 3} \\ \Lambda^2_{\ 0} & \Lambda^2_{\ 1} & \Lambda^2_{\ 2} & \Lambda^2_{\ 3} \\ \Lambda^3_{\ 0} & \Lambda^3_{\ 1} & \Lambda^3_{\ 2} & \Lambda^3_{\ 3} \end{pmatrix} \begin{pmatrix} x^0 \\ x^1 \\ x^2 \\ x^3 \end{pmatrix} + \begin{pmatrix} b^0 \\ b^1 \\ b^2 \\ b^3 \end{pmatrix} \tag{34.16}$$

With the column vector $x = (x^{\beta})$ and the matrix $\Lambda = (\Lambda^{\alpha}_{\ \beta})$ this becomes $x' = \Lambda x + b$; here the upper index of $\Lambda^{\alpha}_{\ \beta}$ is the row index, the lower one the column index. Since Λ and b do not depend on the coordinates, the differential form of (34.15) reads

$$dx'^{\beta} = \Lambda^{\beta}_{\ \alpha}\, dx^{\alpha} \tag{34.17}$$

Herewith, we evaluate the invariance $ds^2 = ds'^2$,

$$ds'^2 = \eta_{\alpha\beta}\, dx'^{\alpha}\, dx'^{\beta} = \eta_{\alpha\beta}\, \Lambda^{\alpha}_{\ \gamma}\, \Lambda^{\beta}_{\ \delta}\, dx^{\gamma} dx^{\delta} \stackrel{!}{=} ds^2 = \eta_{\gamma\delta}\, dx^{\gamma} dx^{\delta} \tag{34.18}$$

Since the invariance holds for any dx, it follows:

$$\Lambda^{\alpha}_{\ \gamma}\, \Lambda^{\beta}_{\ \delta}\, \eta_{\alpha\beta} = \eta_{\gamma\delta} \quad \text{or} \quad \Lambda^{\mathrm{T}} \eta\, \Lambda = \eta \tag{34.19}$$

This corresponds to the condition $\alpha^{\mathrm{T}} \alpha = 1$ for orthogonal transformations.

Special Lorentz Transformation

For rotational, spatial, or temporal displacements there are no differences between Galilean and Lorentz transformations. The principle of relativity of Galilei and that of Einstein equally imply the isotropy and the homogeneity of space and the homogeneity of time; hence, for example, the equivalence of inertial systems with differently oriented axes. Therefore, we restrict ourselves in the following to the relative motion between two IS. We refer to the special arrangement of Figure 34.1 and determine the corresponding (special) Lorentz transformation (LT).

At time $t = 0$, the systems IS and IS$'$ in Figure 34.1 coincide with each other; both clocks are set to zero ($t' = t = 0$). The event given by this coincidence has the coordinates $(ct, x, y, z) = (0, 0, 0, 0)$ and $(ct', x', y', z') = (0, 0, 0, 0)$, respectively. We set these coordinates to (34.15) and get

$$b^{\alpha} = 0 \tag{34.20}$$

In IS, at any x we can add an axis parallel to the y-axis (or z-axis). Thus, always one of these parallel axes can be brought into congruence with the current y'-axis so that

$$y' = y, \quad z' = z \tag{34.21}$$

The only thing we are looking for is the transformation between x, t and x', t'. This transformation cannot depend on the y- or z-coordinates, since the arrangement is homogeneous with respect to these directions. The Λ of the special Lorentz transformation (LT) is therefore of the form

$$\Lambda = \left(\Lambda_{\alpha}^{\beta}\right) = \begin{pmatrix} \Lambda_0^0 & \Lambda_1^0 & 0 & 0 \\ \Lambda_0^1 & \Lambda_1^1 & 0 & 0 \\ 0 & 0 & 1 & 0 \\ 0 & 0 & 0 & 1 \end{pmatrix} \tag{34.22}$$

This Λ must satisfy (34.19). Furthermore, the origin of IS$'$, $x' = 0$, must have the x-coordinate $x = vt$ in IS. The position of the origin of IS$'$ can be understood as a sequence of events. The coordinates of these events must be connected by the LT we are looking for:

$$\left(x'^0 = ct',\ x'^1 = 0\right) \overset{\text{LT}}{\longleftrightarrow} \left(x^0 = ct,\ x^1 = vt\right) \tag{34.23}$$

The coordinates $x'^2 = x'^3 = 0$ and $x^2 = x^3 = 0$ were not written down.

We first evaluate the condition (34.19). For this, we display $\Lambda^{\mathrm{T}}\eta\Lambda = \eta$ for (34.22) in the relevant subspace:

$$\begin{pmatrix} \Lambda^0_0 & \Lambda^1_0 \\ \Lambda^0_1 & \Lambda^1_1 \end{pmatrix} \begin{pmatrix} 1 & 0 \\ 0 & -1 \end{pmatrix} \begin{pmatrix} \Lambda^0_0 & \Lambda^0_1 \\ \Lambda^1_0 & \Lambda^1_1 \end{pmatrix} = \begin{pmatrix} 1 & 0 \\ 0 & -1 \end{pmatrix} \tag{34.24}$$

Multiplied out, these are four conditions, but two of them are equal. Hence, we are left with three conditions:

$$\left(\Lambda^0_0\right)^2 - \left(\Lambda^1_0\right)^2 = 1, \quad -\left(\Lambda^1_1\right)^2 + \left(\Lambda^0_1\right)^2 = -1, \quad \Lambda^0_0\,\Lambda^0_1 - \Lambda^1_0\,\Lambda^1_1 = 0 \tag{34.25}$$

Without the restriction of generality, one can set $\Lambda^1_0 = -\sinh\psi$ and $\Lambda^0_1 = -\sinh\varphi$. Then from the first condition follows $\Lambda^0_0 = \pm\cosh\psi$, from the second $\Lambda^1_1 = \pm\cosh\varphi$. The transformation is supposed to contain the limiting case of the identical transformation; therefore we allow in each case only the plus sign. From the third condition in (34.25), it follows $\varphi = \psi$. Thus, we obtain

$$\begin{pmatrix} \Lambda^0_0 & \Lambda^0_1 \\ \Lambda^1_0 & \Lambda^1_1 \end{pmatrix} = \begin{pmatrix} \cosh\psi & -\sinh\psi \\ -\sinh\psi & \cosh\psi \end{pmatrix} \tag{34.26}$$

One may compare this with the well-known form of a special orthogonal transformation (21.11): the signs in the quadratic form of ds^2 lead to hyperbolic functions instead of trigonometric functions. The condition (34.23) gives

$$x'^1 = 0 = \Lambda^1_0\, c\,t + \Lambda^1_1\, x^1 = \left(\Lambda^1_0\, c + \Lambda^1_1\, v\right) t \tag{34.27}$$

From this follows $\Lambda^1_0/\Lambda^0_0 = -\tanh\psi = -v/c$ or

$$\psi = \operatorname{artanh}\frac{v}{c} \quad \text{rapidity} \tag{34.28}$$

The quantity ψ is called *rapidity*. The relation between the rapidity ψ and the velocity v is shown in Figure 34.2. Some relations in the SRT take a particularly simple form if one uses the rapidity instead of the velocity.

For any finite ψ holds

$$v < c \quad \text{for the relative velocity } v \text{ between IS and IS}' \tag{34.29}$$

This restriction follows from Einstein's principle of relativity. There is no IS′, which moves relative to another one with the speed of

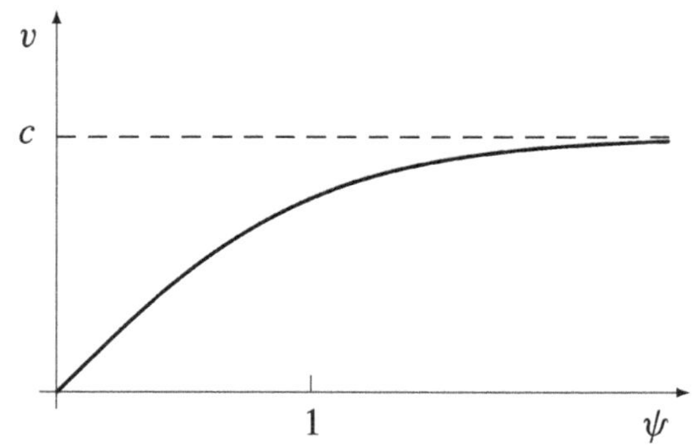

Figure 34.2 Relation between the velocity v and the rapidity ψ.

light (or faster). The practical impossibility of such a reference frame (e.g. of a rocket laboratory) follows from the equations of motion (Chapter 38). By the restriction $v < c$ the concept of the IS is modified. We have introduced IS as reference systems, which move with respect to the fixed-star sky with constant velocity. In the principle of relativity of Galilei, "all IS" include also those with $v > c$. In the relativity principle of Einstein, it remains with the formulation "all IS", but there are just none with $v \geq c$.

The derived special LT can be written as

$$\begin{pmatrix} ct' \\ x' \end{pmatrix} = \begin{pmatrix} \cosh\psi & -\sinh\psi \\ -\sinh\psi & \cosh\psi \end{pmatrix} \begin{pmatrix} ct \\ x \end{pmatrix} \qquad \text{Lorentz transformation} \tag{34.30}$$

with $\psi = \operatorname{artanh}(v/c)$. We introduce a quantity γ by

$$\gamma = \frac{1}{\sqrt{1 - v^2/c^2}} = \cosh\psi \tag{34.31}$$

Then $\sinh\psi = \gamma\, v/c$. (This γ has nothing to do with the occasionally used index γ.) The special LT can now also be written as

$$\begin{pmatrix} ct' \\ x' \end{pmatrix} = \begin{pmatrix} \gamma & -\gamma\, v/c \\ -\gamma\, v/c & \gamma \end{pmatrix} \begin{pmatrix} ct \\ x \end{pmatrix} \tag{34.32}$$

This is equivalent to

$$x' = \frac{x - vt}{\sqrt{1 - v^2/c^2}}, \quad y' = y, \quad z' = z, \quad ct' = \frac{ct - xv/c}{\sqrt{1 - v^2/c^2}} \quad (34.33)$$

For $v \ll c$ this reduces to the special Galilean transformation (34.2). The general LT will be given in Chapter 36.

Exercises

34.1 *Inverse Lorentz Transformation*

The special Lorentz transformation $x' = x'(x, t)$ and $t' = t'(x, t)$ between two inertial frames is assumed to be known. What is the reverse transformation? Express the result alternatively by the velocity v or the rapidity ψ

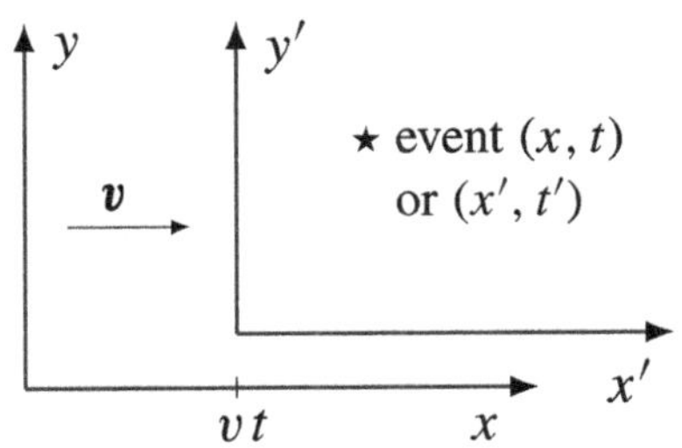

34.2 *Matrix notation for line element*

Write the Lorentz transformation

$$dx'^{\alpha} = \Lambda^{\alpha}_{\beta}\, dx^{\beta}$$

in matrix notation. In this notation, give the condition $ds^2 = ds'^2$ for the Minkowski line element.

Chapter 35

Length and Time Measurement

The measurement of a velocity is a combined length and time measurement. The independence of the speed of light from the inertial system of the experimenter has therefore consequences for the measurement of these quantities. We discuss in particular the length contraction, the time dilation, and the meaning of simultaneity.

The transition from Galilei to Lorentz transformation (LT) was forced by experimental findings. The realization of an experiment can answer the question "what results from a time or length measurement"; only in this restricted and well-defined sense, we treat the questions "what is time" and "what is length". This restriction in itself is nothing new; physical quantities are in general defined by measurement prescriptions. In view of the paradoxical consequences of the Lorentz transformation, it is appropriate to emphasize this point. Besides, these consequences, as well as at first sight the constancy of the speed of light itself, are not compatible with "natural" space–time presuppositions. This is not surprising, because these conceptions — consciously or unconsciously — are abstracted from everyday experiences and therefore refer to velocities $v \ll c$.

Resting Rulers and Clocks

We begin with measurements made by experimenters within an inertial system (IS). For the length definition, a rod or ruler is declared as the standard rod, and its length is defined as 1 unit of length (historically, one refers to a specific standard meter stick kept in Paris, *mètre des archives*). By adapting to the standard ruler, an arbitrary number of equally long single-unit rulers is produced. In this calibration, the congruent rulers are at rest relative to each other. Within an IS, a coordinate net can be realized by resting rods (Figure 35.1). For this purpose, imagine a laboratory room or a train, in which a grid of rigid rods forms the coordinate net. All spatial IS coordinates are thus defined as lengths.

For the measurement of time by suitable clocks, we recall the discussion in Chapter 1. Within an IS we can place clocks at different places, which all show the same IS- time, which are thus *synchronized*. The synchronization of clocks is done by the exchange of signals: As the standard clock, we choose the clock at the origin. Then at the time t, a signal is sent to another clock. The signal is sent back immediately and arrives at time $t + \Delta t$ again at the standard clock. Then at an arbitrary time t_i a signal with the information "on the reception it is $t_i + \Delta t/2$" is sent to the clock to be synchronized. Due to the isotropy of the space and the homogeneity of the time, the travel time is the same for the forth and back directions. Any uniform motion may serve as a signal, for example, a uniformly moving body or a light signal.

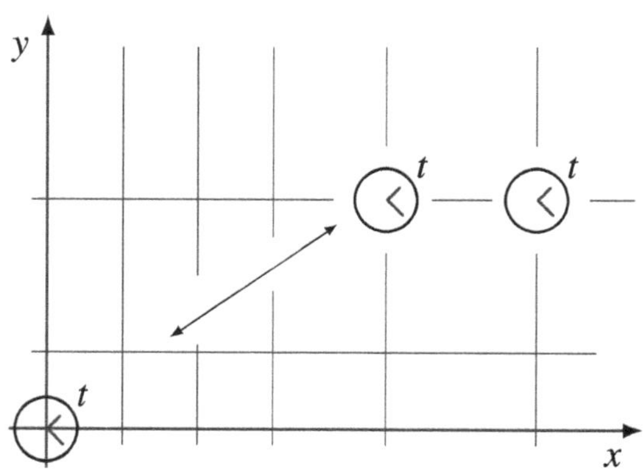

Figure 35.1 The coordinate net of an IS can be defined by resting, calibrated rulers. Furthermore resting identical clocks are placed at arbitrary positions. By an exchange of signals (indicated by a double arrow) they can be synchronized. This defines the length and time measurements within the considered IS.

Moving Ruler

In an IS, the length of an arbitrary rod at rest is determined by comparing it to calibrated resting rulers. The rod can then be given an inscription indicating its *proper length* or rest length l_0. This quantity l_0 is then independent of the IS; it is, therefore, a Lorentz scalar. To summarize: the proper length of an object is its length as measured in the object's rest frame. The considered object shall be in the following a rod (a bar or a stick).

We consider the systems IS and IS′ (Figure 35.2) and a rod at rest in IS′ with proper length l_0. We investigate which length l experimenters in IS measure for this rod. It will turn out that $l \neq l_0$ (length contraction).

The rod rests in IS′ between x_1' and x_2', so that

$$x_2' - x_1' = l_0 = \text{proper length} \tag{35.1}$$

To determine the length of the rod in IS, two observers in IS mark the position of the beginning and the end of the rod on the x-axis at the same IS time t. This time we arbitrarily choose $t = 0$. The positions of the marks are denoted by x_1 and x_2. The passage of the beginning and end of the bar corresponds to two events, which we denote by 1 and 2:

1: Rod end passes an observer in IS at $x_1 = 0$ at time $t_1 = 0$.
2: Rod beginning passes an observer in IS at x_2 at time $t_2 = 0$.

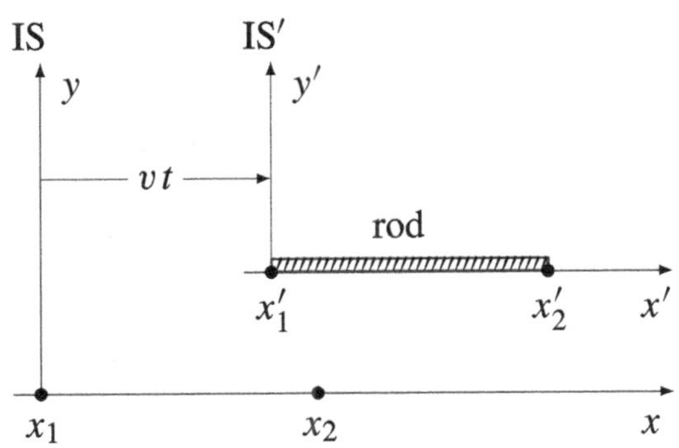

Figure 35.2 In IS the length of a rod moving with v is to be measured. For this purpose, IS-observers mark at time $t = 0$ the points x_1 and x_2, where the rod's end and beginning are located.

The IS and IS$'$ coordinates of the two events are:

$$\text{Event 1:} \quad \begin{cases} x_1 = 0, \; t_1 = 0 \\ x_1' = 0, \; t_1' = 0 \end{cases} \qquad \text{Event 2:} \quad \begin{cases} x_2, \; t_2 = 0 \\ x_2' = l_0, \; t_2' \end{cases} \tag{35.2}$$

For the special LT (34.32) under consideration, the coordinate systems coincide at time $t_1 = t_1' = 0$. The dashed and undashed coordinates of event 1 trivially satisfy the LT. In IS$'$, the rod has the length l_0 (thus $x_2' - x_1' = l_0$). We apply the special Lorentz transformation at time zero to x_2':

$$x_2' = \gamma\,(x_2 - v\,t_2) = l_0 \quad \stackrel{t_2=0}{\longrightarrow} \quad x_2 = l_0/\gamma = l_0\,\sqrt{1 - v^2/c^2} \tag{35.3}$$

The IS observer measures the length $l = x_2 - x_1$

$$l = l_0\,\sqrt{1 - \frac{v^2}{c^2}} \quad \text{length contraction} \tag{35.4}$$

The shortening with respect to the proper (or rest) length is called *length contraction*. Due to the isotropy of space, the result does not change if we rotate both, the rod and the velocity orientation by some angle. Therefore, (35.4) with $v^2 = \boldsymbol{v}^2$ is also valid for a general LT, as long as rod and $\boldsymbol{v}$ are parallel.

We consider the special LT as in Figure 35.2, but a rotated rod with length $l_{0\perp}$ in the y-direction and $l_{0\|}$ in the x-direction; it is $l_0^2 = l_{0\perp}^2 + l_{0\|}^2$. The LT then says $y' = y$, i.e. the length perpendicular to $\boldsymbol{v} = v\,\boldsymbol{e}_x$ does not change; for the length parallel to $\boldsymbol{v}$ the above derivation is valid. Thus, we obtain

$$l_\| = l_{0\|}\,\sqrt{1 - \frac{\boldsymbol{v}^2}{c^2}}, \quad l_\perp = l_{0\perp} \quad \text{(length contraction)} \tag{35.5}$$

This result is valid also for a general Lorentz transformation.

If the rod moves with the time-dependent velocity $v = v(t)$ the measurement of the length l in IS also gives (35.4). For this purpose one considers at a certain time t_0 an IS$'$ which moves with the (constant) velocity $v(t_0)$ with respect to IS; this system is called instantaneously co-moving IS$'$ or momentary rest system of the rod. At the time $t = t_0$ the rod is momentarily resting in IS$'$ and the given derivation applies. At a later time, one had to use another instantaneous rest system IS$''$. As a result, the formula (35.4) remains valid also with $v = v(t)$.

A rod of proper length l_0 at rest in IS is measured in IS$'$ also with $l = l_0/\gamma < l_0$; because the effect (35.4) does not depend on the sign of v. Whether the discussed length contraction applies to the rulers in IS or IS$'$, depends on the measurement made. It is the measurement itself, which leads to an asymmetry between IS and IS$'$. For the measurement described above, the two events must be simultaneous in IS. However, these events are not simultaneous in IS$'$, because

$$t'_1 = 0, \quad t'_2 = \gamma\left(t_2 - \frac{x_2\, v}{c^2}\right) = -\frac{l_0}{c}\,\frac{v}{c} \tag{35.6}$$

Thus, the two events (35.2) cannot be used in IS$'$ in order to measure the length of a rod that is resting in IS.

Finally, it should be emphasized once more that the length contraction is a statement that refers to a specific measurement prescription. However, this prescription is not arbitrary, but the obvious and reasonable definition[1] for the quantity *length*. However, the result depends on whether the measurement is made in IS or IS$'$. The formulation "moving bars are shorter" represents the result only incompletely; such formulations invite misinterpretations. The events belonging to a measurement must be defined unambiguously (like in (35.2)) so that the Lorentz transformation can be applied to them.

In order to explain the Michelson–Morley experiment, Lorentz considered in 1892 a length contraction which at that time was considered real. So, in a historic context, the terminus "Lorentz contraction" may be found in connection with a real contraction

[1]Consider the following example: A train, which is running with constant speed is so long that the measurement of its length by application of measuring rods is impossible. One could reasonably determine its length in this way: Many observers are positioned along the railroad track. At an agreed time (the observers have synchronized clocks) the two observers, where the beginning or the end of the train is located, mark the track. Afterwards the length between the two track marks can be determined by using stationary measuring rods. This measurement prescription for the length of the train corresponds to the one given in Figure 35.2.

(compression) of the rod. Today, however, the terms *length contraction* and *Lorentz contraction* are used synonymously; both mean the result (34.4). For this length contraction, the rod is, of course, not compressed. The length contraction is just the result of an exactly and reasonably defined length measurement.

Moving Clock

Figure 35.3 shows an IS$'$ clock that is moving with velocity v relative to IS. We compare the time t' shown by this clock with the IS-time t.

At the positions x_1 and x_2, there are two observers with resting, synchronized clocks, which display the IS time t. The observers read the IS$'$ clock as it passes by them. We arbitrarily place the first IS observer at $x_1 = 0$. The passage of the moving IS$'$-clock at the two IS-clocks defines two events:

1: IS$'$ clock passes observer in IS at $x_1 = 0$ at time $t_1 = 0$.
2: IS$'$ clock passes observer in IS at $x_2 = vt_2$ at time t_2.

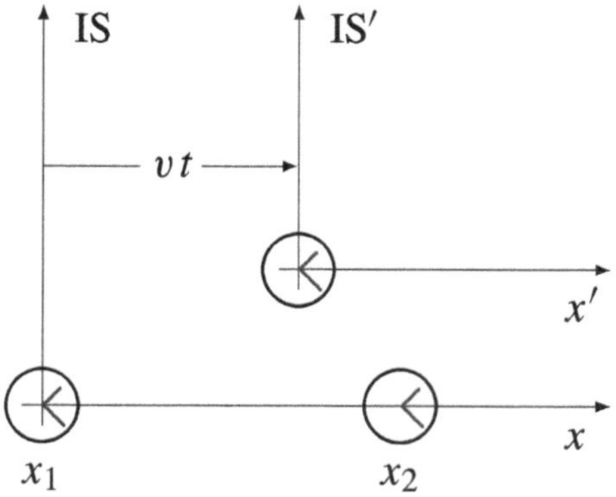

Figure 35.3 A clock moves with velocity v relative to IS. We make the position of this clock to the origin of an IS$'$; the clock then rests at $x' = 0$ and shows the IS$'$ time t'. The t' display is read off in IS when the clock passes the positions x_1 and x_2. Subsequently, the time differences (as measured in IS and in IS') are compared.

The coordinates of these two events are:

$$\text{Event 1:} \begin{cases} x_1 = 0, \ t_1 = 0 \\ x'_1 = 0, \ t'_1 = 0 \end{cases} \qquad \text{Event 2:} \begin{cases} x_2 = v t_2, \ t_2 \\ x'_2 = 0, \ t'_2 \end{cases} \tag{35.7}$$

For event 1, the coordinates of IS and IS′ are trivially connected by the Lorentz transformation (LT). Event 2 is given by the specification that the moving IS′ clock passes the position $x_2 = vt$. We use the special Lorentz transformation (34.32) for t'_2:

$$t'_2 = \gamma \left(t_2 - \frac{x_2 \, v}{c^2} \right) = t_2 \sqrt{1 - v^2/c^2} \tag{35.8}$$

At event 1, the IS clock and the IS′ clock are both at zero. At event 2, the moving clock shows

$$t_0 = t'_2 - t'_1 = t'_2 = \text{IS}'\text{–time interval between 1 and 2} \tag{35.9}$$

When the clock passes x_2 in IS, the IS observer reads the time

$$t = t_2 - t_1 = t_2 = \text{IS–time interval between 1 and 2} \tag{35.10}$$

From the last three equations we get

$$t = \frac{t_0}{\sqrt{1 - v^2/c^2}} \quad \text{(time dilation)} \tag{35.11}$$

The result $t_0 < t$ means that the moving clock is lagging with respect to the IS clocks. This effect is called *time dilation* (or relativistic time dilation).

The IS′-clock shall now move with $\boldsymbol{v}$ in any direction. By rotation, we can put the x- and x'-axis of both inertial systems always in the direction of $\boldsymbol{v}$. With this we reach again the arrangement in Figure 35.3. Therefore, (35.11) with $v^2 = \boldsymbol{v}^2$ also holds for a general LT,

$$t = \frac{t_0}{\sqrt{1 - \boldsymbol{v}^2/c^2}} \quad \text{time dilation} \tag{35.12}$$

The abbreviated formulation of this result as “a moving clock goes slower” is problematic. Without the background given here one could conclude that "the IS′-clock goes slower than the IS clock" and (since from the IS′ point of view IS is moving) "the IS clock goes slower than the IS′ clock"; thus one would have constructed a contradiction. However, it is the measurement (Figure 35.3), which introduces an asymmetry between IS and IS′: for the measurement described here, two clocks in IS are necessary, and one clock resting in IS′ moves past. For this measurement, the formula (35.12) is a well-defined, verifiable statement.

The described measurement refers to a clock that moves relative to IS with constant velocity; in this case, at least three clocks are required for the clock comparison. However, one could think of the following experiment for only two clocks:

1. Both clocks are started at the same position.
2. The clocks are moved relative to each other.
3. They are brought back to their original position.

Which clock shows less now? This cannot be answered from the given information. The forth and back movements necessarily imply accelerations for at least one of the clocks (the other clock could keep resting at the starting position, or it might be accelerated, too). The above discussion (using IS) is therefore not transferable without further ado. In the next section, we determine the time of an arbitrarily moving clock. After that also the result of the experiment with only two clocks can be calculated.

Proper time

If a clock moves relative to an IS with a time-dependent velocity $\boldsymbol{v}(t)$, then it can no longer be made the origin of an IS′. The time actually displayed by the clock can be calculated as follows: For a given time $t = t_0$ we introduce an IS′ which moves with the (constant) velocity $\boldsymbol{v}(t_0)$ relative to IS. In this IS′, the clock then rests momentarily (at the time t_0). Therefore, the time interval displayed by the clock between t_0 and $t_0 + dt$ is equal to the corresponding time interval dt' in IS′:

$$d\tau = dt' \overset{(35.12)}{=} dt\,\sqrt{1 - \frac{v(t_0)^2}{c^2}} \tag{35.13}$$

If we decompose the time between two events 1 and 2 into infinitesimal intervals, we obtain from (35.13) the total time displayed by the moving clock:

$$\tau = \int_{t_1}^{t_2} dt \, \sqrt{1 - \frac{v(t)^2}{c^2}} \quad \text{proper time} \tag{35.14}$$

The quantity τ is called *proper time* because it is independent of the IS. The independence from the IS can be seen like this: Let event 1 be the starting of the clock, event 2 the stopping. Thus, a certain pointer position of the clock (thus τ) is reached, which every observer can read off, independently of his state of motion. An IS$'$ observer would calculate the time with a different time coordinate in $\int dt' \, \sqrt{1 - v'(t')^2/c^2}$, but he will get the same result τ.

The independence of the inertial system follows also from the invariance of the line element ds^2 under LT, by noting that $d\tau$ equals the line element ds of the clock:

$$\left(ds^2\right)_{\text{clock}} = \left(c^2\, dt^2 - d\boldsymbol{r}^2\right)_{\text{clock}} = c^2 \left(1 - \frac{\boldsymbol{v}(t)^2}{c^2}\right) dt^2 = c^2\, d\tau^2 \tag{35.15}$$

With ds, also $d\tau$ is invariant under Lorentz transformations, so it is independent of the chosen IS.

We return to the experiment outlined at the end of the last section. At a certain position, there are two clocks which are turned on at the same time. They are then somehow moved independently of each other, brought together again, and turned off. The clocks then display the times τ_1 and τ_2, respectively; using (35.14) these times can be calculated. The proper times τ_1 and τ_2 will be in general different; they depend on the corresponding time courses of the clocks' velocities. An application to this is the twin paradox discussed in Chapter 39.

An example of the relation between the proper time and the IS time is the lifetime of fast muons from cosmic radiation (Exercise 35.1). Experimentally, the time dilation has been confirmed for clocks in airplanes and satellites. In these cases, the influence of the gravitational field on the clock's rate must be taken into account (Exercise 35.3).

Simultaneity

The length measurement of the moving rod or ruler (Figure 35.2) was executed by marking the beginning and end of the rod at the same IS-time. As seen from IS, the rod lies then just over the length $l = l_0/\gamma < l_0$. Seen from IS$'$ this does not mean a measurement of the IS length l, because the two events (the coincidence of the rod ends with the markings) are not simultaneous in IS$'$. Rather, in IS$'$ these events are separated by the time difference $(l_0/c)(v/c)$ (according to (35.6)).

This shows that simultaneity is a relative notion; it is a statement that depends on the IS of the observer. This raises the question to what extent the temporal order of events is arbitrary. Is causality (cause before effect) possibly violated?

We consider two events that we can define without loss of generality by

$$\text{Event 1:} \quad \begin{cases} x_1 = 0, \ t_1 = 0 \\ x_1' = 0, \ t_1' = 0 \end{cases} \qquad \text{Event 2:} \quad \begin{cases} x_2 = x, \ t_2 = t \\ x_2' = x', \ t_2' = t' \end{cases} \tag{35.16}$$

The interval between the two events is classified according to

$$s_{12}^2 = c^2 t^2 - x^2 = c^2 t'^2 - x'^2 \begin{cases} = 0 & \text{lightlike} \\ < 0 & \text{spacelike} \\ > 0 & \text{timelike} \end{cases} \tag{35.17}$$

as *lightlike*, *spacelike*, or *timelike*. This classification is independent of the chosen inertial system (because ds^2 is invariant under LT). In the x–t diagram in Figure 35.4, different areas are assigned to these cases.

Let the two events be connected by a fictitious object which moves with the velocity $V = x/t$ along the x-axis. We order the two events so that $V \geq 0$. From $s_{12}^2 = (c^2 - V^2)\, t^2$ follows:

$$\text{lightlike: } V = c, \quad \text{spacelike: } V > c, \quad \text{timelike: } 0 \leq V < c \tag{35.18}$$

For the temporal interval between the two events (t in IS and t' in IS$'$) we obtain

$$t' = \gamma\left(t - \frac{v\,x}{c^2}\right) = \gamma t\left(1 - \frac{v\,V}{c^2}\right) \tag{35.19}$$

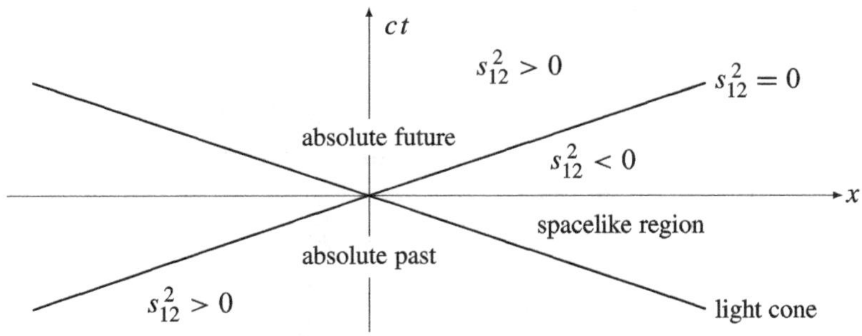

Figure 35.4 The event 1 is represented by the origin ($x = 0, t = 0$). The event 2 is represented by some other point with coordinates $(x^0, x^1) = (ct, x)$. For $s_{12}^2 > 0$ the event 2 lies in the absolute future or past (as seen from 1). For $s_{12}^2 < 0$ the two events are absolutely distant from each other; their temporal order is not defined. On the light cone $s_{12}^2 = 0$ are events that can be connected with 1 by a photon.

The first equality follows from the LT (34.33), the second by inserting $x = Vt$ for the considered fictitious particle.

Spacelike interval

For a spacelike interval, i.e. for $V > c$, the events are not causally connected: If event 1 shall influence event 2, then something has to get from 1 to 2. As we will see in Chapter 38, material particles (particles with a mass not equal to zero) can only move with a velocities $V < c$. Massless particles (like photons) always move with $V = c$. Therefore, c is also the maximum velocity with which any effect can propagate.

Since of $V/c > 1$, the last bracket in (35.19) can be positive or negative (depending on the size of v), i.e.

$$1 - \frac{v}{c}\frac{V}{c} \lesseqgtr 0 \tag{35.20}$$

Thus, the temporal order of the two events depends on the chosen inertial frame. Since the two events cannot be causally connected with each other, their temporal order has no specific physical meaning. An example of a spacelike interval is the two events defined for the length measurement (Figure 35.2).

Lightlike or timelike interval

For lightlike or timelike intervals, that is, for $V \le c$, the events may be causally connected; the earlier event could be the cause of the later one. Therefore, the temporal order is physically relevant; it is the same for every observer. The last bracket in (35.19) is positive (because of $0 \le V \le c$ and $|v/c| < 1$). Since

$$\text{sign}\; t' = \text{sign}\; t \tag{35.21}$$

the two events have the same temporal order in all IS.

Summary

We summarize: The temporal sequence of events, which might have influenced each other causally, has an absolute meaning, i.e. it is independent of the IS. Everything that we call the past in everyday language lies in the area of the absolute past in Figure 35.4. Only for events that cannot have influenced each other, the temporal order depends on the reference system; but this order is without practical importance.

As an example, we consider the statement made in Berlin "At 12 o'clock on January 1st of the year 2000 a solar eruption takes place". Hereby, the simultaneity of the two events "Berlin, 12 o'clock at noon, January 1st in 2000" and "sun eruption" is expressed; we refer to an IS, in which the solar system as a whole rests. For the astronaut of a fast-passing spaceship (system IS$'$ with relative velocity v) the two events are not simultaneous. For the astronaut the temporal interval results from (35.19) to $t' = -\gamma v x/c^2 = -8\gamma(v/c)\,\text{min}$; here the interval Berlin-Sun with $x = 8$ light minutes was inserted. The time interval t' can be positive or negative depending on the sign of v.

However, any consequences of the eruption on the earth are an absolute future with respect to the eruption. In IS-Berlin, the consequences occur at the earliest after the light travel time, i.e. at about 12.08 o'clock. For the astronaut, however, these consequences can occur shortly after the eruption, namely when for him the Earth–Sun distance goes to zero due to the length contraction.

House Stick Experiment

Finally, we discuss the house stick experiment outlined in Figure 35.5 (sometimes also described as a barn-ladder experiment or paradox). Here, the length contraction appears to lead to a contradictory result. Can one close both doors in this experiment without jamming the stick?

The relevant events are:

1: The end of the stick passes the left front door.
2: The beginning of the stick passes the right front door.

In Figure 35.5, center, event 1 has already occurred, but event 2 has not yet occurred. For the IS-time t we get therefore $t_1 < t < t_2$ and $t_1 < t_2$.

In Figure 35.5, right, event 2 has already occurred, but event 1 has not yet occurred. For the IS′-time t' we get therefore $t'_2 < t' < t'_1$ and $t'_2 < t'_1$.

Thus, the order of the two events is different in the considered reference frames IS and IS′. This implies that the interval of events 1 and 2 is spacelike, and that the two events cannot causally influence each other.

We specify the experiment in such a way that the right door is closed immediately before event 2, and that the left door is closed at

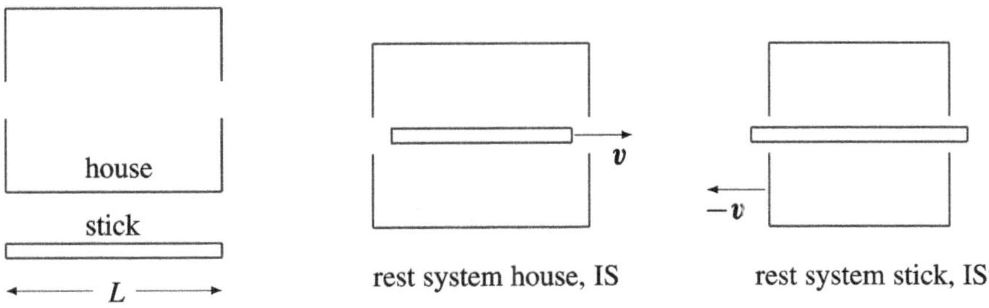

Figure 35.5 A house with two open doors has the same proper length L as a stick (left). The stick now moves with the velocity v through the house. Seen from the IS of the house, the stick has a shortened length (center); therefore it seems possible to close both doors while the stick is in the interior of the house. Seen from the rest system IS′ of the stick, the house moves with the velocity $-\boldsymbol{v}$ and has a shortened length (right); closing the doors without jamming does not seem possible.

the same IS time. Thus, it is clear to the observer in IS (Figure 35.5 middle) that both doors can be closed.

We now discuss this experiment from the point of view of an observer moving along with the stick (i.e., from IS′, Figure 35.5 right). Immediately after the closing of the right-hand door the stick bumps into this door. Let the door be so solid that the stick is stopped here. If the stick were a *rigid body* then at this stop the left end of the now resting rod would be inside the left door opening, and the left door could not be closed anymore. In this case, we would have constructed a contradiction; the experiment would depend on the reference frame.

The reason for this contradiction is the assumption of a *rigid* stick. The condition *rigid body* is incompatible with special relativity. *Rigid* means constant lengths between different parts of the body. But this implies an infinitely large speed of action: If one end of a rigid stick moves, then the other end must move simultaneously so that the distance between the beginning and the end remains the same. According to the special theory of relativity, the movement of one end can be noticed at the other end at the earliest after the time L/c. (In reality, the time L/c_{S} is needed, where c_{S} is the sound velocity of the stick material.)

Back to the house-stick experiment: As soon as the stick reaches the right, closed door, it crumbles (if the door is stable enough). The end of the rod is not affected by this. This still holds for event 1, i.e. when the end of the rod passes the left door; because the interval of events 1 and 2 is spacelike. Seen from IS′, event 1 occurs after event 2. Nevertheless, the left door can be closed, because the end of the stick is not aware of the crash at the right.

The result of the experiment is therefore, as it must be, independent of the chosen reference frame. The (broken) stick can be locked up in the house.

The example makes clear that with the assumption of a *rigid body* contradictions can be constructed. This is especially so when accelerations occur and deformations of the considered body are unavoidable.[2] But this is not due to a contradiction within the special relativity theory, but because the assumption of a rigid body is in contradiction to the special relativity theory (and to the reality).

Exercises

35.1 *Lifetime of muons*

Muons are produced at a height of about $h \approx 30\,\mathrm{km}$ by cosmic rays. In their rest system, the muons have a lifetime $\tau \approx 2 \cdot 10^{-6}\,\mathrm{s}$. Despite this height and their short lifetime ($c\tau \approx 600\,\mathrm{m}$) they still reach the earth's surface to a large part.

Consider the deviation $\varepsilon = (c - v)/c \ll 1$ of the speed of muons from the speed of light. Under which condition for ε the muons may reach the earth's surface? What does an observer in the rest system of the muon measure for the height h?

35.2 *Snapshot of a passing sphere*

In its rest system IS′, a body forms a sphere with diameter D. The body moves with the relativistic velocity $\boldsymbol{v} = v\,\boldsymbol{e}_x$ in an inertial system IS. An IS observer takes a photograph of the object. The observer is so far away ($L \to \infty$) that the light rays reaching him are parallel to the y-axis (figure below). What shape (sphere? ellipsoid?) appears in the photograph? Which parts of the sphere are shown?

An equator of a sphere moving with v appears as an ellipse because of the length contraction in IS. For the stationary sphere, the point P on the equator would be a just visible point. Due to the aberration, the light ray in IS′ would have to be directed into the inside of the sphere to be visible in IS in the direction $-\boldsymbol{e}_y$. The point opposite to P can be seen without further ado. On this side one can see even further: A ray emanating tangentially downward from A forms a certain angle with $-\boldsymbol{e}_{y'}$. If this angle is equal to φ_{A}, then

[2]For another example refer to W. Rindler, Length contraction paradox, *American Journal of Physics* 29 (1961) 365,

A is just visible. The great circle marked by A and B separates the visible and invisible (hatched) parts of the sphere.

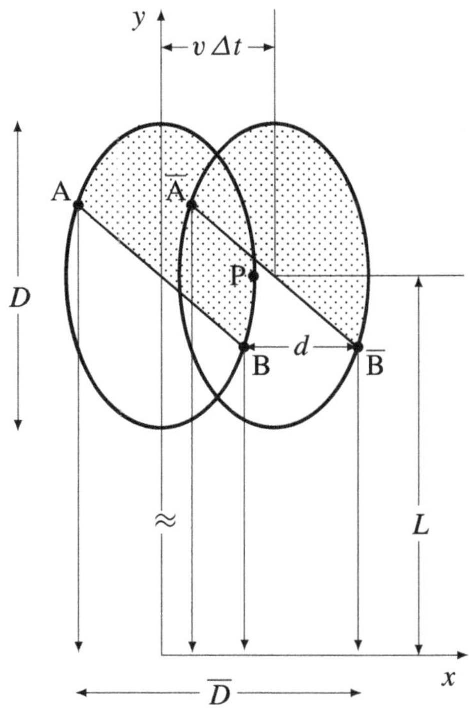

Notes: For a light ray to travel in $-\boldsymbol{e}_y$-direction in IS, it must travel in the moving system IS′ under an aberration angle φ_{A} relative to direction $-\boldsymbol{e}_{y'} = -\boldsymbol{e}_y$. This angle is given by (we refer to (22.40) in [2] or to (14.20) in [5])

$$\tan\varphi_{\mathrm{A}} = \frac{v/c}{\sqrt{1 - v^2/c^2}} \stackrel{!}{=} \frac{dx'}{dy'} \tag{35.22}$$

So in IS′ this light ray must have the slope dy'/dx'.

Calculate the coordinates of A and B from the ellipse equation and from the condition that the elliptic tangent has the angle φ_{A} relative to $-\boldsymbol{e}_y$.

The camera registers at a certain time t the light from A and B. Due to the different light travel times, light from B has to be sent at a time Δt later than that of A. During this time Δt the left ellipse has moved to the position of the right one, and B has moved to $\overline{\mathrm{B}}$. On the photo then A and $\overline{\mathrm{B}}$ mark the diameter $\overline{D}$ of the object.

35.3 *Time offset for satellites*

A satellite (mass m) moves on a circular orbit (radius r_0) in the gravitational potential

$$V(r) = -\frac{G M_{\mathrm{E}}\, m}{r} = m\,\Phi(r) \tag{35.23}$$

Here, G is the gravitational constant and M_{E} is the mass of the earth. A clock in the satellite shows the time t_{S}. A clock, which is resting at $r = \infty$ shows the time t_∞. Determine the time difference due to relativistic time dilation in the form $t_{\mathrm{S}}/t_\infty = 1 + \delta$ in the lowest, nonvanishing order in v/c. Express δ by $\Phi(r_0)$.

In addition, the gravitational field influences the rate of the clock:

$$\frac{t_{\mathrm{S}}}{t_\infty} = 1 + \delta + \frac{\Phi(r_0)}{c^2} \tag{35.24}$$

A clock in the laboratory on the earth shows the time $t_{\mathrm{L}} \approx t_\infty (1 + \Phi(R)/c^2)$; where R is the earth radius, and where the (small) velocity due to the earth rotation may be neglected. Determine the relative time shift $(t_{\mathrm{L}} - t_{\mathrm{S}})/t_{\mathrm{L}}$ between laboratory and satellite as a function of r_0/R for $\Phi/c^2 \ll 1$. What are the size and the sign of this effect for a near-earth and for a geostationary satellite (one may use the results of Exercise 17.3)?

Chapter 36

Lorentz Group

We generalize the special Lorentz transformation (LT) from Chapter 34 to the general LT. The LT form a group. The evaluation of two successive LT leads to the addition formula for velocities.

General Lorentz Transformation

According to Galileo's principle of relativity, any two inertial systems (IS and IS′) are connected by the transformation (5.9),

$$x'_i = \sum_{j=1}^{3} \alpha_{ij}\, x_j - v_i\, t - a_i, \quad t' = t - t_0 \tag{36.1}$$

This transformation depends on 10 parameters: The parameters a_i and t_0 describe a constant displacement in time and space, the orthogonal matrix $\alpha = (\alpha_{ij})$ depends on three rotation angles and describes the relative position of the spatial coordinate axes of IS and IS′, the three parameters v_i give the relative velocity $\boldsymbol{v}$ between IS and IS′. These transformations form a group which is called the Galilean group.

The general Lorentz transformation is of the form (34.15),

$$x'^{\beta} = \Lambda^{\beta}_{\delta}\, x^{\delta} + b^{\beta} \quad \text{or} \quad x' = \Lambda\, x + b \tag{36.2}$$

One distinguishes between the *homogeneous* ($b = 0$) and the *inhomogeneous* ($b \neq 0$) Lorentz transformations.

For mere displacements and rotations (i.e. for $v^i = 0$) there are no differences to the Galilei transformation: From the comparison of

(36.2) and (36.1) we see that b^β describes displacements in space and time,

$$b = \left(b^\beta\right) = (-c\,t_0, -a_i) \tag{36.3}$$

A rotation of the Cartesian axes of IS relative to those of IS$'$ is mediated by the LT with

$$\Lambda(\alpha) = \left(\Lambda^\gamma_{\ \beta}\right) = \begin{pmatrix} 1 & 0 & 0 & 0 \\ 0 & \alpha_{11} & \alpha_{12} & \alpha_{13} \\ 0 & \alpha_{21} & \alpha_{22} & \alpha_{23} \\ 0 & \alpha_{31} & \alpha_{32} & \alpha_{33} \end{pmatrix} \tag{36.4}$$

The condition (34.19) for an LT, $\Lambda^{\rm T}\eta\Lambda = \eta$, becomes here $\alpha^{\rm T}\alpha = 1$.

Differences between Galilei and Lorentz transformations arise only for a relative motion of IS and IS$'$ with the constant velocity $\boldsymbol{v}$. According to (34.32), the special LT for $\boldsymbol{v} = v\,\boldsymbol{e}_1$ is given by

$$\Lambda(v) = \left(\Lambda^\beta_{\ \delta}\right) = \begin{pmatrix} \gamma & -\gamma\, v/c & 0 & 0 \\ -\gamma\, v/c & \gamma & 0 & 0 \\ 0 & 0 & 1 & 0 \\ 0 & 0 & 0 & 1 \end{pmatrix} \tag{36.5}$$

The generalization for arbitrary $\boldsymbol{v}$ will be given below.

The Lorentz transformations form a group, that is two successive LT result in one LT again. We consider an LT from IS to IS$'$ and an LT from IS$'$ to IS$''$ (in matrix notation)

$$x' = \Lambda x + b, \quad x'' = \Lambda' x' + b' \tag{36.6}$$

From this follows

$$x'' = \Lambda'' x + b'' \quad \text{with} \quad \Lambda'' = \Lambda'\Lambda, \quad b'' = b' + \Lambda' b \tag{36.7}$$

For this to be an LT again, Λ'' must satisfy the condition (34.19):

$$\Lambda''^{\rm T}\,\eta\,\Lambda'' = (\Lambda'\Lambda)^{\rm T}\,\eta\,\Lambda'\Lambda = \Lambda^{\rm T}\big(\Lambda'^{\rm T}\,\eta\,\Lambda'\big)\Lambda = \Lambda^{\rm T}\,\eta\,\Lambda = \eta \tag{36.8}$$

Since this is the case, two successive LTs yield one LT again. We still show the remaining group properties: The associativity $\Lambda(\Lambda'\Lambda'') = (\Lambda\Lambda')\Lambda''$ follows from the associativity of the matrix multiplication. The unit matrix $\Lambda = 1$ is the 1-element. By $\alpha \to \alpha^{\rm T}$ and $\boldsymbol{v} \to -\boldsymbol{v}$ we get the inverse element. The group properties of the shifts are evident. The group of inhomogeneous LT is called the *Poincaré group*, and that of the homogeneous LT is called the *Lorentz group*.

We generalize (36.5) for an arbitrary direction of $\boldsymbol{v}$. Let the origins of the systems IS and IS′ are connected by the vector $\boldsymbol{v}t$; and let the axes be parallel and $b = 0$. Thus, we have the situation as in Figure 34.1, except that now $\boldsymbol{v}$ is not parallel to $\boldsymbol{e}_1 = \boldsymbol{e}'_1$. We can, however, go from the system IS by a suitable $\Lambda(\alpha)$ to a system IS″ with $\boldsymbol{e}''_1 \parallel \boldsymbol{v}$, then with $\Lambda(v)$ from (36.5) to an IS‴, and from IS‴ with $\Lambda(\alpha^{\mathrm{T}})$ to IS′. The searched transformation matrix

$$\Lambda(\boldsymbol{v}) = \Lambda(\alpha^{\mathrm{T}})\,\Lambda(v)\,\Lambda(\alpha) \tag{36.9}$$

is composed of already-known cases. For the multiplication of $\Lambda(v)$ with $\Lambda(\alpha)$ and $\Lambda(\alpha^{\mathrm{T}})$ we note: The element $\Lambda^0_0 = \gamma$ remains unaffected. The Λ^0_i and Λ^i_0 change like a 3-vector, and the Λ^i_j like a 3-tensor of the second rank. From these observations and the limiting case (36.5), one can determine the transformations matrix for $\boldsymbol{v} = \sum v_i\,\boldsymbol{e}_i$ without explicit reference to α. Since (Λ^0_i) behaves like a vector and equals $(-\gamma\, v/c,\ 0,\ 0)$ for $(v_i) = (v, 0, 0)$, it must be $(\Lambda^0_i) = (-\gamma\, v_i/c)$. The quantity

$$\gamma = \frac{1}{\sqrt{1 - \boldsymbol{v}^2/c^2}} \tag{36.10}$$

behaves like a 3-scalar. One easily verifies that the (Λ^i_j) in

$$\Lambda(\boldsymbol{v}) = \begin{pmatrix} \gamma & -\gamma\, v_1/c & -\gamma\, v_2/c & -\gamma\, v_3/c \\ -\gamma\, v_1/c & & & \\ -\gamma\, v_2/c & & \delta_{ij} + \dfrac{v_i\, v_j\,(\gamma - 1)}{v^2} & \\ -\gamma\, v_3/c & & & \end{pmatrix} \tag{36.11}$$

behave like a second rank tensor for rotations (i.e. a 3-tensor), and for $(v_i) = (v, 0, 0)$ it agrees with (36.5). This matrix is symmetric, $\Lambda(\boldsymbol{v}) = \Lambda(\boldsymbol{v})^{\mathrm{T}}$. However, this symmetry does not apply for $\Lambda(\alpha)$ from (36.4), and thus it does not apply for a general Λ either.

The general homogeneous LT results for an arbitrary relative velocity $\boldsymbol{v}$ and for arbitrarily rotated axes; it can be described by

$\Lambda = \Lambda(\alpha)\,\Lambda(\boldsymbol{v})$. For the general LT, we add a constant space–time shift by b:

$$x' = \Lambda(\alpha)\,\Lambda(\boldsymbol{v})\,x + b \qquad \begin{array}{c}\text{(general Lorentz}\\ \text{transformation)}\end{array} \tag{36.12}$$

The rotation (α) depends on three angles, the relative velocity on the three values v_i. Thus the elements of the Lorentz group (homogeneous LT) depend on 6 parameters. With four additional parameters b^β we get the Poincaré group.

From $\Lambda^{\mathrm{T}}\eta\,\Lambda = \eta$ follows $(\det\Lambda)^2 = 1$ and $(\Lambda^0_{\ 0})^2 = 1 + \sum_i(\Lambda^i_{\ 0})^2$; this gives $\det\Lambda = \pm 1$, and furthermore $\Lambda^0_{\ 0} \geq 1$ or $\Lambda^0_{\ 0} \leq -1$. By

$$\det\Lambda = 1, \quad \Lambda^0_{\ 0} \geq 1 \tag{36.13}$$

we exclude a change in time direction (such as $t' = -t$) and spatial reflections (such as $x' = -x$). The so-restricted transformations are called *proper* LT. They are distinguished by the fact that a continuous transition (velocity, angle of rotation, and displacement towards zero) leads to the identity transformation.

Successive Lorentz Transformation

We consider Figure 36.1 with the reference frames IS and IS′ moving relative to each other with velocity $\boldsymbol{v}_1$. In IS′, a particle moves with velocity $\boldsymbol{v}(t)$. With which velocity $\boldsymbol{V}$ does this particle move in IS?

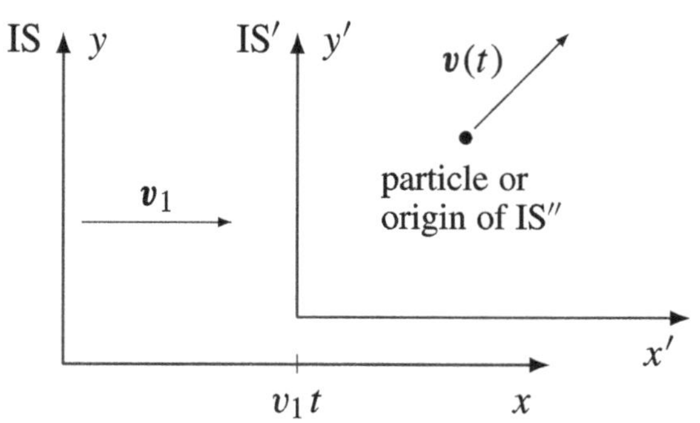

Figure 36.1 In IS′ a particle moves with the velocity $\boldsymbol{v}(t)$. Which velocity has the particle than in IS?

We introduce an IS$''$ moving relative to IS$'$ with constant velocity $\boldsymbol{v}_2 = \boldsymbol{v}(t_0)$; let the coordinate axes of IS, IS$'$ and IS$''$ be parallel. At time $t = t_0$, the particle is at rest in IS$''$. Then its velocity in IS is equal to the velocity $\boldsymbol{V}$ of IS$''$ in IS. According to (36.7), the LT from IS to IS$''$ is given by

$$\Lambda(\boldsymbol{V}) = \Lambda(\boldsymbol{v}_2)\,\Lambda(\boldsymbol{v}_1) \tag{36.14}$$

Since the axes are parallel, $\Lambda(\boldsymbol{V})$, $\Lambda(\boldsymbol{v}_1)$, and $\Lambda(\boldsymbol{v}_2)$ are all of the form (36.11). By calculating the right-hand side of (36.14) and comparison with (36.11) one obtains the sought velocity

$$\boldsymbol{V} = \boldsymbol{V}(\boldsymbol{v}_1, \boldsymbol{v}_2) \tag{36.15}$$

We start with the case of *parallel* velocities, $\boldsymbol{v}_1 \parallel \boldsymbol{v}_2$; the general form of $\boldsymbol{V}(\boldsymbol{v}_1,\, \boldsymbol{v}_2)$ will be given below. We rotate the coordinate axes of IS and IS$'$ so that $\boldsymbol{v}_1$ and $\boldsymbol{v}_2$ are parallel to the x- and x'-axes. Then the two Λ's in (36.14) are of the form (34.22) with (34.26). In the relevant subspace, we write down the matrix multiplication (36.14):

$$\Lambda(\boldsymbol{V}) = \begin{pmatrix} \cosh\psi_2 & -\sinh\psi_2 \\ -\sinh\psi_2 & \cosh\psi_2 \end{pmatrix} \begin{pmatrix} \cosh\psi_1 & -\sinh\psi_1 \\ -\sinh\psi_1 & \cosh\psi_1 \end{pmatrix} \tag{36.16}$$

We perform the matrix multiplication:

$$\begin{aligned} \Lambda(\boldsymbol{V}) &= \begin{pmatrix} \cosh(\psi_1+\psi_2) & -\sinh(\psi_1+\psi_2) \\ -\sinh(\psi_1+\psi_2) & \cosh(\psi_1+\psi_2) \end{pmatrix} \\ &= \begin{pmatrix} \cosh\psi & -\sinh\psi \\ -\sinh\psi & \cosh\psi \end{pmatrix} \end{aligned} \tag{36.17}$$

Thus, the rapidity $\psi(\boldsymbol{V})$ simply results from the addition

$$\psi = \psi_1 + \psi_2 \tag{36.18}$$

One can compare this with the multiplication of two orthogonal matrices describing two rotations around the same axis. The parallelism of the velocity corresponds to the same axis of rotation, and the addition of the rapidities corresponds to the addition of the angles of rotation. For parallel velocities the LT commute with each other,

$$\Lambda(\boldsymbol{v}_1)\,\Lambda(\boldsymbol{v}_2) = \Lambda(\boldsymbol{v}_2)\,\Lambda(\boldsymbol{v}_1) \quad (\boldsymbol{v}_1 \parallel \boldsymbol{v}_2) \tag{36.19}$$

If the velocities are not parallel (or the axes of rotation are not equal), then the transformations must not be interchanged.

Velocity Addition Formula

We evaluate (36.18) explicitly for the velocities. This yields the addition formula for the velocities. Subsequently, this formula is generalized for non-parallel velocities.

For $\tanh(\psi_1 + \psi_2)$ we use the addition formula for the tangent hyperbolicus

$$\tanh \psi = \tanh(\psi_1 + \psi_2) = \frac{\tanh \psi_1 + \tanh \psi_2}{1 + \tanh \psi_1 \tanh \psi_2} \tag{36.20}$$

With $\tanh \psi = V/c$ and $\tanh \psi_i = v_i/c$ this becomes

$$V = \frac{v_1 + v_2}{1 + v_1 v_2/c^2} \quad \text{for} \quad \boldsymbol{v}_1 \parallel \boldsymbol{v}_2 \tag{36.21}$$

This is called the *velocity addition formula* (or addition formula for short). Obviously, the relation (36.18) is simpler than (36.21). In SRT, therefore, the rapidities are a preferred measure for the velocities, because for these some relations (like the addition formula) are simpler. From (36.21) follows specifically

$$V \begin{cases} \approx v_1 + v_2 & (v_1 \ll c \ \text{ and } \ v_2 \ll c) \\ \to c & (v_1 \to c \ \text{ or } \ v_2 \to c) \end{cases} \tag{36.22}$$

The addition of the velocities is illustrated in Figure 36.2.

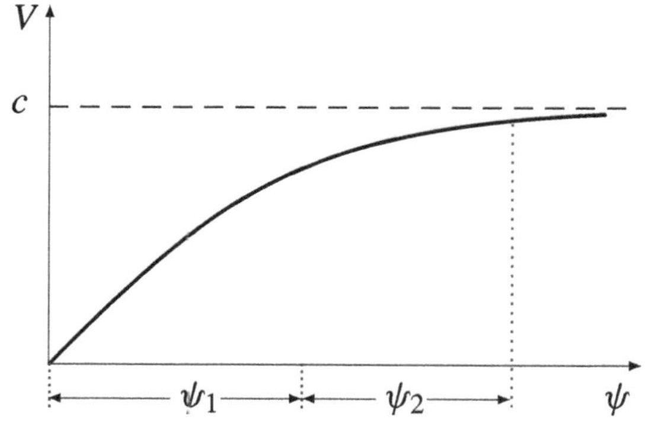

Figure 36.2 Velocity-addition formula: two parallel velocities add up according to $\psi = \psi_1 + \psi_2$. The resulting velocity is then $V = c\tanh(\psi_1 + \psi_2)$. The depicted graph is that of the tangens hyperbolicus.

Since the derivation of (36.21) was done via LT, the velocities are constrained by $v_1 < c$ and $v_2 < c$. If the particle considered in Figure 36.1 is a photon, its velocity in each IS is equal to c (because the velocity of light is the same in all IS). Since the formula (36.21) for $v_2 = c$ also yields $V = c$, we can include this limiting case in the addition formula. Due to the symmetry with respect to v_1 and v_2, we can also allow for $v_1 = c$.

General case

We now determine the addition formula $\boldsymbol{V} = \boldsymbol{V}(\boldsymbol{v}_1, \boldsymbol{v}_2)$ for non-parallel velocities $\boldsymbol{v}_1$ and $\boldsymbol{v}_2$. For this purpose, we compute the $0j$-element of $\Lambda(\boldsymbol{V})$ in (36.14):

$$\begin{aligned} -\gamma\,\frac{V_j}{c} &= \Lambda^0_j(\boldsymbol{V}) = \Lambda^0_\beta(\boldsymbol{v}_2)\,\Lambda^\beta_j(\boldsymbol{v}_1) = \Lambda^0_0(\boldsymbol{v}_2)\,\Lambda^0_j(\boldsymbol{v}_1) \\ &\quad + \sum_{i=1}^{3} \Lambda^0_i(\boldsymbol{v}_2)\,\Lambda^i_j(\boldsymbol{v}_1) \\ &= -\gamma_1\gamma_2\,\frac{v_{1,j}}{c} - \sum_{i=1}^{3} \gamma_2\,\frac{v_{2,i}}{c}\left(\delta_{ij} + \frac{v_{1,i}\,v_{1,j}}{v_1^2}\,(\gamma_1 - 1)\right) \end{aligned} \tag{36.23}$$

where $\gamma = \gamma(V)$, $\gamma_1 = \gamma(v_1)$, and $\gamma_2 = \gamma(v_2)$. We multiply (36.23) by $-c/\gamma$ and use the vector notation:

$$\boldsymbol{V} = \frac{\gamma_1\gamma_2}{\gamma}\left[\boldsymbol{v}_1 + \frac{\boldsymbol{v}_2}{\gamma_1} - \boldsymbol{v}_1\,\frac{\boldsymbol{v}_1\cdot\boldsymbol{v}_2}{v_1^2}\left(\frac{1}{\gamma_1} - 1\right)\right] \tag{36.24}$$

We determine the 00 element:

$$\begin{aligned} \gamma &= \Lambda^0_0(\boldsymbol{V}) = \Lambda^0_0(\boldsymbol{v}_2)\,\Lambda^0_0(\boldsymbol{v}_1) + \sum_{i=1}^{3} \Lambda^0_i(\boldsymbol{v}_2)\,\Lambda^i_0(\boldsymbol{v}_1) \\ &= \gamma_1\gamma_2 + \sum_{i=1}^{3} \gamma_1\gamma_2\,\frac{v_{1,i}\,v_{2,i}}{c^2} = \gamma_1\gamma_2\left(1 + \frac{\boldsymbol{v}_1\cdot\boldsymbol{v}_2}{c^2}\right) \end{aligned} \tag{36.25}$$

From this, we read off the prefactor $\gamma_1\gamma_2/\gamma = 1/(1 + \boldsymbol{v}_1\cdot\boldsymbol{v}_2/c^2)$ in (36.24). We split the velocity $\boldsymbol{v}_2 = \boldsymbol{v}_{2\parallel} + \boldsymbol{v}_{2\perp}$ into the parts parallel and perpendicular to $\boldsymbol{v}_1$. Thus, (36.24) becomes

$$\boldsymbol{V} = \frac{\boldsymbol{v}_1 + \boldsymbol{v}_{2\parallel} + \boldsymbol{v}_{2\perp}\sqrt{1 - \boldsymbol{v}_1^2/c^2}}{1 + \boldsymbol{v}_1 \cdot \boldsymbol{v}_2/c^2} \tag{36.26}$$

The result is symmetric in $\boldsymbol{v}_1$ and $\boldsymbol{v}_2$ only for $\boldsymbol{v}_1 \parallel \boldsymbol{v}_2$. In general, the LT is (like rotations) non-commutative

$$\Lambda(\boldsymbol{v}_1)\,\Lambda(\boldsymbol{v}_2) \neq \Lambda(\boldsymbol{v}_2)\,\Lambda(\boldsymbol{v}_1) \quad (\text{for } \boldsymbol{v}_1 \times \boldsymbol{v}_2 \neq 0) \tag{36.27}$$

Chapter 37

Lorentz Tensor

Just like the 3-tensors are defined by their behavior under orthogonal transformations (Chapter 21), the Lorentz tensors or 4-tensors are defined by their behavior under Lorentz transformations. The calculation rules for Lorentz tensors are established. These rather formal developments are not presupposed in the following; however, the first two pages of this chapter should be read in any case.

We consider the four-dimensional space, which is spanned by Cartesian coordinate axes for the quantities

$$\left(x^{\alpha}\right) = \left(x^0, x^1, x^2, x^3\right) = (ct,\, x,\, y,\, z) \tag{37.1}$$

The squared line element

$$ds^2 = \eta_{\alpha\beta}\, dx^{\alpha}\, dx^{\beta} \tag{37.2}$$

is invariant under Lorentz transformations (LT). We refer to the 4-dimensional space with this line element as the *Minkowski space.*

This can be compared to the three-dimensional space defined by the coordinate axes for

$$\left(x_i\right) = \left(x_1, x_2, x_3\right) = (x,\, y,\, z) \tag{37.3}$$

The squared line element

$$dl^2 = \sum_{i=1}^{3} \left(dx_i\right)^2 \tag{37.4}$$

is invariant under orthogonal transformations.

In Chapter 21, we defined the tensors of the three-dimensional space as indexed quantities, which transform under orthogonal transformations componentwise like the coordinates. Here, we analogously define tensors of Minkowski space as indexed quantities that transform componentwise under LT like the coordinates. For distinction, we use the notations 3-tensor and 4-tensor.

The significance of these formal definitions is the following: under the assumption of the isotropy of the three-dimensional space, physical laws must not depend on the orientation of the Cartesian coordinate system. Therefore, the laws must not change their form under orthogonal transformations. For this purpose the laws are defined as 3-tensor equations, for example, $L_i = \sum \Theta_{ij}\, \omega_j$. According to the principle of relativity, all IS are equivalent. Therefore, the form of physical laws must not depend on the IS. According to Einstein's principle of relativity, this means that the laws must keep their form under LT. For this purpose, the laws are formulated as 4-tensor equations. This formal condition is often sufficient to find the relativistic generalization of a known non-relativistic law.

We consider a quantity V^α with one index. This is called a *Lorentz vector* if it transforms like the coordinates x^α,

$$V'^\beta = \Lambda^\beta_\alpha\, V^\alpha \tag{37.5}$$

Synonymously this is designated as a 4-vector or a *Lorentz tensor of first rank.* We recall that we sum over two equal indices (one at the top, one at the bottom), and that Greek indices can take the values 0, 1, 2, 3. Under *tensor*, we always understand the totality of the indexed quantities V^α. For a specific α, however, V^α is just a number (maybe multiplied by a physical unit).

A *Lorentz tensor* or *4-tensor* of Nth rank is an N-fold indexed quantity that transforms like

$$T'^{\alpha_1 \ldots \alpha_N} = \Lambda^{\alpha_1}_{\beta_1} \ldots \Lambda^{\alpha_N}_{\beta_N}\, T^{\beta_1 \ldots \beta_N} \tag{37.6}$$

A *Lorentz scalar* or tensor of 0th rank is then an LT-invariant quantity. Examples of Lorentz scalars are the proper length l_0 of a rod, the line element ds^2, and the proper time $d\tau$ of a clock. Examples of Lorentz vectors are x^α and dx^α. Examples of a second-rank Lorentz tensor are the product $x^\alpha dx^\beta$ or $\eta_{\alpha\beta}$.

Calculation Rules

To simplify the notation, we define for each 4-vector V^α by

$$V_\beta = \eta_{\beta\alpha}\, V^\alpha \tag{37.7}$$

an associated indexed quantity. In particular,

$$(x_\alpha) = (x_0, x_1, x_2, x_3) = (ct, -x, -y, -z) \tag{37.8}$$

This assignment is done accordingly for all tensors, for example,

$$T_{\alpha\beta} = \eta_{\alpha\alpha'}\,\eta_{\beta\beta'}\,T^{\alpha'\beta'} \qquad T^\alpha{}_\beta = \eta_{\beta\beta'}\,T^{\alpha\beta'}, \qquad T_\alpha{}^\beta = \eta_{\alpha\alpha'}\,T^{\alpha'\beta} \tag{37.9}$$

The $T^{\alpha\beta\ldots}$ are called *contravariant* components of the tensor, the $T_{\alpha\beta\ldots}$ *covariant* components.[1] We may also have mixed components like in $T^\alpha{}_\beta$. As a simplification, we call the indexed quantity itself a tensor, or a co- or contravariant tensor. For the mixed components, one has to pay attention to the sequence of the indices, because in general $T^\alpha{}_\beta \neq T_\beta{}^\alpha$. For a symmetric tensor, the indices might be written on top of each other, $T^\alpha{}_\beta = T_\beta{}^\alpha = T^\beta_\alpha$.

The indices of the transformation matrix Λ^α_β can be written on top of each other because this quantity is not a tensor. To satisfy the tensor definition, Λ would have to be defined first in IS and IS$'$. But Λ is a quantity that is assigned to the transition between IS and IS$'$. Notwithstanding this restriction, Λ^α_β behaves similarly to a tensor; for this, see the discussion of the 3-tensor properties of (36.11).

The definition (37.6) leads immediately to some rules for the construction of tensors. If S and T are tensors, then the following holds:

1. Addition: $a\,S^{\alpha_1\ldots\alpha_N} + bT^{\alpha_1\ldots\alpha_N}$ is a tensor of rank N; where a and b are numbers.
2. Multiplication: $S^{\alpha_1\ldots\alpha_N}\,T^{\beta_1\ldots\beta_M}$ is a tensor of rank $N+M$.
3. Contraction: $\eta_{\beta\gamma}\,S^{\alpha_1\ldots\beta\ldots\gamma\ldots\alpha_N} = S^{\alpha_1\ldots\beta\ldots}{}_\beta{}^{\ldots\alpha_N}$ is a tensor of rank $(N-2)$. In particular, $ds^2 = dx^\alpha\,dx_\alpha$ and $S^\alpha T_\alpha$ are Lorentz scalars.
4. Tensor equations: If $S^\alpha = U^{\alpha\beta}T_\beta$ holds in all IS, then $U^{\alpha\beta}$ is a second rank tensor.

[1] The terminus *covariant* is used in two ways: it is used as a synonym for *form invariant*, or it is used for the lower indices of a tensor.

Points 1 to 3 follow directly from the tensor definition, and the last statement can be proven as in (21.19)–(21.21).

The matrix $(\eta_{\alpha\beta})$ was defined by the number assignment in (34.11). By the same assignment, we define a matrix $(\eta^{\alpha\beta})$:

$$\left(\eta^{\alpha\beta}\right) = \left(\eta_{\alpha\beta}\right) = \begin{pmatrix} 1 & 0 & 0 & 0 \\ 0 & -1 & 0 & 0 \\ 0 & 0 & -1 & 0 \\ 0 & 0 & 0 & -1 \end{pmatrix} \tag{37.10}$$

The matrices $\left(\eta^{\alpha\beta}\right)$ and $\left(\eta_{\alpha\beta}\right)$ are inverses of each other:

$$\eta^{\alpha\beta}\,\eta_{\beta\gamma} = \delta^{\alpha}_{\gamma} = \begin{cases} 1 & (\alpha = \gamma) \\ 0 & (\alpha \neq \gamma) \end{cases} \tag{37.11}$$

The four-dimensional line element can be written in the following forms:

$$ds^2 = \eta_{\alpha\beta}\, dx^\alpha dx^\beta = dx^\alpha dx_\alpha = \eta^{\alpha\beta}\, dx_\alpha\, dx_\beta \tag{37.12}$$

According to the listed calculation rules, $dx^\alpha dx^\beta$ is a second rank tensor, which becomes a scalar ds^2 by contraction. The form $V^\alpha W_\alpha$ is the scalar product of two Lorentz vectors. In the next section, it is formally shown that $\eta_{\alpha\beta}$ can be regarded as a Lorentz tensor, too. Therefore, ds^2 can also be considered as the scalar obtained by twofold contraction of the tensor $\eta^{\alpha\beta}\, dx^\gamma\, dx^\delta$.

We calculate the transformation properties of the covariant tensors. We multiply (37.7) with $\eta^{\alpha\beta}$ and use (37.11):

$$V^\alpha = \eta^{\alpha\beta}\, V_\beta \tag{37.13}$$

The quantities $\eta_{\alpha\beta}$ are defined by the number assignment (37.10), and do not depend on the IS. Therefore, the covariant vector V^α transforms according to

$$V'_\alpha = \eta_{\alpha\beta}\, V'^{\beta} = \eta_{\alpha\beta}\, \Lambda^{\beta}_{\gamma}\, V^\gamma = \eta_{\alpha\beta}\, \Lambda^{\beta}_{\gamma}\, \eta^{\gamma\delta}\, V_\delta = \overline{\Lambda}{}^{\delta}_{\alpha}\, V_\delta \tag{37.14}$$

In the last step, we introduced $\overline{\Lambda}{}^{\delta}_{\alpha}$ by

$$\overline{\Lambda}{}^{\delta}_{\alpha} = \eta_{\alpha\beta}\, \Lambda^{\beta}_{\gamma}\, \eta^{\gamma\delta} \tag{37.15}$$

We multiply this by $\Lambda^{\alpha}_{\epsilon}$:

$$\overline{\Lambda}^{\delta}_{\alpha}\,\Lambda^{\alpha}_{\epsilon} = \eta_{\alpha\beta}\,\Lambda^{\beta}_{\gamma}\,\eta^{\gamma\delta}\,\Lambda^{\alpha}_{\epsilon} = \eta^{\gamma\delta}\,\eta_{\gamma\epsilon} = \delta^{\delta}_{\epsilon} \tag{37.16}$$

In the penultimate step, (34.19) was used. Analogously

$$\Lambda^{\epsilon}_{\delta}\overline{\Lambda}^{\delta}_{\alpha} = \Lambda^{\epsilon}_{\delta}\,\eta_{\alpha\beta}\,\Lambda^{\beta}_{\gamma}\,\eta^{\gamma\delta} = \eta^{\epsilon\beta}\,\eta_{\alpha\beta} = \delta^{\epsilon}_{\alpha} \tag{37.17}$$

In the penultimate step, $\Lambda^{\epsilon}_{\delta}\,\Lambda^{\beta}_{\gamma}\,\eta^{\gamma\delta} = \eta^{\epsilon\beta}$ was used. This follows from (34.19), $\Lambda^{\mathrm{T}}\eta\,\Lambda = \eta$, if one uses in this the contravariant components of $\eta = (\eta^{\alpha\beta})$.

From (37.16) follow the reverse transformations

$$V^{\gamma} = \delta^{\gamma}_{\beta}\,V^{\beta} = \overline{\Lambda}^{\gamma}_{\alpha}\,\Lambda^{\alpha}_{\beta}\,V^{\beta} = \overline{\Lambda}^{\gamma}_{\alpha}\,V'^{\alpha} \tag{37.18}$$

$$V_{\gamma} = \delta^{\beta}_{\gamma}\,V_{\beta} = \overline{\Lambda}^{\beta}_{\alpha}\,\Lambda^{\alpha}_{\gamma}\,V_{\beta} = \Lambda^{\alpha}_{\gamma}\,V'_{\alpha} \tag{37.19}$$

We summarize: Contravariant vectors are transformed with Λ^{α}_{β}, and covariant ones with $\overline{\Lambda}^{\alpha}_{\beta}$. The respective other Λ mediates the back transformation.

We will briefly discuss the matrix notation. For the matrix $\Lambda = (\Lambda^{\beta}_{\gamma})$, we have agreed with (34.16) that the upper index indicates the row, and the lower one the column. For (37.15) we use the matrix notation

$$\overline{\Lambda} = \left(\eta\Lambda\eta\right)^{\mathrm{T}} \tag{37.20}$$

Now, (37.16) and (37.17) correspond to the matrix equations $\overline{\Lambda}\Lambda = 1$ and $\Lambda\overline{\Lambda} = 1$. From this follows

$$\Lambda^{-1} = \overline{\Lambda} = \left(\eta\Lambda\eta\right)^{\mathrm{T}} \tag{37.21}$$

We discuss this result further using the special cases of a mere rotation (36.4) and a mere velocity transformation (36.11). The respective inverse transformations

$$\Lambda^{-1}(\alpha) = \Lambda(\alpha^{\mathrm{T}}) \tag{37.22}$$

$$\Lambda^{-1}(\boldsymbol{v}) = \Lambda(-\boldsymbol{v}) \tag{37.23}$$

can be read directly from (36.4) and (36.11) (α denotes here the orthogonal 3×3 matrix of rotation). On the right-hand side of (37.20) the multiplication by the η's means that the 00 and ij components remain unchanged, and that the $0i$- components change their sign (i.e. $\boldsymbol{v} \to -\boldsymbol{v}$). The transposition exchanges the ij components; because of $\alpha^{-1} = \alpha^{\mathrm{T}}$, this reverses a rotation.

Minkowski and Levi-Civita tensor

By (37.10) $(\eta_{\alpha\beta})$ and $(\eta^{\alpha\beta})$ are defined as constant matrices. However, we can consider these indexed quantities as tensors, too, and transform them accordingly. This follows from

$$\eta'_{\alpha\beta} \overset{(37.6)}{=} \bar{\Lambda}_\alpha^\gamma\, \bar{\Lambda}_\beta^\delta\, \eta_{\gamma\delta} \overset{(34.19)}{=} \bar{\Lambda}_\alpha^\gamma\, \bar{\Lambda}_\beta^\delta\, \Lambda_\gamma^\mu\, \Lambda_\delta^\nu\, \eta_{\mu\nu} = \eta_{\alpha\beta} \tag{37.24}$$

In the last step, (37.16) and (37.17) were used. The tensor η is called the *Minkowski tensor*. Since

$$\eta^\alpha{}_\beta \overset{(37.9)}{=} \eta^{\alpha\gamma}\, \eta_{\gamma\beta} = \delta_\beta^\alpha \tag{37.25}$$

also, the Kronecker symbol δ_β^α is a 4-tensor. Since η is symmetric, the indices can be written on top of each other, $\eta^\alpha{}_\beta = \eta_\beta{}^\alpha = \eta_\beta^\alpha$.

Also, the totally antisymmetric tensor

$$\epsilon^{\alpha\beta\gamma\delta} = \begin{cases} +1 & \text{if } (\alpha,\beta,\gamma,\delta) \text{ is an even permutation of } (0,1,2,3) \\ -1 & \text{if } (\alpha,\beta,\gamma,\delta) \text{ is an odd permutation of } (0,1,2,3) \\ 0 & \text{otherwise} \end{cases} \tag{37.26}$$

can be considered as a tensor in Minkowski space. For $\det \Lambda = 1$ one easily shows (Exercise 37.1) that this *Levi-Civita symbol* is a tensor, too, i.e.

$$\epsilon'^{\alpha\beta\gamma\delta} = \Lambda_{\alpha'}^\alpha\, \Lambda_{\beta'}^\beta\, \Lambda_{\gamma'}^\gamma\, \Lambda_{\delta'}^\delta\, \epsilon^{\alpha'\beta'\gamma'\delta'} = \epsilon^{\alpha\beta\gamma\delta} \tag{37.27}$$

The covariant components are defined by

$$\epsilon_{\alpha\beta\gamma\delta} = \eta_{\alpha\alpha'}\, \eta_{\beta\beta'}\, \eta_{\gamma\gamma'}\, \eta_{\delta\delta'}\, \epsilon^{\alpha'\beta'\gamma'\delta'} = -\,\epsilon^{\alpha\beta\gamma\delta} \tag{37.28}$$

Tensor fields

For the sake of completeness, we also give the definition of Lorentz tensor fields; we do not need them in mechanics. The functions $S(x)$, $V^\alpha(x)$ and $T^{\alpha\beta}(x)$ are a scalar, vector, or tensor field, respectively, if

$$S'(x') = S(x) \tag{37.29}$$

or

$$V'^{\alpha}(x') = \Lambda^{\alpha}_{\beta}\, V^{\beta}(x) \tag{37.30}$$

or

$$T'^{\alpha\beta}(x') = \Lambda^{\alpha}_{\gamma}\, \Lambda^{\beta}_{\delta}\, T^{\gamma\delta}(x) \tag{37.31}$$

Here, the arguments have to be co-transformed, so $x' = (x'^{\alpha}) = (\Lambda^{\alpha}_{\beta}\, x^{\beta})$.

Tensor fields can be differentiated with respect to their arguments. We show that the partial derivative $\partial/\partial x^{\alpha}$ transforms like a covariant vector. We start with

$$\frac{\partial}{\partial x'^{\alpha}} = \frac{\partial x^{\beta}}{\partial x'^{\alpha}} \frac{\partial}{\partial x^{\beta}} \tag{37.32}$$

Since dx^{β} is a Lorentz vector, (37.18) holds, so $dx^{\beta} = \overline{\Lambda}^{\beta}_{\alpha}\, dx'^{\alpha}$. From this follows

$$\frac{\partial x^{\beta}}{\partial x'^{\alpha}} = \overline{\Lambda}^{\beta}_{\alpha} \tag{37.33}$$

This gives

$$\frac{\partial}{\partial x'^{\alpha}} = \overline{\Lambda}^{\beta}_{\alpha}\, \frac{\partial}{\partial x^{\beta}} \tag{37.34}$$

Therefore,

$$\partial_{\alpha} \equiv \frac{\partial}{\partial x^{\alpha}} \tag{37.35}$$

transforms according to (37.14) and thus is a covariant vector. Accordingly

$$\partial^{\alpha} \equiv \frac{\partial}{\partial x_{\alpha}} \tag{37.36}$$

is a contravariant vector. From the vector property of ∂^{α} and ∂_{α}, it follows that the *d'Alembert operator*

$$\Box \equiv \partial^{\alpha}\, \partial_{\alpha} = \eta^{\alpha\beta}\, \partial_{\alpha}\partial_{\beta} = \frac{1}{c^2}\, \frac{\partial^2}{\partial t^2} - \Delta \tag{37.37}$$

is a Lorentz scalar.

Exercises

37.1 *Levi-Civita tensor in Minkowski space*

Show that the Levi-Civita tensor is a pseudo-Lorentz tensor of rank 4, i.e. that

$$\epsilon'^{\alpha\beta\gamma\delta} = (\det\Lambda)\, \Lambda^{\alpha}_{\alpha'}\, \Lambda^{\beta}_{\beta'}\, \Lambda^{\gamma}_{\gamma'}\, \Lambda^{\delta}_{\delta'}\, \epsilon^{\alpha'\beta'\gamma'\delta'} \tag{37.38}$$

is equal to $\epsilon^{\alpha\beta\gamma\delta}$.

Chapter 38

Equation of Motion

We derive the relativistic generalization of Newton's second law and determine the relativistic energy of a free particle. Thereby, we obtain Einstein's famous formula $E = mc^2$, i.e. the equivalence of mass and energy.

Newton's second law reads

$$m\,\frac{d\boldsymbol{v}}{dt} = \boldsymbol{F}_{\mathrm{N}} \quad \text{(non-relativistic)} \tag{38.1}$$

For the force, we now use the symbol $\boldsymbol{F}_{\mathrm{N}}$ instead of the previous one $\boldsymbol{F}$. There are two reasons for this: In Part IX and Appendix A, $\boldsymbol{F}$ shall denote the spatial component of the relativistic force. Furthermore, the Newtonian force $\boldsymbol{F}_{\mathrm{N}}$ will be defined more precisely in the following.

The inertial frame in which the particle has the velocity $\boldsymbol{v}(t)$, is denoted by IS. We now consider the instantaneous rest system IS$'$, which is moving relative to IS with the constant velocity $\boldsymbol{v}(t_0)$. In IS$'$, the particle is momentarily at rest (at time $t = t_0$). An equation of motion like (38.1) refers to a time and its environment. In this infinitesimal interval ($t_0 - dt \le t \le t_0 + dt$), the velocity of the particle in IS$'$ is arbitrarily small. Newton's second law is well confirmed for velocities $v \ll c$. We, therefore, assume that Newton's second law holds in IS$'$ *exactly*:

$$m\,\frac{d\boldsymbol{v}'}{dt'} = \boldsymbol{F}_{\mathrm{N}} \quad \text{(relativistically valid in IS}'\text{)} \tag{38.2}$$

The dashes refer to the IS$'$ coordinates $(x'^{\alpha}) = (ct', x', y', z')$. The quantities m and $\boldsymbol{F}_{\mathrm{N}}$ are specified below. From (38.2) we can derive the relativistic equation of motion in an arbitrary IS.

As explained in Chapter 2, Newton's second law implies the definition of the mass m and the force $\boldsymbol{F}_{\mathrm{N}}$ as measurable quantities. We retain these definitions, but we now relate them to (38.2). This means

$$m = \text{mass in IS}' = \text{rest mass} \tag{38.3}$$

$$\boldsymbol{F}_{\mathrm{N}} = \text{force in IS}' \tag{38.4}$$

So, we continue to define the force and mass in Newtonian terms, but these *definitions refer now to the momentary rest system*. In Newtonian mechanics, these quantities are independent of the IS, thus $m(\text{IS}) = m(\text{IS}')$ and $\boldsymbol{F}_{\mathrm{N}}(\text{IS}) = \boldsymbol{F}_{\mathrm{N}}(\text{IS}')$; in this context, we exclude frictional forces. In the non-relativistic limit, one can neglect the differences between the respective quantities in (38.2) and in (38.1). In the relativistic case, equation (38.1) is, however, invalid, and the correct equation is (38.2) with the quantities defined as in (38.3, 38.4).

Newton's second law refers to a mass point with the invariable property "mass m". Accordingly, we apply the relativistic equation to particles which have a constant rest mass m. This applies in particular to elementary particles, but often also to complex particles, which are idealized as mass points.

The interaction of the particle with its environment generally depends on the position and the time. The Newtonian force is then a force field $\boldsymbol{F}_{\mathrm{N}}(\boldsymbol{r}, t)$.

From a valid equation in IS$'$ we obtain the corresponding equation in IS by a Lorentz transformation; this follows from Einstein's principle of relativity. So an LT applied to (38.2) leads to the relativistic equation of motion in an arbitrary IS. Actually, we use a somewhat different procedure: We set up a 4-vector equation, which in IS$'$ reduces to (38.2). This equation is then valid in IS$'$; an LT into another IS is no longer required because of its 4-vector property (i.e. covariance under LT).

The trajectory of the mass point in IS can be represented in the following forms:

$$x^i = x^i(t) \quad \text{or} \quad x^\alpha = x^\alpha(\tau) \quad \text{(trajectory)} \tag{38.5}$$

Here, τ is the time of a clock comoving with the mass point. According to (35.13), τ can be expressed by

$$d\tau = dt\,\sqrt{1-\frac{\boldsymbol{v}(t)^2}{c^2}} = \frac{dt}{\gamma} \tag{38.6}$$

The orbital curve $x^i(t)$ specifies $v^i = dx^i/dt$ and thus $t = t(\tau)$ and $x^0 = c\,t(\tau)$ and $x^i = x^i(\tau)$. The four functions $x^\alpha(\tau)$ contain the same information as the three functions $x^i(t)$.

The obvious generalization of the velocity $v^i = dx^i/dt$ is the 4-velocity u^α

$$u^\alpha = \frac{dx^\alpha}{d\tau} \quad \text{(4-velocity)} \tag{38.7}$$

According to (34.17) and (35.15), dx^α is a 4-vector and $d\tau$ is a 4-scalar; thus u^α is a 4-vector. The 4-velocity u^α can be expressed by the 3-velocity v^i:

$$\big(u^\alpha\big) = \gamma\left(\frac{dx^0}{dt},\,\frac{dx^1}{dt},\,\frac{dx^2}{dt},\,\frac{dx^3}{dt}\right) = \frac{(c, v^1, v^2, v^3)}{\sqrt{1-v^2/c^2}} = \gamma\,(c,\,\boldsymbol{v}) \tag{38.8}$$

In the last step, we combined the spatial components into a vector.

The obvious relativistic generalization of the l.h.s. of (38.2) is $m\,du^\alpha/d\tau$. The rest mass m specified in (38.3) is a 4-scalar. Therefore, $m\,du^\alpha/d\tau$ is a 4-vector. The associated force must then also be a 4-vector, denoted by F^α. This results in

$$m\,\frac{du^\alpha}{d\tau} = F^\alpha \quad \text{equation of motion} \tag{38.9}$$

Equivalent formulations to this are

$$\frac{dp^\alpha}{d\tau} = F^\alpha \quad \text{or} \quad m\,\frac{d^2x^\alpha}{d\tau^2} = F^\alpha \tag{38.10}$$

Here, we have introduced the momentum

$$\big(p^\alpha\big) = \big(m u^\alpha\big) = \left(\frac{mc}{\sqrt{1-v^2/c^2}}\,,\;\frac{m\,\boldsymbol{v}}{\sqrt{1-v^2/c^2}}\right) \quad \text{(4-momentum)} \tag{38.11}$$

which is a 4-vector, too.

In (38.9), m, u^α and τ are already defined, while F^α is an as yet unspecified 4-vector. We determine F^α from the condition that (38.9) reduces to (38.2) in IS$'$. To do this, we express $du^\alpha/d\tau$ by $\boldsymbol{v}$ and $d\boldsymbol{v}/dt$,

$$\left(\frac{du^\alpha}{d\tau}\right) = \gamma\left(\frac{d(\gamma c)}{dt},\ \frac{d(\gamma \boldsymbol{v})}{dt}\right) = \frac{\gamma^4}{c^2}\,\boldsymbol{v}\cdot\frac{d\boldsymbol{v}}{dt}\,(c,\ \boldsymbol{v}) + \gamma^2\left(0,\ \frac{d\boldsymbol{v}}{dt}\right) \tag{38.12}$$

In the momentary rest system IS$'$, this becomes

$$\left(\frac{du'^\alpha}{d\tau}\right) = \left(0,\ \frac{d\boldsymbol{v}'}{dt'}\right) \tag{38.13}$$

In this derivation, note that $\boldsymbol{v}' = 0$ holds in IS$'$, but not $d\boldsymbol{v}'/dt' = 0$; therefore, in the calculation of $du'^\alpha/d\tau$ one cannot use $(u'^\alpha) = (c, 0)$. With (38.13) we write (38.9) in IS$'$,

$$(F'^\alpha) = m\left(\frac{du'^\alpha}{d\tau}\right) = m\left(0,\ \frac{d\boldsymbol{v}'}{dt'}\right) \overset{(38.2)}{=} (0, \boldsymbol{F}_{\rm N}) \tag{38.14}$$

Thus, F'^α is fixed in IS$'$. The IS in which the particle moves with $\boldsymbol{v}$ is reached from IS$'$ by a Lorentz transformation with $-\boldsymbol{v}$. Therefore,

$$F^\alpha = \Lambda^\alpha_\beta(-\boldsymbol{v})\, F'^\beta, \qquad \begin{pmatrix} F^0 \\ F^1 \\ F^2 \\ F^3 \end{pmatrix} = \Lambda(-\boldsymbol{v}) \begin{pmatrix} 0 \\ F_{\rm N}^1 \\ F_{\rm N}^2 \\ F_{\rm N}^3 \end{pmatrix} \qquad \text{(Minkowski force)} \tag{38.15}$$

Through this, F^α is defined as a 4-vector. The quantity F^α is called the *Minkowski force*. The force $F_{\rm N}^i$ is the Newtonian force in the momentary rest system.

On the way to (38.9), we have introduced quantities with the remark that this is the "obvious relativistic generalization". The validity of (38.9) is, however, independent of this plausibility argument. It follows from:

1. Equation (38.9) is a 4-vector equation.
2. Equation (38.9) is valid in IS$'$.

We have shown that the equation of motion becomes (38.2) in IS$'$, i.e. it is valid in IS$'$. Then the correct relativistic equation in

IS is obtained by a Lorentz transformation. This LT has not to be performed explicitly, because (38.9) does not change its form under LT.

The physical basis of point 1 is Einstein's relativity principle and of point 2 the validity of the second law in IS′. The relativistic equation of motion leads to predictions that deviate significantly from those of (38.1) and can be checked experimentally.

We evaluate the relation between the Newtonian force $\boldsymbol{F}_{\rm N}$ and the Minkowski force F^α. For the special LT with $\boldsymbol{v} = v\,\boldsymbol{e}_1$ we obtain from (38.15)

$$\big(F^\alpha\big) = \big(F^0, F^1, F^2, F^3\big) = \Big(\gamma\,\frac{v}{c}\,F_{\rm N}^1,\; \gamma F_{\rm N}^1,\; F_{\rm N}^2,\; F_{\rm N}^3\Big) \qquad (38.16)$$

We divide the Newtonian force $\boldsymbol{F}_{\rm N} = \boldsymbol{F}_{{\rm N}\|} + \boldsymbol{F}_{{\rm N}\perp}$ into parallel and perpendicular parts with respect to the velocity $\boldsymbol{v}$. From (38.16) we then read off:

$$\big(F^\alpha\big) = \big(F^0,\, \boldsymbol{F}\big) = \left(\gamma\,\frac{v\,F_{{\rm N}\|}}{c},\; \gamma\boldsymbol{F}_{{\rm N}\|} + \boldsymbol{F}_{{\rm N}\perp}\right) \qquad (38.17)$$

This is the relation between the Newtonian force specified in (38.4) and the Minkowski force.

Some physics books present a relation $F^i = \gamma F_{\rm N}^i$ which contradicts (38.16). The question of the "correct" connection cannot be decided within the Newtonian theory, because there the forces $(\gamma F_{\rm N}^1, F_{\rm N}^2, F_{\rm N}^3)$ and $\gamma\,(F_{\rm N}^1, F_{\rm N}^2, F_{\rm N}^3)$ are equivalent. From the (undisputed) form (38.9), however, it follows $m\,(0, dv'^i/dt') = (0, F'^i)$ in the momentary rest system IS′. In a relativistic theory, this F'^i will be called "Newtonian force" [9]. From this then follows (38.17).

As just mentioned, one finds different statements on the relation between Newton's and Minkowski's force in the literature. Appendix A investigates this point in detail and discusses its practical and logical relevance.

As a simple example of the relativistic equation of motion, we consider a spacecraft that is accelerated in such a way that the astronaut (mass m) feels his usual earth weight mg (by which he is pressed against the spacecraft floor). In his rest system IS′, the astronaut experiences the acceleration g, and the Newtonian force is $F_{\rm N} = mg$. We restrict ourselves to a one-dimensional motion in one-direction. We set $d\tau = dt/\gamma$ and $F^1 = \gamma F_{\rm N}$ from (38.16) into (38.9) and obtain:

$$\frac{d}{dt}\,\frac{m\,v(t)}{\sqrt{1-v(t)^2/c^2}} = m\,g \tag{38.18}$$

The mass m can be canceled on both sides. This equation of motion is solved in Chapter 39. For the spaceship itself, one would have to consider time-dependent mass and force because the mass of the spaceship is reduced by the fuel consumption.

Lorentz Force

In order to set up the Minkowski force F^α concretely, one needs a relativistic theory of the interaction under consideration. From (38.17), F^α cannot be determined in general, because the force $\boldsymbol{F}_{\rm N}$ defined in (38.4) is not known, or not known with sufficient accuracy.

There are forces for which a relativistic generalization is not possible. These include in particular the constraint forces, which keep the intervals $|\boldsymbol{r}_{\mu\nu}|$ within a rigid body constant. As we discussed in the last section of Chapter 35 (house stick experiment), the concept of a rigid body is in contradiction to the theory of relativity.

As an example, for a relativistic theory, we refer to electrodynamics. In this case, we can evaluate the relation (38.17) and give the exact relativistic equation. For this, we must presuppose the knowledge of the LT for the electromagnetic fields, i.e. we go beyond the framework of relativistic mechanics.

We consider a particle with the charge q in an electromagnetic field. Let the electric field be $\boldsymbol{E}$ and the magnetic field $\boldsymbol{B}$; we use Gauß's system (of units), where $\boldsymbol{E}$ and $\boldsymbol{B}$ have the same dimension. In the instantaneous rest system IS$'$, the Newtonian force on the particle is

$$\boldsymbol{F}_{\rm N} = q\,\boldsymbol{E}' \tag{38.19}$$

Here, $\boldsymbol{E}'$ is the electric field strength in IS$'$. In electrodynamics (Chapter 22 in [2]), one derives the relation between $\boldsymbol{E}'$ and the fields $\boldsymbol{E}$ and $\boldsymbol{B}$ in IS:

$$\boldsymbol{E}'_{\parallel} = \boldsymbol{E}_{\parallel}, \quad \boldsymbol{E}'_{\perp} = \gamma \left(\boldsymbol{E}_{\perp} + \frac{\boldsymbol{v}}{c} \times \boldsymbol{B} \right) \tag{38.20}$$

The index $\parallel$ denotes the part of the field that is parallel to $\boldsymbol{v}$, and the index $\perp$ the one perpendicular to it. The charge q is a Lorentz scalar; thus it does not change at the transition from IS′ to IS. We set $\boldsymbol{F}_{\mathrm{N}} = q\left(\boldsymbol{E}'_{\parallel} + \boldsymbol{E}'_{\perp}\right)$ on the right-hand side of (38.17). Using (38.20) we then obtain

$$\left(F^{\alpha}\right) = \gamma \left(q\,\frac{\boldsymbol{v}\cdot\boldsymbol{E}}{c},\; q\left(\boldsymbol{E} + \frac{\boldsymbol{v}}{c} \times \boldsymbol{B}\right)\right) \tag{38.21}$$

We substitute this and $d\tau = dt/\gamma$ into (38.9) and get:

$$\frac{d}{dt}\,\frac{m c^2}{\sqrt{1 - v(t)^2/c^2}} = q\,\boldsymbol{v}\cdot\boldsymbol{E} \tag{38.22}$$

$$\frac{d}{dt}\,\frac{m\,\boldsymbol{v}(t)}{\sqrt{1 - v(t)^2/c^2}} = q\left(\boldsymbol{E}(\boldsymbol{r},t) + \frac{\boldsymbol{v}}{c} \times \boldsymbol{B}(\boldsymbol{r},t)\right) \tag{38.23}$$

These are the valid relativistic equations for a charged particle in an electromagnetic field. The fields are to be taken at the position $\boldsymbol{r} = \boldsymbol{r}(t)$ of the particle. The right-hand side of (38.23) is the *Lorentz force* $\boldsymbol{F}_{\mathrm{L}} = q\,(\boldsymbol{E} + \boldsymbol{v} \times \boldsymbol{B}/c)$.

For $v \ll c$ we may neglect terms of order v^2/c^2 and (38.23) becomes

$$m\,\frac{d\boldsymbol{v}(t)}{dt} = q\left(\boldsymbol{E}(\boldsymbol{r},t) + \frac{\boldsymbol{v}}{c} \times \boldsymbol{B}(\boldsymbol{r},t)\right) \quad (v \ll c) \tag{38.24}$$

Thus, the Lorentz force can also be used in Newtonian mechanics. The corresponding non-relativistic Lagrange function was given in (9.47). Instead of the Lorentz force $\boldsymbol{F}_{\mathrm{L}}$, one could also use the Newtonian force $\boldsymbol{F}_{\mathrm{N}} = q\boldsymbol{E}'$; except for terms of the order v^2/c^2 both forces are equal. Usually one will prefer the Lorentz force because it uses the fields in the considered inertial frame (where the particle has the velocity $\boldsymbol{v}$).

Occasionally, the *relativistic mass* is introduced:

$$m_{\mathrm{rel}}(v) = \frac{m}{\sqrt{1 - v^2/c^2}} \tag{38.25}$$

This quantity describes inertia increasing with velocity: for example, if a constant force acts on a particle as in (38.18), then the non-relativistic acceleration $d\boldsymbol{v}/dt$ behaves as if the inertial mass of the particle becomes larger and larger; and $v \to c$ implies $d\boldsymbol{v}/dt \to 0$. In the context of the relativistic treatment, however, note that m is a 4-scalar, while $m_{\rm rel}(v)$ transforms like the 0 component of a 4-vector. The concept of the relativistic mass is discussed in more detail in appendix A.

Occasionally it is claimed that one obtains the relativistic equation of motion by substituting $m \to m_{\rm rel}(v)$ into Newton's second law $d(m\boldsymbol{v})/dt = \boldsymbol{F}_{\rm N}$ (where also the mass is subjected to the time derivative). This is only conditionally correct: By this substitution, one gets $d(m\gamma\boldsymbol{v})/dt = \boldsymbol{F}_{\rm N}$. This is incorrect of order v^2/c^2, because from (38.9) with (38.17) follows $d(m\gamma\boldsymbol{v})/dt = \boldsymbol{F}_{\rm N\parallel} + \boldsymbol{F}_{\rm N\perp}/\gamma \neq \boldsymbol{F}_{\rm N}$. Provided one defines $\boldsymbol{F}_{\rm N}$ precisely (as in (38.4)), the substitution $m \to m_{\rm rel}$ is thus not valid. Appendix A goes into more detail on this point.

Relativistic Energy

For evaluating (38.22), we assume that the particle is in an electrostatic field $\boldsymbol{E} = -\operatorname{grad}\Phi(\boldsymbol{r})$. The particle then has the potential energy $q\Phi(\boldsymbol{r})$ and the right-hand side of (38.22) can be written as follows:

$$q\,\boldsymbol{v}\cdot\boldsymbol{E} = -q\,\frac{d\boldsymbol{r}}{dt}\cdot\operatorname{grad}\Phi = -q\,\frac{d\Phi(\boldsymbol{r}(t))}{dt} \tag{38.26}$$

Hence, (38.22) becomes

$$\frac{d}{dt}\left(\frac{mc^2}{\sqrt{1-v(t)^2/c^2}} + q\,\Phi(\boldsymbol{r}(t))\right) = 0 \tag{38.27}$$

From this follows $\gamma mc^2 + q\,\Phi = \text{const.}$, or more detailed

$$\underbrace{mc^2}_{\text{rest energy}} + \underbrace{mc^2\,(\gamma-1)}_{\text{kinetic energy}} + \underbrace{q\,\Phi(\boldsymbol{r})}_{\text{potential energy}} = \text{const.} \tag{38.28}$$

We have obtained a conserved quantity from the equations of motion. From the known meaning of $q\Phi$ (potential energy), it follows that this is the *energy* of the particle in an electrostatic potential.

The first two terms in (38.28) together are called *relativistic energy* or *energy* of the free particle:

$$E = \frac{mc^2}{\sqrt{1 - v^2/c^2}} = \text{relativistic energy} \qquad (38.29)$$

In the non-relativistic limit, we obtain from this

$$E = mc^2 + \frac{mv^2}{2} + \mathcal{O}\left(\frac{mv^4}{c^2}\right) \quad (v \ll c) \qquad (38.30)$$

The *kinetic* energy is the energy that is required to bring the particle from rest to the velocity v,

$$E_{\text{kin}} = E(v) - E(0) = \frac{mc^2}{\sqrt{1 - v^2/c^2}} - mc^2 \qquad (38.31)$$

In the non-relativistic limit, this becomes $E_{\text{kin}} \approx mv^2/2$.

Equivalence of mass and energy

The equations of motion (38.1), (38.2), and (38.9) refer to a mass point with $m = \text{const}$. For these equations, the *rest mass energy*

$$E_0 = mc^2 \quad \text{(rest mass energy)} \qquad (38.32)$$

constitute an insignificant shift of the energy zero point. This statement remains correct as long as we restrict ourselves to processes in which the mass points are preserved. However, the relation (38.32) has a much deeper meaning, which we explain in the following.

It is the energy E *including the rest mass*, which is conserved during processes in closed systems; this follows in a more general context from the homogeneity of time. For a system of N independent particles, this means that

$$E = \sum_{\nu=1}^{N} E_\nu = \sum_{\nu=1}^{N} E_{\text{kin},\nu} + \sum_{\nu=1}^{N} m_\nu c^2 = \text{const.} \qquad (38.33)$$

The kinetic energy $\sum_\nu E_{\text{kin},\nu}$ and the rest of energy $\sum_\nu m_\nu c^2$ each by itself are in general not conserved.

As an example, consider N neutrons (mass m_{n}) and Z protons (mass m_{p}), which initially have no interaction with each other.

We assume now that these nucleons come together to form an atomic nucleus (mass M_{K}). The conservation of energy is valid for this process. Using (38.33), we equate the (conserved) energy of the initial state and of the final state:

$$\left(N m_{\mathrm{n}} + Z m_{\mathrm{p}}\right) c^2 = M_{\mathrm{K}} c^2 + \Delta E' \tag{38.34}$$

According to (38.33), the contribution $\Delta E'$ accounts for the kinetic energies of the particles, in general, however, for all other forms of energy (except the rest mass energies). This could be, for example, the energy of the radiation emitted during the process. Shortened (but at the same time more general) we formulate the result as

$$\Delta E' = \Delta m\, c^2 \tag{38.35}$$

This statement is called the *equivalence of mass and energy*. Any change in mass implies an associated change of all other energy contributions.

If we combine $N + Z$ nucleons to an atomic nucleus, then an energy of about 8 MeV per nucleon is released (on average). The total mass of the nucleus is therefore not $Nm_{\mathrm{n}} + Zm_{\mathrm{p}}$, but it is smaller by the *mass defect* $\Delta m \approx 8(N+Z)\,\mathrm{MeV}/c^2$. The corresponding energy $\Delta E'$ is set free during the process, for example, as kinetic energy or as radiation. The mass defect per nucleon is largest for medium-heavy nuclei (around iron); therefore, in the nuclear fission of heavy nuclei, as well as in the nuclear fusion of light nuclei, energy is released.

The statement (38.35) is often also given in the form $E = mc^2$. The mass–energy equivalence is of particular physical relevance for transformation processes; for this the notation $\Delta E' = \Delta m c^2$ in (38.35) is adequate.

We consider an example for $E = mc^2$: Assume we have an electron and a positron, both with the mass m_{e}. They can annihilate each other by forming two (or more photons). Then $E = mc^2$ means that the mass $m = 2m_{\mathrm{e}}$ is transformed into photons of the energy E. In (38.35), the energy of the photons are denoted by $\Delta E'$, and the mass change by $\Delta m = 2m_{\mathrm{e}}$. In a complete description, we must account also for the kinetic energies of the electron and the positron. For sufficiently high kinetic energies also other particles might be created (see the following chapter, the section about the production of a Z^0-particle).

Maximum particle velocity

We consider the relativistic energy (38.29). From this expression, we see that

$$E = \frac{mc^2}{\sqrt{1 - v^2/c^2}} \stackrel{v \to c}{\longrightarrow} \infty \tag{38.36}$$

This means that the velocities of particles with mass $m \neq 0$ are restricted by

$$v < c \quad \text{(particles with } m \neq 0\text{).} \tag{38.37}$$

This also implies the impossibility of a concrete IS (e.g. a rocket laboratory) with $v \geq c$. Such IS had already been excluded in (34.29).

For a particle with finite energy E it follows from (38.29) $m \to 0$ for $v \to c$. In fact, there are massless particles (photons, neutrinos[1]) that move with the speed of light. From the basic assumptions of the SRT follows, that these particles have the velocity of light in every IS

$$v = c \quad \text{(particle with } m = 0\text{)} \tag{38.38}$$

There are no particles that move with a superluminal velocity, i.e. faster than light.

Relativistic Momentum

The equation of motion (38.23) can be expressed in the form

$$\frac{d\boldsymbol{p}}{dt} = q\left(\boldsymbol{E} + \frac{\boldsymbol{v}}{c} \times \boldsymbol{B}\right) \tag{38.39}$$

---

[1]Neutrinos were long assumed to be massless. Observed neutrino oscillations indicate that neutrinos do have some small mass (dependent of the neutrino type). Experimentally, one can establish upper limits for these masses (about eV/c^2 for the electron–neutrino). — All experiments are consistent with a zero photon mass. Experimentally, however, one can only establish (very small) upper limits for the photon mass (about $10^{-18}\,\mathrm{eV}/c^2$).

where we use the (relativistic) *momentum*

$$\boldsymbol{p} = \frac{m\boldsymbol{v}}{\sqrt{1-v^2/c^2}} \tag{38.40}$$

In the non-relativistic limit case, $\boldsymbol{p}$ becomes the Newtonian momentum $\boldsymbol{p}_{\mathrm{N}} = m\boldsymbol{v}$,

$$\boldsymbol{p} = m\boldsymbol{v} + \mathcal{O}\left(\frac{v^2}{c^2}\right) \approx \boldsymbol{p}_{\mathrm{N}} \quad (v \ll c) \tag{38.41}$$

This justifies the name *momentum* for $\boldsymbol{p}$ in (38.40). The deeper reason for this designation is (similar to the energy), that in processes in a closed system the total momentum $\sum \boldsymbol{p}_\nu$ is conserved. Formally this conservation law follows from the homogeneity of space.

The components of the 4-momentum (38.11) consist of the (relativistic) energy E and the (relativistic) momentum $\boldsymbol{p}$,

$$(p^\alpha) = \left(\frac{mc}{\sqrt{1-v^2/c^2}}, \frac{m\boldsymbol{v}}{\sqrt{1-v^2/c^2}}\right) = \left(\frac{E}{c}, \boldsymbol{p}\right) \tag{38.42}$$

Energy–Momentum Relation

From (38.42) follows

$$E^2 = m^2c^4 + c^2\boldsymbol{p}^2 \tag{38.43}$$

This relation can also be derived from the line element ds^2 and the proper time (33.15):

$$ds^2 = c^2 d\tau^2 = \eta_{\alpha\beta}\, dx^\alpha dx^\beta \quad \longrightarrow \quad c^2 = \eta_{\alpha\beta}\, u^\alpha u^\beta \tag{38.44}$$

From this follows

$$\eta_{\alpha\beta}\, p^\alpha p^\beta = m^2c^2 \tag{38.45}$$

Since $(p^\alpha) = (E/c, \boldsymbol{p})$ this is equivalent to (38.43).

We consider the energy–momentum relation in the non-relativistic and in the relativistic case:

$$E = \sqrt{m^2c^4 + c^2\boldsymbol{p}^2} \approx \begin{cases} mc^2 + \boldsymbol{p}^2/2m & (|\boldsymbol{p}| \ll mc) \\ c\,|\boldsymbol{p}| & (|\boldsymbol{p}| \gg mc) \end{cases} \tag{38.46}$$

The relation (38.43) holds also for particles with vanishing rest mass, like photons or neutrinos,

$$E = c\,|\boldsymbol{p}| \quad (m = 0) \tag{38.47}$$

Exercises

38.1 *Constancy of* $u^\alpha u_\alpha$

The initial condition for $u^\alpha(0)$ satisfies the condition $u^\alpha u_\alpha = c^2$. Show that then the solution $u^\alpha(\tau)$ of the equation of motion $m\, du^\alpha/d\tau = F^\alpha$ also satisfies this condition. Thereby the Minkowski force

$$\big(F^\alpha\big) = \left(\gamma\,\frac{v\,F_{\mathrm{N}\|}}{c},\, \gamma\boldsymbol{F}_{\mathrm{N}\|} + \boldsymbol{F}_{\mathrm{N}\perp}\right), \quad \gamma = \frac{1}{\sqrt{1 - v^2/c^2}} \tag{38.48}$$

is given by the Newtonian force $\boldsymbol{F}_{\mathrm{N}}$.

Chapter 39

Applications

We calculate the relativistic kinetic energy of two colliding particles in the laboratory and in the center-of-mass system (LS and CS). We calculate which part of this energy may be used for the production of new heavy particles. Then we solve the relativistic equation of motion for a constant force. On the basis of this solution, we discuss the so-called twin paradox.

Kinetic Energy: Laboratory Versus Center-of-Mass System

In order to bring a particle at rest to the velocity v we require the kinetic energy

$$E_{\text{kin}} = E(v) - E(0) = \frac{mc^2}{\sqrt{1 - v^2/c^2}} - mc^2 \qquad (39.1)$$

We consider two equal particles ($m_1 = m_2 = m$) which are shot onto each other (Figure 39.1). The sum

$$K = E_{\text{kin}}(v_1) + E_{\text{kin}}(v_2) \qquad (39.2)$$

of their kinetic energies will be calculated in the CS and in the LS.

In the closed system, the center-of-mass momentum $\boldsymbol{P} = \boldsymbol{p}_1 + \boldsymbol{p}_2$ is constant; the center of mass moves uniformly. Therefore, we can choose an inertial system in which the center of mass rests. In this CS, the relativistic momenta of the two particles add up to zero:

$$\boldsymbol{p}_1(t) + \boldsymbol{p}_2(t) = 0 \quad \text{(CS)} \qquad (39.3)$$

For $m_1 = m_2$ this implies $\boldsymbol{v}_1(t) = -\boldsymbol{v}_2(t)$.

Figure 39.1 The collision of two particles, 1 and 2, is observed in the laboratory system (LS, left) and in the center-of-mass system (CS, right). The figure shows the situation before the collision. For equal masses, the velocities in CS are oppositely equal, $v_1 = -v_2 = v$.

As long as the two particles do not interact, the velocities $\boldsymbol{v}_1(t)$ and $\boldsymbol{v}_2(t)$ are constant. The velocity of particle 2 in CS before the collision is denoted by $\boldsymbol{v}_2(-\infty)$. By a Lorentz transformation with velocity $-\boldsymbol{v}_2(-\infty)$, we pass from the CS to another IS in which particle 2 is resting before the collision. This *laboratory system* LS is defined by

$$\boldsymbol{v}_2'(-\infty) = 0 \quad \text{(LS)} \tag{39.4}$$

The quantities of the LS are indicated by a dash. The system with (39.4) is called laboratory system LS. In LS, a particle beam (particle 1) is directed on a target at rest (particle 2).

The kinetic energy in the CS is available for the excitation of other degrees of freedom or for the production of new particles. Since the momentum $\boldsymbol{P} = \boldsymbol{p}_1 + \boldsymbol{p}_2$ is conserved, the energy $\boldsymbol{P}^2/2M$ is not available for the production of new particles. This "reserved" energy is minimal in CS (namely zero). The relevant condition for the production of a particle of mass M reads therefore

$$K_{\text{CS}} > Mc^2 \quad \begin{pmatrix}\text{condition for production} \\ \text{of a particle of mass } M\end{pmatrix} \tag{39.5}$$

In addition, the rest mass energies of the colliding particles may contribute to the available energy. For very high energies, this is of minor importance.

During the collision, the angular momentum is also conserved, so that the corresponding kinetic energy is not available (not even in the CS). The maximum energy K_{CS} therefore is available for central collisions only (i.e. zero impact parameter and vanishing angular momentum).

In the following, we refer to such a central collision. Before the collision all velocities are parallel or antiparallel to the direction of $\boldsymbol{v}_1 = -\boldsymbol{v}_2$; therefore we consider only the components in this direction (Figure 39.1). The CS-velocities are

$$\text{CS}: \quad v_1 = v, \quad v_2 = -v, \quad v_{\text{LS}} = -v \quad (t = -\infty) \tag{39.6}$$

where $v = v_1(-\infty) = -v_2(-\infty)$. The origin of LS coincides for $t = -\infty$ with particle 2; it follows $v_{\text{LS}} = -v$. A Lorentz transformation with $-v$, therefore, leads from the CS to the LS. In the LS the following holds:

$$\text{LS}: \quad v_1' = V, \quad v_2' = 0, \quad v_{\text{CS}} = v \quad (t = -\infty) \tag{39.7}$$

Here, $v_2' = 0$ follows from the definition of LS, and $v_{\text{CS}} = v$ from the relative velocity given in (39.6) between CS and LS.

Particle 1 has the LS velocity V, which may be viewed as the CS velocity v of LS plus the CS velocity of particle 1. For this addition of velocities, we use the velocity addition formula (36.21),

$$V = \frac{v + v}{1 + v^2/c^2} \tag{39.8}$$

The corresponding rapidities are

$$\psi(V) = 2\,\psi(v) \tag{39.9}$$

Using (39.1), we calculate the sum of the kinetic energies K of both particles in the LS and CS, and consider the non-relativistic and the relativistic limit:

$$K_{\text{LS}} = \frac{mc^2}{\sqrt{1 - V^2/c^2}} - mc^2 \approx \begin{cases} mV^2/2 & (v \ll c) \\ mc^2 \cosh\psi(V) & (\psi \gg 1) \end{cases} \tag{39.10}$$

$$K_{\text{CS}} = \frac{2mc^2}{\sqrt{1 - v^2/c^2}} - 2mc^2 \approx \begin{cases} 2\,(mv^2/2) & (v \ll c) \\ 2mc^2 \cosh\psi(v) & (\psi \gg 1) \end{cases} \tag{39.11}$$

In the non-relativistic limit case, $V \approx 2v$ holds and

$$K_{\text{LS}} = 2\,K_{\text{CS}} \quad (v \ll c). \tag{39.12}$$

For the relativistic limit case, we use

$$\cosh(2x) = \frac{\exp(2x) + \exp(-2x)}{2} \approx \frac{\exp(2x)}{2} \tag{39.13}$$
$$\approx 2\left(\frac{\exp(x) + \exp(-x)}{2}\right)^2 = 2\cosh^2 x \qquad (x \gg 1)$$

With $\psi(V) = 2\psi(v)$ this becomes $\cosh\psi(V) \approx 2\cosh^2\psi(v)$. We use this for the relativistic limits in (39.10) and (39.11) and obtain

$$K_{\rm LS} \approx \frac{(K_{\rm CS})^2}{2\,mc^2} \qquad (\psi(v) \gg 1) \tag{39.14}$$

This approximation (39.13) is already usable for $\psi(v) > 2$.

Producing a Z^0 particle

As the following example shows, the differences between $K_{\rm LS}$ and $K_{\rm CS}$ can be quite large. We consider the production of a Z^0-particle in a collision of electrons with positrons or of protons with antiprotons,

$$e^+ + e^- \to Z^0 \qquad \text{or} \qquad p + \bar{p} \to Z^0 \tag{39.15}$$

A theory by Glashow, Salam, and Weinberg, developed in the 1960s, postulated the existence of the Z^0-particle. It should have a mass of about $M(Z^0) \approx 90\ \text{GeV}/c^2$ and have the same quantum numbers as a photon. The production of the Z^0 is possible in a colliding beam experiment if the particles in each beam have at least an energy of about $E_{\rm kin} = 50$ GeV. In this *colliding beam* experiment, the energy

$$K_{\rm CS} = 2\,E_{\rm kin} = 100\,\text{GeV} \tag{39.16}$$

is available for the production of new particles. With such an experiment (proton-antiproton collision), the Z^0-particles were produced for the first time at CERN in 1983.

If one wanted to perform the same experiment in the LS,[1] i.e. for instance to direct a particle beam onto matter with resting particles,

---

[1]The CS and LS systems are defined in Figure 39.1. Deviating from this definition one could also understand by "laboratory system" the system in which the experimental setup is at rest.

then the particles in the incident beam would have to possess the energy

$$K_{\rm LS} = \frac{(K_{\rm CS})^2}{2mc^2} = \frac{(100\ \text{GeV})^2}{2mc^2} \approx \begin{cases} 10^7\ \text{GeV} & (\text{e}^+ - \text{e}^-) \\ 5\cdot 10^3\ \text{GeV} & (\text{p} - \overline{\text{p}}) \end{cases} \tag{39.17}$$

We used $m_{\rm e}c^2 \approx 0.5$ MeV and $m_{\rm p}c^2 \approx 1$ GeV. In the LS, therefore, the production of Z^0-particles is excluded in the foreseeable future.

Hyperbolic Motion

We write down the spatial part of (38.9), considering $\gamma = dt/d\tau$, $(F^\alpha) = (F^0, \boldsymbol{F})$, and $(u^\alpha) = \gamma(c, \boldsymbol{v})$:

$$\frac{d}{dt}\frac{m\boldsymbol{v}(t)}{\sqrt{1-v^2/c^2}} = \frac{\boldsymbol{F}}{\gamma} \tag{39.18}$$

For a constant force, we consider a one-dimensional motion in the direction of the force

$$\frac{d}{dt}\frac{m\,v(t)}{\sqrt{1-v(t)^2/c^2}} = F_{\rm N} = \text{const.} \quad (\boldsymbol{v}\parallel \boldsymbol{F}) \tag{39.19}$$

Since $\boldsymbol{v}\parallel\boldsymbol{F}$, here $\boldsymbol{F} = \gamma \boldsymbol{F}_{\rm N}$ holds, (38.17). This equation of motion could describe the following cases:

- A particle with charge q is accelerated in a homogeneous and constant electric field E. From (38.23) one sees that the right-hand side in (39.19) is equal to qE. The Newtonian force is given by $F_{\rm N} = qE' = qE$ because $E' = E$ holds for $\boldsymbol{v}\parallel\boldsymbol{E}$, (22.6) in [2].
- A spacecraft is accelerated in such a way that the astronaut (mass m) is pressed against the spaceship floor with his earth weight mg. In his momentary rest system IS′, the astronaut experiences the acceleration g. Hence, the Newtonian force is

$$F_{\rm N} = mg \tag{39.20}$$

For these examples, (39.19) becomes

$$\frac{d}{dt}\frac{v(t)}{\sqrt{1-v(t)^2/c^2}} = \begin{cases} \dfrac{qE}{m} & \text{charge in field} \\ g & \text{spaceship} \end{cases} \tag{39.21}$$

We continue the calculation with the bottom line of (39.21); for the other case, g would have to be replaced by qE/m.

With the initial condition

$$v(0) = 0 \tag{39.22}$$

we integrate (39.21) to

$$\frac{v(t)}{\sqrt{1 - v(t)^2/c^2}} = g\,t \tag{39.23}$$

We resolve this to $v(t)$

$$v(t) = \frac{g\,t}{\sqrt{1 + g^2\,t^2/c^2}} \approx \begin{cases} g\,t & (t \ll c/g) \\ c & (t \gg c/g) \end{cases} \tag{39.24}$$

For small times we get the Newtonian result $v = g\,t$, for large times v approaches the speed of light (Figure 39.2, left).

We put the x-axis in the direction of the force, so $\boldsymbol{e}_{\|} = \boldsymbol{e}_x$ and $v(t) = dx/dt$. From (39.24) and the initial condition $x(0) = 0$ then follows

$$x(t) = \frac{c^2}{g}\left(\sqrt{1 + \frac{g^2\,t^2}{c^2}} - 1\right) = \begin{cases} g\,t^2/2 & (t \ll c/g) \\ c\,t - c^2/g & (t \gg c/g) \end{cases} \tag{39.25}$$

In the x–t diagram (Figure 39.2, right), this is a *hyperbola* in contrast to the non-relativistic parabola $x = g\,t^2/2$. Therefore, this section

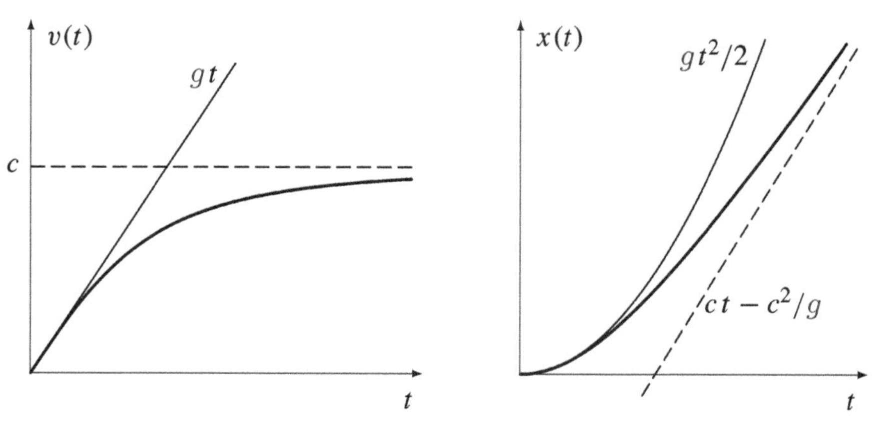

Figure 39.2 Velocity and position as a function of time for a particle under the influence of the constant force mg. The velocity (left) initially increases linearly with time, $v = g\,t$, then deviates significantly from this linearity, and approaches asymptotically the speed of light (for $t \gtrsim c/g$). The classical parabolic motion in the x–t-diagram (right) is replaced by a hyperbola.

has been titled *hyperbolic motion*. While the velocity approaches asymptotically c, the energy keeps increasing:

$$E = \frac{mc^2}{\sqrt{1 - v^2/c^2}} = mc^2 \sqrt{1 + \frac{g^2 t^2}{c^2}} \tag{39.26}$$

Twin Paradox

A spacecraft is accelerated according to (39.21) with $g = 10\,\mathrm{m/s^2}$. The acceleration g takes place relative to an IS$'$ in which the spacecraft is momentarily at rest so that the astronaut feels his familiar weight. This acceleration is maintained until 5 years have passed in the spacecraft. During the next 5 years of spaceship time, it is decelerated with $-g$. Afterwards, the spaceship returns to earth in an analogous way (another 5 years with $-g$, and 5 years with g). A twin brother of the astronaut remains on earth. The astronaut is 20 years older after his return but the twin who stayed behind is much older.

The basis for the following calculation of the times was given in the section "Moving clock" in Chapter 35. In the experiment considered here, the twin remaining on earth is (approximately) in an IS, while the spacecraft represents a non-IS. The result (different ages of the twins) seems to be paradoxical (hence the name "twin paradox") only, if one does not take into account these different reference frames.

We perform the calculation in an IS where the sun is at rest. Since of $v_{\mathrm{E}}/c \approx 10^{-4} \ll 1$ the earth time t_{E} is approximately equal to the IS time, $t_{\mathrm{E}} \approx t$. In the IS under consideration, the spacecraft performs a hyperbolic motion (39.25) during the first five spacecraft years. At the launch of the spacecraft, we start an earth clock and the spacecraft clock, so $\tau = t = 0$. With this initial condition for $\tau = \tau(t)$ we integrate

$$d\tau = dt \sqrt{1 - \frac{v(t)^2}{c^2}} \overset{(39.26)}{=} \frac{dt}{\sqrt{1 + g^2 t^2/c^2}} \tag{39.27}$$

to

$$\tau = \frac{c}{g} \,\mathrm{arsinh}\, \frac{gt}{c} \quad \text{or} \quad t = \frac{c}{g} \sinh \frac{g\tau}{c} \tag{39.28}$$

For $g = 10\,\mathrm{m/s^2}$ we get

$$\frac{c}{g} \approx 0.97 \text{ a} \tag{39.29}$$

Here, "a" is the unit for one year (the first letter of the Latin word annus). We evaluate (39.28) for $\tau = 5\,\mathrm{a}$,

$$t \approx 84\,\mathrm{a} \quad \text{for} \quad \tau = 5\,\mathrm{a} \tag{39.30}$$

During the four stages of the journey, the distortion factor $(1 - v(t)^2/c^2)^{1/2}$ between dt and $d\tau$ adopts the same values in equal time intervals. Therefore, upon return

$$t \approx 336\,\mathrm{a} \quad \text{and} \quad \tau = 20\,\mathrm{a} \tag{39.31}$$

Thus, the twin on earth is about 316 years older than the returning astronaut.

From (39.25) follows the spatial distance of the spacecraft at time $\tau = 5\,\mathrm{a}$

$$x(\tau = 5\,\mathrm{a}) = \frac{c^2}{g}\left(\sqrt{1 + \frac{g^2 t^2}{c^2}} - 1\right) \approx 83\,\mathrm{ly} \tag{39.32}$$

The maximum distance from earth is about 166 light-years (ly). This means that the average velocity of the spacecraft is close to c. As seen from the earth, the velocity $v \approx c$ is reached after about one year. After this, the continuous acceleration can no longer enhance substantially the spaceship's velocity; but it continues to increase the ratio t/τ.

Experimentally the relation between t and τ was tested, for example, with airplanes. For the corresponding small effect, one must take into account also (i) that the gravitational field influences the rate of the clocks (Exercise 35.3) and (ii) that a clock on earth is not (with the accuracy necessary here) in an IS. Taking into account these effects, the experiments verify the relation between dt and $d\tau$.

Accelerated reference frame

The designation *paradox* explains itself from the following question: Seen from the astronaut the earth moves with $-v(t)$. Thus, $v(t)^2$

runs through the same values as in (39.27). Wouldn't the clocks on the earth go slower than the clocks in the spacecraft?

The answer is: This is not so, because the spaceship does not represent an IS. In the reference frame connected with the spaceship, the laws of the SRT are not valid (at least not in their usual form). In particular, the relation $d\tau = (1 - v^2/c^2)^{1/2}\, dt$ holds only in IS, but not in the reference system of the spacecraft.

The binding of laws to reference systems is not surprising; it does not appear here for the first time either. In particular, Newton's laws are valid only in IS, but not in an accelerated reference system S′. An attempt of the astronaut to use $d\tau = (1 - v^2/c^2)^{1/2}\, dt$ for calculating the earth time is analogous to the attempt of a billiard player trying to calculate the path of the ball on a merry-go-round using Newton's laws. Such an attempt fails: The billiard ball (without spin) does not describe a linear path, as the (inadmissible) application of Newton's first law would suggest.

But the billiard player on the carousel can do the following: He starts from the valid IS laws and applies a transformation from IS to his accelerated system S′. This results in different laws; inertia forces (like centrifugal and Coriolis forces) will appear. The laws modified in this way are valid in S′; they can describe the motion of the billiard ball on the carousel. Only in this specific sense, Newton's laws may be used in an accelerated S′. This applies analogously to relativistic mechanics and shall be briefly sketched here.

We consider the modifications in the relativistic laws which result from a transformation from the inertial system IS to an accelerated system S′. The starting point of relativistic mechanics was the four-dimensional line element, which has the form $ds^2 = \eta_{\alpha\beta}\, dx^\alpha dx^\beta$ in IS. We examine the form of ds^2 after a transformation from the IS with coordinates x^α to an accelerated S′ with coordinates x'^α. We specifically consider a system S′ obtained by the following transformation,

$$t = t', \quad x = x' + \frac{g\,t'^2}{2}, \quad y = y', \quad z = z' \tag{39.33}$$

The origin $x' = 0$ of S′ moves in IS with $x = gt^2/2$. For $t \ll c/g$ this is a possible transformation into the spacecraft's system S′.

We calculate ds^2 as a function of the coordinates of S′,

$$ds^2 = \eta_{\alpha\beta}\, dx^\alpha dx^\beta = c^2 dt'^2 \left(1 - \frac{g^2\, t'^2}{c^2}\right) - 2g\, t' dx' dt' - dx'^2 - dy'^2 - dz'^2 \tag{39.34}$$

We can express this as

$$ds^2 = g_{\alpha\beta}(x')\; dx'^\alpha\, dx'^\beta \tag{39.35}$$

Such a form is also obtained for an *arbitrary* coordinate transformation $x'^\alpha(x)$.

The new coordinates $x'^\alpha = x'^\alpha(x^0, x^1, x^2, x^3)$ in general do not have the meaning of IS coordinates (referring to resting rods and clocks). Thus, t' in S′ is only a *time parameter*. The connection of t' with the time of clocks resting in S′ has still to be established. In this respect, the choice of the time parameter t' is largely arbitrary. A similar arbitrariness exists also for the spatial coordinates of a general coordinate transformation $x'^\alpha = x'^\alpha(x^0, x^1, x^2, x^3)$. The new spatial coordinates could be, for example, spherical coordinates; then $x'^3 = \phi$ is an angle and not a length. The meaning of the new coordinates is determined by the form (39.35) of the line element.

According to (39.35), the Minkowski tensor $\eta_{\alpha\beta}$ is replaced by the coordinate, dependent quantities $g_{\alpha\beta}(x)$ in S′. Thereby the relativistic laws are modified. This is, as already said, not surprising; it is comparable to the transition from Newton's second law to the equation of motion in the rotating system. Whether one wants to use the new (generally more complicated) laws in S′ is a question of expediency.

As an example of the application of relativistic laws in accelerated systems, we calculate with (39.34) the times of clocks that are at rest either in S′ or in IS. According to (35.15), the display of a clock shows

$$c\, d\tau = ds_{\text{clock}} = \sqrt{ds^2_{\text{clock}}} \tag{39.36}$$

We can determine the rate of an arbitrary clock by either the IS or the S′-form of ds^2:

$$d\tau = \begin{cases} \dfrac{1}{c}\,\sqrt{\eta_{\alpha\beta}\; dx^\alpha_{\text{clock}}\, dx^\beta_{\text{clock}}} & \text{(calculation in IS)} \\[2ex] \dfrac{1}{c}\,\sqrt{g_{\alpha\beta}(x'_{\text{clock}})\; dx'^\alpha_{\text{clock}}\, dx'^\beta_{\text{clock}}} & \text{(calculation in S}'\text{)} \end{cases} \tag{39.37}$$

In the following, we use (39.34) for the S′-calculation. We consider an IS clock at rest on earth, and a S′ clock resting in the spacecraft. For the coordinate differentials of these clocks, the following holds:

$$\begin{array}{lcc} & \text{IS-clock} & \text{S}'\text{-clock} \\ \text{calculation in IS:} & dx = 0 & dx = g\,t\,dt \\ \text{calculation in S}'\text{:} & dx' = -g\,t'\,dt' & dx' = 0 \end{array} \tag{39.38}$$

For both clocks $dy = dz = dy' = dz' = 0$ holds. With this, we can calculate the line element (i.e. the displayed proper time) of both clocks either in IS or in S′. With the coordinate differentials (39.38) inserted into (39.37), we obtain

$$\begin{array}{lcc} & \text{IS-clock} & \text{S}'\text{-clock} \\ \text{calculation in IS:} & d\tau = dt & d\tau = \sqrt{1 - g^2\,t^2/c^2}\;dt \\ \text{calculation in S}'\text{:} & d\tau = dt' & d\tau = \sqrt{1 - g^2\,t'^2/c^2}\;dt' \end{array} \tag{39.39}$$

When calculating the S′-time, the factor $(1 - v^2/c^2)^{1/2}$ (with $v = v_{\text{clock}}$) in the IS-calculation equals to the factor $(g_{00})^{1/2}$ in the S′ calculation. In his system S′, the astronaut can therefore calculate the various clock times and finds in particular, that his twin brother, who stayed back on earth ages "faster".

Relativistic theory of gravitation

In Chapter 6, we discussed that due to the equivalence of inertial and gravitational masses, the gravitational forces can be eliminated by the transition into a suitably accelerated system S′. Such systems S′ are in particular a freely falling elevator and a satellite laboratory orbiting the earth.

One can now take the following standpoint: In an orbiting (free-falling) satellite laboratory, the processes take place as if no gravitational field were present. Therefore, the known relativistic laws apply without gravity, in particular $ds^2 = \eta_{\alpha\beta}\,dx^\alpha dx^\beta$. In this specific sense, the satellite laboratory is a *local* IS. A laboratory on earth is accelerated relative to the satellite laboratory. The transformation to the earth, therefore, leads to the form (39.35) for ds^2. As seen from the earth, there is a gravitational field at the position of the satellite, which is described by the fields $g_{\alpha\beta}(x)$.

The train of thought outlined here leads to a relativistic theory of gravitation, the *general theory of relativity* of Einstein.

Exercises

39.1 *Kinetic energy in the center of mass and laboratory system*

The collision of two identical particles is described in the CS and in the LS:

$$\text{LS:}\quad v_1 = V, \quad v_2 = 0$$
$$\text{CS:}\quad v_1 = v, \quad v_2 = -v$$

The corresponding kinetic energies are

$$K_{\rm LS} = \frac{mc^2}{\sqrt{1 - V^2/c^2}} - mc^2, \quad K_{\rm CS} = \frac{2\,mc^2}{\sqrt{1 - v^2/c^2}} - 2\,mc^2$$

Calculate the exact relationship between $K_{\rm LS}$ and $K_{\rm CS}$.

39.2 *Relativistic motion in the electric field*

A particle with charge q moves in a homogeneous, constant electric field $\boldsymbol{E} = E_0\,\boldsymbol{e}_x$. Solve the relativistic equations of motion (38.23) for the initial conditions $\boldsymbol{r}(0) = 0$ and $\boldsymbol{v}(0) = v_0\,\boldsymbol{e}_y$. Which trajectory curve results in the x–y-plane?

39.3 *Time in an accelerated system*

In an inertial system IS (with coordinates t, x, y, z) the position of a clock oscillates according to $\boldsymbol{r}_{\rm clock} = \boldsymbol{e}_x\, a \sin(\omega t)$; it holds $a\omega \ll c$. At time $t = 0$, the clock is synchronized with an IS clock (a clock resting at $\boldsymbol{r} = 0$).

After a half oscillation ($t = t_0 = \pi/\omega$) the moving clock is again at $\boldsymbol{r} = 0$, and is compared with the IS-clock. Which time periods Δt and $\Delta t'$ are shown by the IS clock and the moving clock? Calculate these time spans first in the IS. Then use a suitable transformation from IS to the S$'$ of the moving clock, and calculate the times in this system. The relativistic effects shall be taken into account in leading order only.

Chapter 40

Lagrange Function

We formulate the Hamiltonian principle for the relativistic motion of a mass point. Specifically, we consider the force-free motion and the motion in an electromagnetic field.

The trajectory of a mass point can be described by $x^i(t)$ or $x^\alpha(\tau)$. The respective expressions for the relativistic Lagrange functions are

$$\mathcal{L}_1 = \mathcal{L}_1(\boldsymbol{r}, \boldsymbol{v}, t) \quad \text{or} \quad \mathcal{L}_2 = \mathcal{L}_2(x, u) \tag{40.1}$$

In the argument of $\mathcal{L}_2$, x stands for x^0, x^1, x^2, x^3 and u for u^0, u^2, u^2, u^3. Via $x^0 = ct$, also $\mathcal{L}_2(u, x)$ may depend on the time.

The functions $\mathcal{L}_1$ and $\mathcal{L}_2$ are not physical quantities; therefore they do not have to be equal. However, they must yield the same equations of motion, i.e. Hamilton's principle $\delta S = 0$ must be the same. We restrict the freedoms for $\mathcal{L}$ in such a way that besides δS also the actions S are the same, i.e. that

$$S = \int_{t_1}^{t_2} dt\ \mathcal{L}_1(\boldsymbol{r}, \boldsymbol{v}, t) = \int_{\tau_1}^{\tau_2} d\tau\ \mathcal{L}_2(x, u) \tag{40.2}$$

First, we consider the Lagrange function $\mathcal{L}_1 = \mathcal{L}_1(\boldsymbol{v}, \boldsymbol{r}, t)$ for a *free* particle. For this system the general space–time symmetries are valid. i.e.

- Homogeneity of space: $\mathcal{L}_1$ does not depend on $\boldsymbol{r}$.
- Homogeneity of time: $\mathcal{L}_1$ does not depend on t.
- Isotropy of space: $\mathcal{L}_1$ depends only on $\boldsymbol{v}^2$.

This gives us

$$\mathcal{L}_1 = f(\boldsymbol{v}^2) \tag{40.3}$$

We now compare the associated Lagrange equation with the known relativistic equation of motion in the force-free case:

$$\frac{d}{dt}\frac{\partial \mathcal{L}_1}{\partial \boldsymbol{v}} = 0 \quad \overset{\text{comparison}}{\longleftrightarrow} \quad \frac{d}{dt}\frac{m\,\boldsymbol{v}(t)}{\sqrt{1-v(t)^2/c^2}} = 0 \tag{40.4}$$

This determines the Lagrange function up to constants:

$$\mathcal{L}_1(\boldsymbol{v}) = \text{const.}\cdot\sqrt{1-\frac{\boldsymbol{v}^2}{c^2}} + \text{const.} = -mc^2\sqrt{1-\frac{\boldsymbol{v}^2}{c^2}} \tag{40.5}$$

The constants were chosen such that the non-relativistic limit yields the known $\mathcal{L}_1 = m\boldsymbol{v}^2/2$ for $v \ll c$ (up to an additive constant). Note that $\mathcal{L}_1$ from (40.5) is not equal to the kinetic energy (39.1); the relation $\mathcal{L} = T - U$ holds only in the non-relativistic case.

Now, we consider the alternative form $\mathcal{L}_2(u,x)$ of the relativistic Lagrange function. For a free particle, the general space–time symmetries imply

- Homogeneity of space: $\mathcal{L}_2$ does not depend on x^i.
- Homogeneity of time: $\mathcal{L}_2$ does not depend on $x^0 = ct$.
- Isotropy of space and relativity of space-time: $\mathcal{L}_2$ can depend on $u^\alpha u_\alpha$ only.

According to these conditions, $\mathcal{L}_2(x,u)$ may be written as

$$\mathcal{L}_2 = f\big(u^\alpha u_\alpha\big) = f\big(\eta_{\alpha\beta}\,u^\alpha u^\beta\big) \tag{40.6}$$

Here, we use Lorentz vectors with lower (covariant) indices given by

$$u_\alpha = \eta_{\alpha\beta}\,u^\beta, \quad u^\alpha = \eta^{\alpha\beta}\,u_\beta \tag{40.7}$$

with $(\eta^{\alpha\beta}) = (\eta_{\alpha\beta})$ and $\eta^{\alpha\beta}\,\eta_{\beta\gamma} = \delta^\alpha_\gamma$. The Euler–Lagrange equations of the Hamiltonian principle $\delta\!\int d\tau\,\mathcal{L}_2 = 0$ are then

$$\frac{d}{d\tau}\frac{\partial \mathcal{L}_2}{\partial u^\alpha} = \frac{\partial \mathcal{L}_2}{\partial x^\alpha}, \quad \text{leading to} \quad \frac{d}{d\tau}\left(2f'(\eta_{\alpha\beta}\,u^\alpha u^\beta)\,u_\alpha\right) = 0 \tag{40.8}$$

According to (38.44), the following holds for the 4-velocities

$$u^\alpha u_\alpha = c^2 \tag{40.9}$$

Therefore, $2f'(\eta_{\alpha\beta}\, u^\alpha u^\beta) = \text{const.}$, so that this term can be omitted. Thus, (40.8) becomes $du_\alpha/d\tau = 0$ or

$$\frac{du^\alpha(\tau)}{d\tau} = 0 \quad (\alpha = 0, 1, 2, 3) \tag{40.10}$$

These equations of motion are to be supplemented by the condition (40.9).

The function f in (40.6) has no influence on the equations of motion. We fix this function via (40.2): for (40.2) to hold for arbitrary time intervals, the integrands $\mathcal{L}_1\, dt$ and $\mathcal{L}_2\, d\tau$ in (40.2) must agree:

$$\begin{aligned} \mathcal{L}_1(\boldsymbol{v})\, dt &= -mc^2 \sqrt{1 - \frac{\boldsymbol{v}^2}{c^2}}\; dt = -mc \sqrt{\left(\frac{d(ct)}{d\tau}\right)^2 - \left(\frac{d\boldsymbol{r}}{d\tau}\right)^2}\; d\tau \\ &= -mc\, \sqrt{u^\alpha\, u_\alpha}\; d\tau = \mathcal{L}_2(u)\; d\tau \end{aligned} \tag{40.11}$$

Thus, the Lagrangian function $\mathcal{L}_2$ of the relativistic, free particle reads

$$\mathcal{L}_2 = -mc\, \sqrt{u^\alpha\, u_\alpha} \tag{40.12}$$

The condition $u^\alpha u_\alpha = c^2$ must not be used in $\mathcal{L}_2(u)$ itself, because for the variation δS also orbits $u^\alpha + \delta u^\alpha$ are allowed, which do not fulfill this condition. But the condition is valid for all actual trajectories. Therefore, it can be inserted into any physical quantity or into the equation of motion (here r.h.s. of (40.8)).

For the Lagrangian function (40.12), the action is equal to the path integral of the particle's trajectory,

$$\begin{aligned} S &= \int_{\tau_1}^{\tau_2} d\tau\; \mathcal{L}_2 = -mc \int_{\tau_1}^{\tau_2} d\tau\; \sqrt{u^\alpha\, u_\alpha} \\ &= -mc \int_1^2 \sqrt{dx^\alpha\, dx_\alpha} = -mc \int_1^2 ds \end{aligned} \tag{40.13}$$

This action is the simplest possible Lorentz scalar for the trajectory of a mass point.

As already mentioned several times, the choice of a Lagrange function is not unique. The form (40.12) was derived here via the connection (40.11), where $\mathcal{L}_1$ was chosen so that in the non-relativistic limit the familiar $\mathcal{L}_1 \to m\boldsymbol{v}^2/2$ is obtained. The choice

(40.12) has the minor disadvantage that the generalized momentum $\partial\mathcal{L}_1/\partial u_\alpha$ is equal to $-p^\alpha$ and not equal to $p^\alpha = mu^\alpha$. To avoid this, one could choose the sign in (40.12) differently. But then the connection (40.11) is lost (unless one accepts $\mathcal{L}_1 \to -m\boldsymbol{v}^2/2$), and the sign in the coupling term in (40.19) must also be changed.

Conservation of energy and momentum

The system of a free particle is invariant with respect to spatial and temporal translations. For the Lagrangian function $\mathcal{L}_2$ we derive from this symmetry the corresponding conservation laws.

The symmetry transformation is

$$\begin{aligned} x^\alpha &\;\to\; x^{*\alpha} = x^\alpha + \epsilon\,\delta^\alpha_\beta, & \psi^\alpha &= \delta^\alpha_\beta \\ \tau &\;\to\; \tau^* = \tau, & \varphi &= 0 \end{aligned} \tag{40.14}$$

For each value $\beta = 0, 1, 2, 3$, this is a one-parameter transformation. The transformation is analogous to (15.8) and (15.9) so that we can take over the results from Chapter 15. The invariance condition (15.12) is trivially satisfied because $\mathcal{L}_2$ does not depend on x^α. According to (15.19), we obtain the conservation laws

$$Q = \frac{\partial\mathcal{L}_2}{\partial u^\alpha}\,\psi^\alpha = -\frac{mc}{\sqrt{u^\gamma u_\gamma}}\,u_\alpha\,\delta^\alpha_\beta = -m\,u_\beta = \text{const.} \tag{40.15}$$

Therefore, from the spatial and temporal translational invariance follows the conservation of the 4-momentum,

$$\big(p^\alpha\big) = m\,\big(u^\alpha\big) = \big(E/c, \boldsymbol{p}\big) = \text{const.} \tag{40.16}$$

Motion in an electromagnetic field

In order to account for forces in $\mathcal{L}_2$ we need a relativistic theory of the underlying force fields. The standard example is electrodynamics, in which it is shown, that

$$\big(A^\beta(x)\big) = (\varPhi, A_x, A_y, A_z) = \big(\varPhi(\boldsymbol{r},t), \boldsymbol{A}(\boldsymbol{r},t)\big) \tag{40.17}$$

is a 4-vector field; here x stands for $(x^\alpha) = (\boldsymbol{r}, t)$. The potentials $\varPhi$

and $\boldsymbol{A}$ determine the electromagnetic fields,

$$\boldsymbol{E}(\boldsymbol{r},t) = -\operatorname{grad}\Phi(\boldsymbol{r},t) - \frac{1}{c}\,\frac{\partial \boldsymbol{A}(\boldsymbol{r},t)}{\partial t}\,, \qquad \boldsymbol{B}(\boldsymbol{r},t) = \operatorname{curl}\boldsymbol{A}(\boldsymbol{r},t) \tag{40.18}$$

For the motion of a particle in the field $A^\alpha(x)$ there should be an additional term in (40.12) depending on $A^\alpha(x)$. Since the electromagnetic forces are proportional to the fields $\boldsymbol{E}$ and $\boldsymbol{B}$, the potential $A^\alpha(x)$ should enter linearly in the Lagrange function. From the principle of relativity, it follows that the additional term must be a Lorentz scalar. The only Lorentz scalars which are linear in $A^\alpha(x)$, and which can be formed with the arguments of $\mathcal{L}_2(u,x)$ are $A_\alpha(x)\,u^\alpha$ and $A_\alpha(x)\,x^\alpha$. The homogeneity of space and time excludes the term $A_\alpha(x)\,x^\alpha$. The only remaining possibility is

$$\mathcal{L}_2(x,u) = -mc\,\sqrt{u^\alpha u_\alpha} \;-\; \frac{q}{c}\,A_\alpha(x)\,u^\alpha \tag{40.19}$$

The $A_\alpha(x)$ are external fields; therefore they do not occur as an argument in $\mathcal{L}_2(x,u)$. Via the arguments $x = (x^\alpha)$ of $A_\alpha(x)$, the Lagrange function $\mathcal{L}_2$ now depends explicitly on the position and on the time. The prefactor of the additional term is chosen in such a way that the correct equations of motion are obtained.

We evaluate the Euler–Lagrange equations for (40.19),

$$\begin{aligned}\frac{d}{d\tau}\,\frac{\partial \mathcal{L}_2}{\partial u^\beta} &= \frac{d}{d\tau}\left(\frac{-mc\,u_\beta}{\sqrt{u^\alpha u_\alpha}} - \frac{q}{c}\,A_\beta(x)\right) = -m\,\frac{du_\beta}{d\tau} - \frac{q}{c}\,\frac{\partial A_\beta}{\partial x^\alpha}\,\frac{dx^\alpha}{d\tau}\\ &= -m\,\frac{du_\beta}{d\tau} - \frac{q}{c}\,\frac{\partial A_\beta}{\partial x^\alpha}\,u^\alpha = \frac{\partial \mathcal{L}_2}{\partial x^\beta} = -\frac{q}{c}\,\frac{\partial A_\alpha}{\partial x^\beta}\,u^\alpha \end{aligned} \tag{40.20}$$

We equate the first and the third expression of the second line

$$m\,\frac{du_\beta}{d\tau} = \frac{q}{c}\left(\frac{\partial A_\beta}{\partial x^\alpha} - \frac{\partial A_\alpha}{\partial x^\beta}\right)u^\alpha = \frac{q}{c}\,F_{\beta\alpha}\,u^\alpha \tag{40.21}$$

The occurring derivatives of the 4-potential were combined to the *field strength tensor* $F_{\alpha\beta}$:

$$(F_{\alpha\beta}) = \left(\frac{\partial A_\beta}{\partial x^\alpha} - \frac{\partial A_\alpha}{\partial x^\beta}\right) = \begin{pmatrix} 0 & E_x & E_y & E_z \\ -E_x & 0 & -B_z & B_y \\ -E_y & B_z & 0 & -B_x \\ -E_z & -B_y & B_x & 0 \end{pmatrix} \tag{40.22}$$

The spatial components of (40.21) are

$$\frac{d\boldsymbol{p}}{dt} = q\left(\boldsymbol{E} + \frac{\boldsymbol{v}}{c} \times \boldsymbol{B}\right) \tag{40.23}$$

The time derivative of the relativistic momentum $\boldsymbol{p} = m\gamma\boldsymbol{v}$ is equal to the Lorentz force. Thus, we have formally derived (38.23).

We determine the Lagrange function $\mathcal{L}_1(\boldsymbol{r}, \boldsymbol{v}, t)$ which corresponds to $\mathcal{L}_2$ of (40.19). We look again in (40.2) and rewrite the potential term in $\mathcal{L}_2\, d\tau$ so that it fits into $\mathcal{L}_1\, dt$,

$$-\frac{q}{c}\, A^\beta\, u_\beta\, d\tau = -\left(-q\,\Phi + \frac{q}{c}\, \boldsymbol{v}\cdot\boldsymbol{A}\right) dt \tag{40.24}$$

From this follows

$$\mathcal{L}_1(\boldsymbol{r}, \boldsymbol{v}, t) = -mc^2\sqrt{1 - \frac{\boldsymbol{v}^2}{c^2}} - q\,\Phi(\boldsymbol{r}, t) + \frac{q}{c}\,\boldsymbol{v}\cdot\boldsymbol{A}(\boldsymbol{r}, t) \tag{40.25}$$

The Euler–Lagrange equations again give (40.23). From (40.25) follows the generalized momentum

$$\boldsymbol{p} = \frac{\partial\mathcal{L}_1}{\partial\boldsymbol{v}} = \frac{m\boldsymbol{v}}{\sqrt{1 - v^2/c^2}} + \frac{q}{c}\,\boldsymbol{A} \tag{40.26}$$

The Hamiltonian function $H = \boldsymbol{p}\cdot\boldsymbol{v} - \mathcal{L}_1$ belonging to (40.25) is then

$$H(\boldsymbol{r}, \boldsymbol{p}, t) = c\sqrt{m^2c^2 + \left(\boldsymbol{p} - \frac{q}{c}\,\boldsymbol{A}(\boldsymbol{r}, t)\right)^2} + q\,\Phi(\boldsymbol{r}, t) \tag{40.27}$$

Other fields

For a 4-vector field A^α the simplest approach (40.19) leads to a successful physical theory. We supplement these considerations with the simplest approaches for the motion of a particle in a Lorentz tensor field of the zeroth or second rank.

If we had a 4-scalar field Ψ instead of a 4-vector field A^α, the most simple approach is

$$\mathcal{L}_2(u, x) = -mc\,\sqrt{u^\alpha u_\alpha} - q\,\Psi(x) \tag{40.28}$$

For a 4-tensor field $g_{\alpha\beta}(x)$, the simplest scalar is $g_{\alpha\beta}\, u^\alpha u^\beta$. Such an additional term is equivalent to substituting $\eta_{\alpha\beta}$ into

$\mathcal{L}_2 = -mc\,(\eta_{\alpha\beta}\, u^\alpha u^\beta)^{1/2}$ by $g_{\alpha\beta}$, i.e.

$$\mathcal{L}_2(u,x) = -mc\,\sqrt{g_{\alpha\beta}(x)\; u^\alpha\, u^\beta} \tag{40.29}$$

According to the equivalence principle (in the theory of general relativity), the gravitational potentials are described by $g_{\alpha\beta}(x)$ (last section in Chapter 39). Therefore, (40.29) is the Lagrangian function for the relativistic motion of a particle in the gravitational field.

Exercises

40.1 *Conserved quantities of the relativistic Lagrangian function*

Determine the conservation laws resulting from the invariance of the relativistic Lagrangian function

$$\mathcal{L} = -\,mc^2\,\sqrt{1 - v^2/c^2}$$

with respect to spatial and temporal shifts.

40.2 *Relativistic Lagrange function for a particle in an electromagnetic field*

Determine the Euler-Lagrange equation for the relativistic Lagrange function

$$\mathcal{L}(\boldsymbol{r},\boldsymbol{v},t) = -\,mc^2\,\sqrt{1 - \frac{\boldsymbol{v}^2}{c^2}} - q\,\Phi(\boldsymbol{r},t) + \frac{q}{c}\,\boldsymbol{v}\cdot\boldsymbol{A}(\boldsymbol{r},t) \tag{40.30}$$

of a charged particle in the electromagnetic field.

40.3 *Relativistic Hamilton function for a particle in an electromagnetic field*

Derive the relativistic Hamiltonian function $H = \boldsymbol{p}\cdot\boldsymbol{v} - \mathcal{L}$ from the Lagrangian function (40.25), and establish the canonical equations of motion.

Appendix A

Newtonian Force and Minkowski Force

In every mechanics book one finds Newton's equation of motion and the corresponding Newtonian force[1] $\boldsymbol{F}_{\mathrm{N}}$. *A modern theoretical mechanics must contain the special relativistic generalization. In the relativistic equation of motion, the Newtonian force is replaced by the Minkowski force* $(F_\alpha) = (F_0, \boldsymbol{F})$. *This inevitably raises the question of the connection between the Newtonian force* $\boldsymbol{F}_{\mathrm{N}}$ *and the spatial part* $\boldsymbol{F}$ *of the Minkowski force.*

For this connection, one finds different statements in textbooks. We investigate the possible reasons or derivations for these different statements. The Lorentz force serves as an instructive example.

In concrete applications, all authors eventually use the correct equations of motion. In this respect, the question discussed here could be dismissed as merely academic, i.e. without practical relevance. However, the question of the correct connection is related to the logic of establishing physical laws by means of symmetry principles.

Different Statements in the Literature

The dynamics of a mass point is determined in the non-relativistic case by the Newtonian force $\boldsymbol{F}_{\mathrm{N}}$, and in the relativistic case by the

[1]In the non-relativistic parts, I–VIII, of this book, the Newtonian force was denoted by $\boldsymbol{F}$. In the relativistic treatment (Part IX and this appendix), however, we use $\boldsymbol{F}$ for the spatial part of the Minkowski force and $\boldsymbol{F}_{\mathrm{N}}$ for the Newtonian force. In the non-relativistic limit case ($v/c \to 0$), the differences between $\boldsymbol{F}_{\mathrm{N}}$ and $\boldsymbol{F}$ vanish.

Minkowski force $(F^\alpha) = (F^0, \boldsymbol{F})$. In the literature, one finds two different specifications for the relation between these forces. For the spatial components, these alternatives are:

$$\text{Version A:} \qquad \boldsymbol{F} = \gamma\, \boldsymbol{F}_{\text{N}} \tag{A.1}$$

$$\text{Version B:} \qquad \boldsymbol{F} = \gamma\, \boldsymbol{F}_{\text{N}\|} + \boldsymbol{F}_{\text{N}\perp} \tag{A.2}$$

Here, the fractions parallel and perpendicular to the velocity $\boldsymbol{v}$ of the particle were distinguished, and $\gamma = 1/\sqrt{1 - v^2/c^2}$. The characteristic features of these two versions[2,3] are the same (A) and the different (B) treatment of the force components. The following discussion examines alternatives A and B in terms of their possible justification and their practical and logical relevance.

Formal Derivation

We derive the relation between Newtonian force and Minkowski force formally. The basic consideration for the establishment of relativistic laws is: In the limiting case $v \to 0$ (instantaneous inertial rest frame)

---

[2]Some sources for version A are:

E. Schmutzer, *Grundlagen der Theoretischen Physik*, BI-Verlag 1989, equation (6.7.4).

H. Goenner, *Spezielle Relativitätstheorie*, Elsevier-Spektrum 2004, equation (4.29).

R. M. Dreizler und C. S. Lüdde, *Theoretische Physik 2: Elektrodynamik und spezielle Relativitätstheorie*, Springer 2005, equation (8.55).

H. Günther, *Die Spezielle Relativitätstheorie*, Springer Spektrum 2013, equation (439).

F. Kuypers, *Klassische Mechanik*, 10th edition, John Wiley & Sons 2016, equation (25.2-12).

W. Nolting, *Theoretical Physics 4: Special Theory of Relativity*, Springer 2017, equation (2.47).

[3]Some sources for version B are:

S. Weinberg, *Gravitation and Cosmology*, John Wiley 1972, equation (2.3.5).

T. Fliessbach, *Mechanics for Physicists: An Introduction, including Special Relativity* (this book), or *Mechanik*, 8th edition, equation (38.17).

F. Scheck, *Mechanics*, 6th edition, Springer 2018, equation (4.80).

E. Rebhan, *Theoretische Physik: Relativitätstheorie und Kosmologie*, Spektrum Akademischer Verlag 2012, equation (4.18).

Newton's equations are also *relativistically* valid. From these valid equations, a Lorentz transformation leads to the relativistic equations in an arbitrary inertial frame.

This reasoning is used as follows: One sets up the Lorentz invariant equations which for $v \to 0$ reduces to the Newtonian limit case. The Lorentz invariant equations

$$m\,\frac{du^\alpha}{d\tau} = F^\alpha \tag{A.3}$$

(with $(u^\alpha) = \gamma(c, \boldsymbol{v})$, $d\tau = dt/\gamma$ and the mass m) lead in the limiting case $v \to 0$ to the Newtonian form

$$m\,\frac{dv'^i}{dt'} = F'^i \quad \text{in IS}' \tag{A.4}$$

In the instantaneous rest frame IS$'$, Newton's law is relativistically valid. So the force on the right side is equal to Newton's force:

$$\big(F'^\alpha\big) = \big(0, F'^i\big) = \big(0, \boldsymbol{F}_\mathrm{N}\big) \tag{A.5}$$

The spatial components F'^i constitute the Newtonian force $\boldsymbol{F}_\mathrm{N}$. The Minkowski force F^α in (A.3) results from this with the help of the well-known Lorentz back transformation $F^\alpha = \Lambda^\alpha_\beta(-\boldsymbol{v})\, F'^\beta$ (for details, see Chapter 38). The result is (38.17)

$$\big(F^\alpha\big) = \big(F^0,\, \boldsymbol{F}\big) = \left(\gamma\,\frac{v\,F_{\mathrm{N}\|}}{c},\ \gamma\,\boldsymbol{F}_{\mathrm{N}\|} + \boldsymbol{F}_{\mathrm{N}\perp}\right) \tag{A.6}$$

For both versions, A and B, we write down the spatial components of the equation of motion:

$$\frac{d}{dt}\,\frac{m\,\boldsymbol{v}(t)}{\sqrt{1 - v^2/c^2}} = \frac{\boldsymbol{F}}{\gamma} = \begin{cases} \boldsymbol{F}_\mathrm{N} & \text{Version A} \\[2ex] \boldsymbol{F}_{\mathrm{N}\|} + \dfrac{\boldsymbol{F}_{\mathrm{N}\perp}}{\gamma} & \begin{cases} \text{Version B} \\ \text{Weinberg}^3 \\ \text{(A.6)} \end{cases} \end{cases} \tag{A.7}$$

The equation of motion (left part) follows from (A.3) with $(F^\alpha) = (F^0, \boldsymbol{F})$, $(u^\alpha) = \gamma(c, \boldsymbol{v})$ and $d\tau = dt/\gamma$.

The formal derivation has led to (A.6), i.e., to the second line in (A.7). Nevertheless, version A (first line) is quite common (see footnote 2). In concrete applications, however, even these authors (see footnote 2) use the correct equations of motion. The reasons for this show up most clearly in the example of the Lorentz force.

Lorentz Force

A possible reason for version A starts from the motion in the electromagnetic field (electric field $\boldsymbol{E}$ and magnetic field $\boldsymbol{B}$). For a particle (mass m, charge q) is acted upon by the Lorentz force, Newtons's equation of motion reads

$$m\,\frac{d\boldsymbol{v}(t)}{dt} = q\left(\boldsymbol{E} + \frac{\boldsymbol{v}}{c} \times \boldsymbol{B}\right) \quad (v \ll c) \tag{A.8}$$

The exact relativistic equation can be written in the form

$$\frac{d}{dt}\,\frac{m\,\boldsymbol{v}(t)}{\sqrt{1 - v(t)^2/c^2}} = q\left(\boldsymbol{E} + \frac{\boldsymbol{v}}{c} \times \boldsymbol{B}\right) \tag{A.9}$$

The comparison with (A.7) gives the Minkowski force:

$$\boldsymbol{F} = \gamma\, q\left(\boldsymbol{E} + \frac{\boldsymbol{v}}{c} \times \boldsymbol{B}\right) \tag{A.10}$$

If now the right-hand side of Newton's equation (A.8) is identified with $\boldsymbol{F}_{\text{N}}$, then (A.10) becomes $\boldsymbol{F} = \gamma \boldsymbol{F}_{\text{N}}$. Thus, one has apparently derived equation (A.1), i.e. version A.

The argumentation in favor of version A given here has the following problem: Equation (A.8) is a legitimate Newtonian equation, but it is not the Newtonian limiting case $v \to 0$. In the Newtonian limit case,

$$m\,\frac{d\boldsymbol{v}'}{dt'} = q\boldsymbol{E}' = \boldsymbol{F}_{\text{N}} \quad v' \to 0 \tag{A.11}$$

the vector $\boldsymbol{E}'$ is the electric field in the instantaneous rest frame IS$'$ of the particle under consideration. It holds[4] $\boldsymbol{E}' = \boldsymbol{E} + (\boldsymbol{v}/c) \times \boldsymbol{B} + \mathcal{O}(v^2/c^2)$. The forces in (A.8) and (A.11) thus differ in terms of relative magnitude v^2/c^2. In the framework of Newtonian mechanics, such forces are equivalent. But a relation of the kind (A.1) or (A.2) requires an exact definition of the Newtonian force. For this, the Newtonian limit is the unambiguous and adequate choice. Therefore, we identify the force $q\boldsymbol{E}'$ in (A.11) as Newtonian force $\boldsymbol{F}_{\text{N}}$.

The transformation of the fields from the instantaneous rest frame IS′ to the frame IS, in which the particle is moving with velocity $\boldsymbol{v}$, is known[4]

$$\boldsymbol{E}' = \boldsymbol{E}'_{\parallel} + \boldsymbol{E}'_{\perp} = \boldsymbol{E}_{\parallel} + \gamma\left(\boldsymbol{E}_{\perp} + \frac{\boldsymbol{v}}{c} \times \boldsymbol{B}\right) \tag{A.12}$$

Here, we distinguished between the parts parallel and perpendicular to the velocity $\boldsymbol{v}$. In order to get from the Newtonian force $\boldsymbol{F}_{\rm N} = q\boldsymbol{E}'$ to the Minkowski force $\boldsymbol{F} = \gamma q(\boldsymbol{E} + (\boldsymbol{v}/c) \times \boldsymbol{B})$, the parallel part must be multiplied by γ. This gives Equation (A.2), i.e. version B.

The importance of the exact specification of the Newtonian limit case results from two points. On the one hand, the result of the relation between Minkowski's force and Newton's force would be arbitrary without such a specification. On the other hand, such a specification follows from the logic with which physical laws are generalized with the help of symmetry principles.

Relativistic Mass

Occasionally the *relativistic mass*

$$m_{\rm rel}(v) = \frac{m}{\sqrt{1 - v^2/c^2}} \tag{A.13}$$

is introduced. Writing (A.8) in the form $d(m\boldsymbol{v})/dt = \cdots$ and replacing m with $m_{\rm rel}(v)$, one obtains the valid relativistic equations (A.9). Therefore, this substitution is often given as a recipe for the relativistic generalization of Newton's second law. However, the validity of this recipe correlates with that of version A.

A possible way to version A is the recipe $m \to m_{\rm rel}$ for the transition from Newton's second law to the relativistic equation of motion:

$$\frac{d}{dt}\big(m\boldsymbol{v}(t)\big) = \boldsymbol{F}_{\rm N} \quad \xrightarrow{m \to m_{\rm rel}(v)} \quad \frac{d}{dt}\,\frac{m\boldsymbol{v}(t)}{\sqrt{1 - v^2/c^2}} = \boldsymbol{F}_{\rm N} \tag{A.14}$$

The recipe $m \to m_{\rm rel}$ is always to be understood in such a way that the time derivative also acts on the mass as shown on the left.

[4]T. Fließbach, *Elektrodynamik*, 7th edition Springer, Spektrum, 2022, Chapter 22.

According to (A.14), the recipe $m \to m_{\text{rel}}$ leads to $\boldsymbol{F} = \gamma \boldsymbol{F}_{\text{N}}$ (version A). Starting from version A, $\boldsymbol{F} = \gamma \boldsymbol{F}_{\text{N}}$, we obtain from the valid relativistic equation (left part in (A.7)) the right part in (A.14). Version A and the prescription $m \to m_{\text{rel}}$ are therefore equivalent for the equation of motion considered here.

As shown in (A.3) to (A.6) above, the correct derivation leads to version B. Therefore, the recipe $m \to m_{\text{rel}}$ cannot be of general validity.

In the special case, $\boldsymbol{F}_{\text{N}} \parallel \boldsymbol{v}$ both lines in (A.7) agree, version A then agrees with the correct form (version B). Therefore, For $\boldsymbol{F}_{\text{N}} \parallel \boldsymbol{v}$, the recipe $m \to m_{\text{rel}}(v)$ leads to the correct relativistic equation of motion.

If we write (A.11) in the form $d(m\boldsymbol{v}')/dt' = \cdots$ and replacing m by $m_{\text{rel}}(v)$, one does *not* obtain the valid relativistic equations (A.9). The substitution $m \to m_{\text{rel}}(v)$ is therefore not universally valid. Apart from that the relativistic mass reflects essential features of the relativistic motion such as a diverging inertia for $v \to c$. All relevant aspects of relativistic mass are discussed in a small booklet[5] by the author.

Symmetry Principles and the Logic of Theoretical Physics

Symmetry principles play a central role in the generalization of physical laws: The more general law must satisfy certain symmetries and must be known to be valid in a special case. We explain this by the example of the symmetry principles *isotropy of space*, *Einstein's principle of relativity*, and *Einstein's equivalence principle.*

1. Due to the isotropy of the space one establishes physical laws in the form of 3-tensor equations. If the law is known in a special coordinate system, then the 3-tensor equation is to be set up such that it reduces to the known law in the special case.
2. Due to Einstein's principle of relativity, one sets up relativistic laws in the form of Lorentz tensor equations. In the non-relativistic case, the law must reduce to the known Newtonian limit.

[5]T. Fließbach, *Die relativistische Masse*, Springer, Spektrum, 2018.

In particular for the motion of a particle: The relativistic equation of motion is to be set up in such a way that it reduces to the known Newtonian limit in the instantaneous rest frame ($v \to 0$).

3. Due to Einstein's equivalence principle, one establishes general relativity laws (GRT law) in the form of Riemann tensor equations (in a generally covariant form). In the limiting case without gravity, the wanted law must reduce to the known special relativity law (SRT law).

   The gravitational field is described by the metric tensor $g_{\alpha\beta}$. In the limiting case $g_{\alpha\beta} \to \eta_{\alpha\beta}$ (concretely in a free-falling satellite laboratory), the GRT law must reduce to the known SRT limiting case.

In (A.3)–(A.6), the correct relation (A.2) between the Newtonian force and the Minkowski force was derived. The basis of this derivation is Einstein's principle of relativity and the logic of theoretical physics in generalizing laws on the basis of symmetry principles. Within the framework of this logic, the known limiting case (here the Newtonian limiting case) must be specified exactly. This exact specification leads to version B, but it is incompatible with version A.

Summary

Academic aspects

If one equates the right-hand side in (A.8) with $\boldsymbol{F}_\mathrm{N}$, then (A.1), $\boldsymbol{F} = \gamma \boldsymbol{F}_\mathrm{N}$, leads to the correct result (A.9). However, this result is obtained by two compensating inaccuracies:

1. First inaccuracy: Newton's equation (A.8) is not adequate in this context because it is not the Newtonian limiting case. Without an exact specification of the Newtonian force, however, the result for the relation to the Minkowski force is arbitrary.
2. Second inaccuracy: the statement $\boldsymbol{F} = \gamma \boldsymbol{F}_\mathrm{N}$ is not valid in general.

In the present case, these two inaccuracies compensate each other.

However, this criticism has also academic features: First of all, (A.8) is a correct Newtonian equation — only not the Newtonian limit case. For concrete calculations (A.8) is the preferred equation; the limiting case (A.11) is rather impractical for this. Relativistic equations are often set up without explicit reference to the Newtonian limit. And concrete applications are independent of questions A or B discussed here. In this respect, the question A or B is without practical relevance. But the question A or B is connected with the logic for establishing physical laws by means of symmetry principles.

Logical aspects

In theoretical physics one uses the following logic for the generalization of laws:

- In a special case, the law is known.
- One knows the transformations from the special case to the general case.

For details, we refer to the above section on "symmetry principles and the logic of theoretical physics".

This appendix has emphasized the correct derivations of the relativistic equations of motion, (A.3)–(A.6): The special case is the Newtonian limit ($v \to 0$), and the transformations are the Lorentz transformations. In practice, one determines the Lorentz invariant equation which in the limiting case agrees with the known law. The underlying symmetry principle is Einstein's relativity principle.

The logic for establishing more general laws based on symmetry principles is fundamental in Theoretical Physics. In this logic, the known special case must be specified exactly. In this respect, the question A or B is of logical relevance for the structure of Theoretical Physics.

Index

aberration, 410
accelerated reference frame, 51–59, 450–453
acceleration, 8–10
action
    functional, 147, 148, 319
active transformation, 46–47
addition formula (velocity), 418–420
amplitude
    of a wave, 345
    of an oscillation, 276, 278, 281, 291
angular momentum, 117
    conservation, 33–34
    definition, 21
    equation of motion, 22, 33–34
    rigid body, 226
angular velocity (rigid body), 216, 222
antisymmetric, total a.s. tensor, 244
Atwood's falling machine, 81–82, 101–102

barometric formula, 363
beam bending, 349–354
Bernoulli's equation, 365, 366, 370
body cone, 255, 258
body-fixed coordinate system, 216–219
Bohr's quantization rule, 313, 315
boundary condition, 131, 139, 142
    beam, 352–354
    string, 341
brachistochrone, 124, 132

calibration of clocks and rulers (SRT), 395
canonical equations, 307–315
canonical transformation, 319–327
Cartesian coordinates, 6–7
catenary, 142–143
center-of-mass, 32–33
    momentum conservation, 32–33, 169, 171
    scattering, 203
center-of-mass system
    collision, 443–446
    scattering, 208
center-of-mass theorem, 32
central force, 22
central potential, 169–209
centrifugal force, 56
chaotic motion, 187–189
clock
    in SRT, 395–396
    time measurement, 5–6
closed system, 33, 47–49, 113–115
complete integral, 330
compressibility, 360
cone
    body, 257
    precession, 257
    space, 257
conservation law, 21–27, 70, 128–163, 376–379
conservation of angular momentum, 21–22, 118
conservative force, 24–25, 36

conserved quantity, 312
constancy of the speed of light, 385
constant of motion, 92
constraint (variation), 133
constraint force, 63, 66, 67
continuity equation, 357
continuum mechanics, 337–379
contraction of tensor indices, 240, 423
contravariant components, 423
coordinate
  cyclic, 94
coordinate system, 3, 10
coordinate transformation
  Galilean transformation, 39–44, 111–112
  Lorentz transformation, 388–393, 413–420
  orthogonal transformation, 235–239
coordinates
  Cartesian, 6–7
  curvilinear, 104
  cyclic, 94
  generalized, 87
  orthogonal, 104
Coriolis force, potential, 56, 59, 175, 181, 200–203
coupled pendulum, 297–299
covariance
  Galilean transformation, 42–49
  Lorentz transformation, 421–456
  orthogonal transformation, 241–242
covariant components (tensor), 423
critically damped, 279
cross section, 196–208
  Rutherford, 208
crystal model, 304–305
current density
  electric, 376
  liquid, 357
  particle beam, 196
curvature, 10
curvilinear coordinates, 8–10, 103–104
cyclic coordinate, 94

d'Alembert operator, 427
d'Alembert principle, 69–70
d'Alembert, solution method, 347
damped oscillator, 28, 273–284
  energy balance, 282–284
degree of freedom, 85
  rigid body, 215
deterministic chaos, 185–189
differential cross section, 196–208
dispersion relation, 304, 369
dissipation
  damped oscillation, 283
  function, Rayleigh's, 96
dissipative force, 24–25, 35–36, 95–96
double pendulum, 87, 102–103, 108, 303
dyad, 227, 245–246
dynamic pressure, 366
dynamic susceptibility, 279
dynamics of a mass point, 3

earth (oblateness), 233
eccentricity, 182
effect and counter effect, 15
eigenfrequency, 277, 285–290
eigenmode, 290
eigenvalue, eigenvector
  eigenvalue equation, 229
  inertia tensor, 228–231
  small oscillations, 289–290
eigenvibration
  string, 345, 346
Einstein
  equivalence principle, 52
  general relativity, 53, 454
  principle of relativity, 383–386
  special relativity, 383, 461
elastic scattering, 208
electromagnetic field, 458–460
electromagnetic forces, 94–95

ellipse
elliptical orbit, 181–183
energy
balance for scattering, 208–209
conservation, 23–25, 34–36, 70–71, 93–94, 115–116, 129, 173, 377
energy-momentum relation (SRT), 440–441
equivalence with mass, 437–439
kinetic, 24–25, 34–36
potential, 24–25, 34–36
relativistic, 436–439
rigid body, 223–224
equal area law, 22, 184
equation of motion
angular momentum, 22, 33, 34
canonical equations, 307–315
Hamiltonian equations, 307–315
Lagrangian equations
1st kind, 63–71
2nd kind, 85–96
relativistic, 429
equation of state, 359–361
equivalence
inertial and gravitational forces, 52
inertial and gravitational mass, 52, 453
mass and energy, 437–439
Euclidean space, 6–7
Euler angle, 220, 222
Euler equation
hydrodynamics, 358–359
rigid body, 249–258
Euler–Lagrange equation, 125–141
event, 384
interval of two events, 386–387

field equation, 338, 343
beam, 351
general form, 374
hydrodynamics, 358–359
string, 343
field strength tensor, 459
field theory, 337–379
figure axis, 254
first integral of motion, 27
force
conservative, 24–25, 36
dissipative, 24–25, 35–36, 95–96
generalized, 90
internal, external, 31
Lorentz, 95, 434–436, 460
measurement, 13
Minkowski, 432–434
Newtonian, 430–433
forced oscillation, 28, 273–284
form invariance (covariance)
Galilean transformation, 42–49
Lorentz transformation, 421–456
orthogonal transformation, 242
Foucault's pendulum, 59
four-dimensional
line element, 388, 421, 424
Minkowski space, 421
four-vector, 422
free oscillation, 294
free rotation, 252–254
friction force, 25
frictional force, 95–96
damped oscillation, 283
functional, 123–125

Galilean
group, 43
invariance, 161–163
principle of relativity, 383–386
transformation, 111–112, 120, 163
gauge transformation
Lagrange function, 151–153
Schrödinger field, 378
general relativity, 53, 454
generalized coordinates, 85–87
generalized force, 90
generalized momentum, 94
generalized velocity, 88
generating function (canonical transformation), 320–324
geodesic line, 135, 143–145

gravitation, 14, 17, 52–53, 175, 181–193
gravitational mass, 14, 52–53
group
  Galilei, 43
  Lorentz, 413, 420
  orthogonal transformations, 238
gyroscope, 216
  force-free symmetric, 254–258
  heavy symmetric, 260–264

Hamilton
  equations (canonical), 307–315
  formalism, 307–333
  function, 309–312
  principle, 147–153, 319–374, 456
Hamilton–Jacobi equation, 329–333
hard sphere, 210
harmonic oscillator, 273–284
  energy balance, 282–284
  Hamiltonian formalism, 324, 325
heavy symmetric gyro, 260–264
holonomic constraint, 64, 65, 134, 140–142
homogeneity of space, 113–117, 455
homogeneity of time, 113–116, 455
Hooke's law, 350
house stick experiment, 407
hydrodynamics, 355–370
hydrostatics, 362–364
hyperbolic motion, 447–449
hyperbolic orbit, 189–190
hyperbolic trajectory, 182

ideal liquid, 356
impact parameter, 198, 199
impulse, 431, 458
inclined plane, 75–100
incompressible, 360
inertia tensor, 223–231
inertial forces, 52
inertial frame, 12, 39–114
inertial mass, 14, 52–53
inertial system, 50
  local, 453

initial condition, 16, 76
integrable system, 188
interval of events, 386–387, 404–409
invariance, 50
  of action, 155–163
  Galilean transformation, 47–49
irrotational flow, 366
isoperimetric constraint, 134, 137–139
isotropy of space, 113–115, 117–118, 455

KAM-Theorem, 189
Kepler
  laws, 22, 183–184
  problem, 181–193
kinematics
  mass point, 3
  rigid body, 213–270
kinetic energy, 24–25, 34–36
  LS and CS, 443–446
  relativistic, 436–439
  rigid body, 223–224
Kronecker symbol, 243, 426

laboratory system
  collision (SRT), 443–446
  scattering, 203, 208
Lagrange
  density, 351, 371–374
  equations
    1st kind, 63–71
    2nd kind, 85–96
    heavy symmetric gyros, 259–264
  equivalence of, 120
  formalism, 63–120
  function, 88–92
  in electromagnetic field, 94–95, 97
  multiplier, 135–137
Laplace equation, 367
lattice oscillations, 304, 305
latus rectum, 182
law of equal areas, 22
length contraction, 397–409

length measurement (SRT), 399, 407, 409
Lenz vector, 191
Levi-Civita symbol, 243, 244, 426
lightlike interval, 404
line element, 103
  4-dimensional, 388, 421, 424
  three-dimensional, 421
Linear chain, 303
linearization, 367
Liouville theorem, 317, 318
local inertial system, 453
longitudinal wave, 343, 369
Lorentz contraction, 400
Lorentz force, 95, 434–436, 460, 466–467
Lorentz group, 413–415, 420
Lorentz tensor (-scalar, -vector), 421–427
Lorentz transformation, 388–393, 413–420
  homogeneous, inhomogeneous, 413
  special, 389–393

Mach principle, 40
major axis transform, 297
many-body problem, 185–189
mass
  defect, 438
  density, 225–226, 339, 349, 357
  inertial and gravitational, 14, 52–53
  measurement, 13
  relativistic, 435
mass point, 3–4
  on circular cone, 104–108
  on rotating rod, 79–81, 100–101
mass-energy equivalence, 437, 439
matrix notation, 227, 228, 414
Maxwell equations, 49–50, 375, 376, 386
Minkowski force, 432–434, 463, 466
Minkowski space, 421
Minkowski tensor, 426
molecule oscillation, 299–301
moment of inertia, 227–229
momentum
  canonical, 309
  center-of-mass, 32, 33
  conservation, 21, 32, 32–33, 116, 117, 458
  generalized, 94
  mass point, 12
  relativistic, 439–440
moving clocks and bars in SRT, 409
Myon (lifetime), 409

natural frequency, 277, 285
natural oscillation (mode), 290
Navier-Stokes equations, 362
Newton's laws, 11–19
Newtonian force, 430–433, 463–466
Newtonian mechanics, 11–59
nodes (standing wave), 346
Noether theorem, 155–163
  field theory, 376, 379
normal coordinates, 290–293
normal mode, 290
nutation, 263, 264

oblateness of earth, 233
orthogonal coordinates, 104
orthogonal transformation, 42, 229–231, 235–239
orthogonality
  eigenvectors, 292
  eigenvectors of inertia tensor, 230
oscillation
  between turning, 19
  critically damped, 279
  forced, 28, 273–284
  overdamped, 279
  pendulum, 297–299
  small, 273
oscillator
  anharmonic, 20
  damped harmonic, 28, 273–284
  spherical, 175, 179, 180

symmetry and conserved quantities, 164

parabolic trajectory, 182
parameter (Kepler ellipse), 182
particular solution, 279
passive transformation, 46–47
pendulum
coupled, 297–299
double, 108, 303
double pendulum, 87, 102–103
plane, 64, 86
perihelion precession, 185, 192
phase, 278, 280, 345
phase shift, 282
phase space, 313–315
volume, 313
plane wave, 370
planetary motion, 181–185, 193
many-body problem, 185–189
Poincaré group, 415
Poincaré section, 188
point mass, 225
poisson bracket, 325–326
polar coordinates, 8–10
polytropic
equation of state, 360
potential, 17, 18, 25–27
effective, 176
many-particle system, 36–38
potential energy, 24–25, 34–36
potential flow, 366–367
power, 23–24
precession
earth axis, 264–270
heavy gyroscope, 263, 264
regular, 257, 258
precession cone, 257
pressure, 356
principal axis transformation, 228–231, 296
principal moment of inertia, 230
principle of least action, 148
principle of virtual work, 69
projectile, 195
proper length, 397
proper time, 402–403
pseudotensor, 244

quantization rule, 313

radial equation (central potential), 171–172
rapidity, 391
Rayleigh's dissipation function, 96
reduced mass, 170
reference frame, 3–4, 11–12, 39–59, 383
rigid body, 216–219
scattering, 203–208
relativistic
energy, 436–439
equation of motion, 429–441
gravitation theory, 453–454
mass, 435
mechanics, 383–461
momentum, 439–440
relativity
general, 454, 469
special, 383, 461
relativity of space-time, 113–115, 119–120, 161–163
relativity principle, 39–40
Einstein, 383–386
Galilei, 383–386
resonance frequency, -curve, 282
response function, 280
rest energy, 436–439
rest mass, 429–430
rheonomic constraint, 66
rigid body, 213–270
in SRT, 408, 409
rigid rotation, 364
rope in gravity field, 133, 142–143
rotating liquid in gravitational field, 364
rotating reference frame, 53–59
rotation
around principal axes, 252–254

rigid body, 213, 270
symmetry, 113–115, 117–118
Rutherford cross section, 200–208

scalar product, 8, 424
scattering, 176, 195–209
Schrödinger equation, 375
scleronomic constraint, 66
simultaneity, 404–409
small oscillation, 273, 305
soap skin, 125, 132
sound waves, 367–370
space cone, 258
space-fixed coordinate system, 216–219
space-time symmetries, 120
space-time-symmetries, 111
spacelike interval, 404
special relativity, 383–461
speed of light, 49, 383–386
maximum particle velocity, 439
speed of sound, 369
spherical oscillator, 164
spherical wave, 370
stability of equilibrium position, 276, 287
standing wave, 346
star equilibrium, 363
state (system), 92
steady state, 281
Steiner's theorem, 231
streamline, 366
string vibration, 337–347
sum convention, 388
superposition principle, 343
superposition principle of forces, 15
susceptibility, 279
symmetry, 111
conservation law, 155–163, 376–379
space-time, 120
system
state, 92, 313, 315

target, 195
tensor
3-tensor, 224
calculation rules, 240–241
definition, 240
tensor equations, 241–242
tensor field, 246–248
4-tensor, 421, 422, 427
calculation rules, 425
definition, 422
tensor equation, 422
tensor field, 426, 427
time coordinate, 3
time dilation, 401
time measurement, 5–6
SRT, 395–396, 400–403
time shift, 41
timelike interval, 404
torque, 21, 34, 249
total antisymmetric tensor, 244, 426
trajectory, 3
in the central field, 174–179
transformation
canonical, 319–327
Galilean, 39–44, 111–112
laboratory and center-of-mass system, 203–208, 443–446
Lorentz, 388–393, 413–420
orthogonal, 42, 229–231, 235–239
passive and active, 46–47
translation symmetry, 113–117
transversal wave, 342
twin paradox, 449–450
two-body problem, 169–179
energy-conservation, 173
reduction to one-body problem, 170–171
reduction to radial equation, 171–172

uncertainty relation, 315

variational principles, 121–165
calculus, 121–165
with constraint, 133–145
without constraint, 121–131

velocity, 8–10
velocity-addition formula, 418, 420
vibration
  small, 305
  string, 337–347
virtual displacement, 69
vortex-free flow, 366, 367

wave
  longitudinal, 369
  plane, 370
  sound, 367–370
  sphere, 370
  standing, 346
  string, 337–346
wave equation
  string, 341–343
  sound, 367–368
wave function, 375
wave number, 345
wave packet, 346, 369
wave velocity, 345
wavelength, 345
work, 23

Printed in the USA
CPSIA information can be obtained
at www.ICGtesting.com
LVHW010818310824
789671LV00002B/15

9 789811 284571